Principles of Wireless Sensor Networks

Wireless sensor networks are an emerging technology with a wide range of applications in military and civilian domains. The book begins by detailing the basic principles and concepts of wireless sensor networks, including information gathering, energy management, and the structure of sensory nodes. It proceeds to examine advanced topics, covering localization, topology, security, and evaluation of wireless sensor networks, highlighting international research being carried out in this area. Finally, it features numerous examples of applications of this technology to a range of domains, such as wireless, multimedia, underwater, and underground wireless sensor networks. The concise but clear presentation of the important principles, techniques and applications of wireless sensor networks makes this guide an excellent introduction for anyone new to the subject, as well as an ideal reference for students, practitioners and researchers.

Mohammad S. Obaidat, recognized around the world for his pioneering and lasting contributions to several areas, including wireless sensor networks, green ICT, wireless and wired networks, performance evaluation of computer systems and networks, and information and network security, is a Professor of Computer Science at Monmouth University, New Jersey, USA. He is the editor-in-chief or editor of many international journals, and has authored over 30 books and over 600 technical papers to date. He has received numerous awards, including a Nokia Research Fellowship, Distinguished Fulbright Scholar Award, McLeod Founder's Award, SCS Presidential Award, and SCS Modeling & Simulation Hall of Fame Award and Best Paper awards in many conferences. He served as SCS President, Advisor to the President Philadelphia University, and Chair of the Department of Computer Science and Software Engineering at Monmouth University. He is a Fellow of the IEEE and the SCS. He has chaired numerous international conferences all over the world and has been invited to give keynote speeches in international conferences. He served as an IEEE Computer Society Distinguished Speaker and is now serving as an ACM and SCS Distinguished Lecturer/speaker.

Sudip Misra is an Associate Professor at the Indian Institute of Technology, Kharagpur. He has authored over 180 scholarly research papers and has edited 6 books. He was awarded the Canadian Government's prestigious NSERC Post-Doctoral Fellowship and the Humboldt Research Fellowship in Germany.

"The book covers the main aspects regarding modern wireless sensor networks, touching hardware and software platforms, networking architectural organization, and communication protocols and applications. It includes treatment of important issues like localization and tracking, topology management, performance evaluation, security, mobility, and multimedia, as well as of two challenging environments ... underwater and underground.

"The material blends theory and applications, and is presented in a form suitable for students, researchers and practitioners. It provides a comprehensive overview and perspective of the field."

Franco Davoli
University of Genoa

Principles of Wireless Sensor Networks

MOHAMMAD S. OBAIDAT

Monmouth University, New Jersey

SUDIP MISRA

Indian Institute of Technology

CAMBRIDGE
UNIVERSITY PRESS

CAMBRIDGE
UNIVERSITY PRESS

Shaftesbury Road, Cambridge CB2 8EA, United Kingdom

One Liberty Plaza, 20th Floor, New York, NY 10006, USA

477 Williamstown Road, Port Melbourne, VIC 3207, Australia

314–321, 3rd Floor, Plot 3, Splendor Forum, Jasola District Centre, New Delhi – 110025, India

103 Penang Road, #05–06/07, Visioncrest Commercial, Singapore 238467

Cambridge University Press is part of Cambridge University Press & Assessment,
a department of the University of Cambridge.

We share the University's mission to contribute to society through the pursuit of
education, learning and research at the highest international levels of excellence.

www.cambridge.org
Information on this title: www.cambridge.org/9780521192477

First published 2014

A catalogue record for this publication is available from the British Library

Library of Congress Cataloging-in-Publication data
Obaidat, Mohammad S. (Mohammad Salameh), 1952–
Principles of wireless sensor networks / Mohammad S. Obaidat, Monmouth University, New Jersey,
Sudip Misra, Indian Institute of Technology.
    pages   cm
Includes bibliographical references and index.
ISBN 978-0-521-19247-7
1. Wireless sensor networks.   I. Misra, Sudip.   II. Title.
TK7872.D48.O25   2014
681'.2–dc23

2013048935

ISBN   978-0-521-19247-7   Hardback

To Our Families

Contents

Preface

Overview and goals

Small low-cost devices powered with wireless communication technologies along with the sensing capabilities are instrumental in the inception of *wireless sensor networks* (WSNs). Recent years have witnessed a sharp growth in research in the area of WSNs. The characteristics of such distributed networks of sensors are that they have the potential for use in various applications in both the civilian and military fields. Enemy intrusion detection in the battlefield, object tracking, habitat monitoring, patient monitoring, and fire detection are some of the numerous potential applications of sensor networks. The ability of an infrastructure-less network setup with minimal reliance on network planning, and the ability of the deployed nodes to self-organize and self-configure without the association of any centralized control are the smart features of these networks. Leveraging the advantages of these features, the network setup is swift in challenging scenarios such as emergency, rescue, or relief operations. The smart features also enable continuous operation of the network without any intervention in case of any failure.

Along with the above-mentioned attractive features possessed by sensor networks, there are several challenges which hinder hassle-free, autonomous, and involuntary operation of these networks. Some of the challenges are attributed to issues relating to scalability, quality-of-service (QoS), energy efficiency, and security. The protocols should be light-weight enough to be suitable for these networks, which consist of small-sized sensor nodes with limited computation power. Sensor networks are often deployed in large-scale and are expected to function through years. Clearly, battery power is an issue in such cases, and can be achieved with the help of energy-efficient or energy-aware protocols. Finally, QoS is also an issue for applications which demand prompt responses.

There exists vast literature on various issues and dimensions of WSNs. This book attempts to provide a comprehensive guide on fundamental concepts, challenges, problems, trends, models, and results in the areas of WSNs. This book has been prepared keeping in mind that it needs to prove itself to be a valuable resource dealing with both the important core and the specialized issues in the areas. We have attempted to offer a wide coverage of topics. We hope that it will be a valuable reference for students, instructors, researchers and practitioners. We believe this is a particularly attractive feature of this book, as the limited selection of books available on sensor networks are written primarily for academicians/researchers. We have attempted to make this book useful for both the academics and the practitioners alike.

Organization and features

The book is broadly divided into three sections – the first part discusses the basics of WSNs, the second part focuses on the networking aspects and protocols of WSNs, and the third part deals with the advanced issues and topics such as localization, topology management, security, modeling, and simulation. There are 14 chapters in the book, of which the first part has three chapters, the second part has three chapters, and the third part has eight chapters.

In the first part, we provide an introduction to WSNs to the readers in Chapter 1. In this chapter, we provide an up-to-date treatment of the fundamental techniques, applications, taxonomy, and challenges of such networks. We also explain the basic components of a wireless sensor node, and classify the sensor networks. Finally, we discuss the differences between WSNs and wireless mesh networks and RFID systems.

Chapter 2 elaborates on components, structure, and operations of a wireless sensor node. We discuss the limitations and the design challenges of WSNs. The hardware architecture and the operating systems of a sensor node are discussed with examples of sensor nodes. This chapter also includes the effects of the infrastructure on the performance evaluation of WSNs. We also discuss the MEMS technology used to manufacture low-power inexpensive sensor nodes.

Chapter 3 reviews the major WSNs applications to various areas including environmental monitoring, health care, intelligent and smart home, homeland security, underwater applications, agriculture and greenhouse monitoring, and military applications.

Chapter 4 is dedicated to discussions about medium access control (MAC) in WSNs. We first discuss the problems of the traditional MAC schemes. In this chapter, the major MAC schemes for WSNs are discussed in detail.

In Chapter 5, we review the aspects, related advantages, and disadvantages, as well as challenges, of routing in WSNs. We classify the existing routing schemes into various categories, and explain a few schemes from each of the categories.

Chapter 6 deals with the transport protocols and quality-of-service (QoS) issues of the WSNs. We first address the transport protocol requirements for WSNs, and discuss the applicability of the Internet transport protocol in WSNs. Finally, the transport protocols are classified into various categories, and schemes from each of the categories are discussed.

Chapter 7, the first chapter of the third part of the book, presents the localization and target tracking schemes of WSNs. First, we discuss the basics of localization and the various distance estimation techniques. Next, the taxonomy of the existing localization schemes is presented with a few schemes from each category investigated in detail. Similarly, the target tracking schemes are also classified into various categories, and we discuss a few existing schemes as well.

In Chapter 8, the aspects and importance of topology management and control are discussed. A taxonomy of the existing schemes is also presented. Finally, we present a few existing schemes from each of the categories.

In Chapter 9, we provide an up-to-date treatment of the techniques that can be used to evaluate the performance of WSN systems. We discuss modeling and simulation techniques for WSNs, which are important when performance evaluation of these networks is needed. The performance metrics and fundamental models associated with performance evaluation are also discussed.

Chapter 10 discusses the security issues related to WSNs. We present a comprehensive study of the challenges, vulnerabilities, attacks, existing solutions, and then compare the major security techniques related to WSNs.

Chapter 11 presents the issues and aspects related to mobile wireless sensor networks. The authors investigate various issues such as coverage, connectivity, and deployment in mobile WSNs.

In Chapter 12, the authors discuss another variant of WSNs named wireless multimedia sensor networks (WMSNs). The challenges and specific applications of WMSNs are also discussed. This chapter also includes the network and node architecture and the communication layers of WMSNs.

Chapter 13 presents the underwater counterpart of WSNs. It is named as Underwater Sensor Networks (UWSNs). We present the challenges and characteristics of UWSNs, and the underwater physics and dynamics associated with UWSNs. The UWSN sensor nodes, their components, the network architectures, and few localization services are also studied in this chapter. We go through each layer of the protocol stack for UWSNs, and briefly discuss the schemes related to each layer.

Chapter 14 deals with another variant of WSNs, the Wireless Underground Sensor Networks (WUGSNs). The applications, challenges, network architectures of WUGSNs are presented. We also shed some light on the protocol stack, communication channels, and routing schemes for such networks.

Target audience

The book is written primarily for the student community. This includes students of all levels – those being introduced to these areas, those with an intermediate level of knowledge of the topics, and those who are already knowledgeable about many of the topics. In order to achieve this goal, we have attempted to design the overall structure and content of the book in a manner that makes it useful at all learning levels.

The secondary audience for this book is the research community, which includes researchers working in academia, industry, or government. To meet the specific needs to this audience group, most chapters of the book also have a section in which attempts have been made to provide directions for future research.

Finally, we have also taken into consideration the needs of those readers, typically from the industry, and those practitioners who wish to gain insight into the practical significance of the topics, i.e. how the spectrum of knowledge and the ideas are relevant for real-life workings of sensor networks.

Acknowledgements

We would like to thank our students for allowing us to try some parts of the book on them, and for their feedback on the materials as well for their help in some aspects of the preparation of the book.

We are also very much thankful to the editorial staff and the production editor at Cambridge University Press, who tirelessly worked with us and alerted us to publication deadlines. We also thank our families, and especially our wives, for bearing with us during the preparation process of the manuscript and for their strong support.

1 Introduction to wireless sensor networks

Wireless sensor networks (WSNs) are a new class of wireless networks that are becoming very popular with a huge number of civilian and military applications. A wireless sensor network (WSN) is a wireless network that contains distributed independent sensor devices that are meant to monitor physical or environmental conditions. A WSN consists of a set of connected tiny sensor nodes, which communicate with each other and exchange information and data. These nodes obtain information on the environment such as temperature, pressure, humidity or pollutant, and send this information to a base station. The latter sends the info to a wired network or activates an alarm or an action, depending on the type and magnitude of data monitored [1–24].

Typical applications include weather and forest monitoring, battlefield surveillance, physical monitoring of environmental conditions such as pressure, temperature, vibration, pollutants, or tracing human and animal movement in forests and borders [1–23]. They use the same transmission medium (which is air) for wireless transmission as wireless local area networks (WLANs). For nodes in a local area network to communicate properly, standard access protocols like IEEE 802.11 are available. However, this and the other protocols cannot be directly applied to WSNs. The major difference is that, unlike devices participating in local area networks, sensors are equipped with a very small source of energy (usually a battery), which drains out very fast. Hence the need arises to design new protocols for MAC that are energy aware. Clearly there is some difference between a traditional WLAN and a WSN, as the latter has limited resources.

The objective of this chapter is to provide an up-to-date treatment of the fundamental techniques, applications, taxonomy, and challenges of wireless sensor networks.

1.1 Background

A wireless sensor network consists of hundreds, if not thousands, of small and inexpensive nodes, which could have a static location or be dynamically deployed to monitor the intended environment. Owing to their miniature size, they have a number of constraints. The function of a WSN is basically monitoring. There are three classes of monitoring that a WSN can observe: (a) entity monitoring, which means monitoring something such as civil structures like bridges, tunnels, highways, and buildings, or the human body, such as monitoring the organs of the body; (b) area monitoring, which includes monitoring the environmental area alarms; and (c) area-entity monitoring, which includes monitoring vehicles on the highway, and monitoring movement of an object [1–15, 23, 24].

The key positive feature of WSNs does not come from the strength of the individual sensor nodes; it comes from the entire array of interconnected sensor nodes. Hence, WSNs are expected to be large in scale from the point of view that they have a lot of nodes and they are apt to be self-configuring, in order to achieve reliability. Since a wireless sensor node is usually inexpensive, we would expect to have a huge number of nodes in a WSN.

Typically, sensor nodes communicate with each other by means of a multi-hop scheme. The flow of information and data stops at particular nodes called base stations or sinks. A sink or base station usually connects the sensor network to a fixed network to distribute the data sensed for further processing. In general, base stations have enhanced capabilities over regular nodes as they should carry out compound processing. This substantiates the actuality that sinks have more advanced processors such as PCs/laptops with more RAM memory, secondary storage, battery and computational power as they are expected to perform more tasks than regular sensor nodes. It is worth noting here that one of the biggest drawbacks of sensor networks is power use, which is really influenced by the interaction between nodes. In order to work out this problem, aggregation points are set up to the network, which reduce the overall communication traffic between nodes and save energy. Typically, collection points are ordinary nodes that get data from nearby nodes, carry out some sort of processing, and then advance the filtered data to the subsequent hop. Sensor nodes are arranged into groups, each group having a "group/cluster head" as the leader. Communication within a group should travel all the way through the cluster head. Then it is advanced to an adjacent group head until it arrives at its destination, which is the sink or base station. A different scheme for saving energy is to let the nodes go into sleep mode, if they are not needed, and to wake them up when they are needed.

The progress of wireless sensor networks was initially provoked by military applications; however, wireless sensor networks are now employed in many civilian applications such as environment monitoring, industrial process monitoring, health care applications, road and highway traffic control, smart homes and cities, and office automation. In health care applications, wireless devices make patient monitoring less invasive, thus improving health care. For utilities applications, wireless sensors provide an inexpensive scheme for collecting system health data to minimize energy usage and enhance management of resources. As for remote monitoring, a wide range of applications are covered where wireless networks can go together with fixed networks and systems by minimizing wiring costs and permitting new sorts of testing and measurement applications. The main applications of remote monitoring are: (a) environmental monitoring of air, soil, and water, (b) building and structural monitoring of bridges, subways, and buildings, (c) process monitoring, (d) machine monitoring, (e) habitat monitoring, (f) intelligent transportation systems, (g) air traffic control, (h) traffic surveillance, (i) video surveillance, and (j) monitoring carbon transfer in rain forests, among others [1–24]. Each node in a wireless sensor network is usually equipped with a radio transceiver, a tiny microcontroller, and a power source (typically a battery). The cost of a sensor node ranges from hundreds of dollars to a quarter of a dollar, depending on the size of the network and the functionality and sophistication required of each node. The size and price restrictions on sensor nodes produce constraints on resources such as energy, memory, computational power, and throughput. In general, a sensor network forms a wireless ad-hoc computer network, which means that each sensor supports a multi-hop routing scheme.

The major components of a wireless sensor network, which include sensors, signal convertors such as analog-to-digital (A/D) and digital-to-analog (D/A) convertors, processors, communication devices, and a power supply, are all becoming more and more inexpensive and smaller. Stringent power expenditure requirements are necessary because the sensor node needs to be reliable and able to run unattended for a long time, which can be years. Among the factors that should be considered in the design of power sources of a WSN are: (a) choice of power harvesting scheme or battery type, and (b) choice of small power electronic design schemes. Companies that produce these devices are now developing small sensor nodes and networks. Moreover, commercial off-the-shelf personal digital assistants (PDAs) or pocket computers contain impressive computing power in a small package. Such devices can easily be used as powerful sensor nodes. Wireless LANs like the popular IEEE 802.11 standards can now offer performance very close to those of wired networks. Moreover, we have now IEEE 802.15 standard that gives specifications for personal area networks (PANs), which can be employed for WSNs as well.

Furthermore, advances in semiconductor technology allow us to have more chip capacity and more processor capabilities. This progress allows a reduction in the energy/bit requirements for both the computing and communication systems. It is expected that advances in micro-electro-mechanical-systems (MEMS) technology will produce more powerful and versatile sensors. MEMO technology integrates mechanical elements, sensors, actuators, and electronics on a common silicon substrate through microfabrication technology, whereas the electronics are fabricated by using integrated circuit (IC) process sequences such as bipolar, CMOS transistors. The micromechanical elements are made up using well-suited micromachining techniques that purposely add new structural layers to create the mechanical and electromechanical devices [1–14].

1.2   Components of a wireless sensor node

The central component of a wireless sensor network is the sensor node. It is a very tiny device that has the ability to sense its immediate environment and map or store the information. Owing to the progress in semiconductor technology, the cost of these devices is decreasing all the time. These tiny devices consist of the following main components [8, 9].

● Microcontroller. This is a computer-on-a-chip which is very tiny in size although capable of doing powerful tasks including controlling the functions of other devices connected to it. In general, a microcontroller consists of a microprocessor,

Table 1.1 The industrial, scientific and medical (ISM) bands

Band	Frequency range
UHF ISM band	902 to 928 MHz
S-band ISM	2.4 to 2.5 GHz
C-band ISM	5.725 to 5.875 GHz

a RAM memory, and associated peripherals. These days, there are other devices available on the market that can be used in place of microprocessors for performing the same actions. Examples include: field programmable gate arrays (FPGAs), application-specific integrated circuits (ASICs), and digital signal processors (DSPs). Each of these devices has its advantages and disadvantages, but microprocessors are the best choice for small scale to very small scale embedded systems owing to their low power consumption and moderate to good computing capabilities.

- Transceiver. This is a transmitter–receiver that is used for communication purposes to send and receive data, and commands. The choice of communication means of WSNs is the radio-frequency. These sensor nodes usually use the industrial, scientific and medical (ISM) frequency bands. The ISM bands are shown in the Table 1.1 [15].

- External memory. Wireless sensor nodes usually use flash [8–14] memories owing to their small size and reasonable storage capacity, which is always increasing. Based on the requirement of the nodes, we can have a user and a program memory. The size of the external memory depends on the application.

- Power source. The sources of power consumption in the nodes are the node programming, sensing and data collecting, data processing, and data communication. Usually, most of the power is needed for transmitting data. Power is stored in the sensor nodes in the form of batteries. The cost of batteries has recently decreased drastically [7–9]. Typically these batteries are for one-time use.

In general, power sources are typically divided into primary and secondary sources. The primary sources cannot be recharged, where secondary sources have to be charged on a regular basis. The major factors of primary and secondary sources are: range, capacity, temperature, current depletion level, and self-discharge characteristics. Fuel cells are expected to come into use as power sources for the sensors of the WSNs [9–14]. Where secondary cells are employed, the charging source may be harvested from the cell's operational environment. The famous example of this is the harvesting of solar energy in order to charge a battery. However, there are other harvesting energy methods that can be used such as wind power, thermal energy, and vibration. In mechanical driven settings, harvesting a battery may not be needed as the harvested movement is constant, for instance in a pipeline. One popular method is solar systems, which necessitate some degree of installation to guarantee the best direction, especially at soaring elevations. We can obtain only about 25% efficiency from the best available silicon solar cell systems.

Some reported accomplishments were found with vibration harvesting used in industrial applications as well as from oil pipelines [12–14].

- Sensors. In general, sensors may be categorized into classes based on their operating principles: (a) physical sensors, (b) thermal sensors, (c) chemical sensors, (d) biological sensors, and (e) electromagnetic, optical, and acoustic sensors.

  Sensors are typically hardware devices that sense the data from the monitored environments and produce some response that is measurable in nature. An analog-to-digital (A/D) convertor is used for converting the analog collected data to the digital form to be processed further by the microcontroller. The sensors in wireless sensor nodes are typically very small sized microelectronic sensing devices which are equipped with a very limited supply of battery power. Examples of some commercial sensors include: BTnode, BEAN, COTS and DOT, MICA and KMote. Sensors can be placed in any kind of environment for days without any attention. The major challenge for a sensor is the life of the battery, which is limited. The battery has usually short life. Thus, schemes are needed to conserve as much energy as possible. If we envisage that such devices are deployed in the battle ground in enemy areas, we can clearly see that it is not possible to recharge or change the battery of these devices. It is true that we may be able to deploy these sensors in enemy territory by the use of aircraft/helicopter, but it may not be possible to invade the enemy territory just to replace the battery. The major source of energy consumption is the communication between the nodes. Moreover, nodes tend to coordinate with each other for some particular tasks [1–24].

1.3 Classification of sensor networks

Owing to the rapid progress in wireless sensor networks, a variety of applications with different needs have emerged. In order to deal with these changing requirements, there are many distinct network designs, in which protocols for distinct layers of the network have been implemented. Although there are many different ways to categorize the sensor network designs, here we show some of the essential differences in sensor networks.

- Data sink(s). One of the most crucial features of sensor networks is the characteristic of data sink(s). In some circumstances, the end user(s) may be entrenched inside the sensor network or may be mobile access points that gather data once in a while. This difference may be crucial, as efficient dispersed data storage methods may be effective in the latter case.
- Sensor mobility. Another classification of sensor networks may be based on the nature of the sensor being organized. Normally, sensors can be interpreted as being stationary; however, some recent sensor networks projects like ZebraNet use mobile sensor nodes. Moreover, in military applications, sensors may be placed on soldiers' bodies or clothes, or on unmanned aerial vehicles (UAVs) to communicate with an

organized sensor network. These sensors can manipulate protocols at the networking layer as well as for the services of localization with the feature of mobility.

- Sensor resources. Sensor nodes may differ with the availability of computer resources. It is apparent that memory and the conditions of processing should affect the implementation of protocols.
- Traffic patterns. Another important feature to be considered is the traffic that is generated in the network. In most of the event-driven applications, sensors may function for the bulk of the time, producing data traffic only when an event of significance is found, whereas in other applications such as environmental monitoring the data have to be produced constantly.

Wireless sensor network taxonomy can be based on the following dimensions [3, 8–13].

(1) Spatial resolution. This is measured in metric units such as centimeters, meters, or millimeters.
(2) Latency. Here, we can classify the network into categories such as negligible, moderate, or high.
(3) Coverage. In this regard we can have the following classes: partial, full, or redundant.
(4) Control. Classes here can be external, central, or distributed.
(5) Temporal resolution. This is usually measured in seconds.
(6) User types. In this case, we can have single, competitive, cooperative, and collaborative classes.
(7) Lifetime. Here, we can have simple with fixed duration, or complex with multiple phase-specific fixed durations.
(8) Bandwidth. This is an important criterion and characteristic. We can have episodic-small, episodic-large, continuous-small, and continuous-large categories. Units of bandwidth can be bytes/episode or byte/second.
(9) Sense of occurrence. We can identify: single discrete-target, multiple discrete-targets, and single distributed phenomena, and multiple distributed phenomena [1–24].

Others classify WSNs based on the following two concepts: (a) network organization or structure, and (b) node fairness and capabilities [13].

The authors in [12] devised a WSN application requirement taxonomy based on two application classes: precision agriculture, and wildfire management. For each dimension, they listed the class for each application. For instance, control is central for precision agriculture while it is distributed for wildfire management; or users can be single for precision agriculture or cooperative for wildfire management and so on.

1.4   Characteristics of wireless sensor networks

Wireless sensor networks have been recognized as one of the most vital technologies of this century. Inexpensive, smart devices with many on-board sensors networked through wireless links and the Internet and deployed in huge numbers present unique prospects for instrumenting and controlling homes, cities, factories, and the environment.

Moreover, networked sensors offer a new means for surveillance and other tactical applications. While sensor networks for various applications may be quite different, they share common characteristics.

Primarily, sensors are electrical, electronic, or electromechanical devices, even though other kinds of sensors exist. In general, a sensor is a type of transducer that converts an input to another, usually electrical, form. Sensors can be direct or paired. An example of a direct sensor is a thermometer or an electrical meter which indicates directly. A paired sensor uses an analog-to-digital (A/D) converter in order to convert an analog signal to a digital signal. Sensors are often used in applications such as medicine, industry, environment, robotics, and military. With the advances in material technology, more and more sensors are being built with Micro-Electro-Mechanic-Systems (MEMS) technology.

A good sensor/transducer should have the following main characteristics [16–24].

(1)   It should be responsive to the considered property.
(2)   It should be insensible to any other property.
(3)   It is desirable that the output signal of the sensor is exactly proportional to the value of the measured characteristic.
(4)   It should have a reasonable lifetime.
(5)   It should not consume much power.

A WSN is made up of hundreds or even thousands of nodes that use sensing devices (sensors) to observe different conditions and environments, such as motion, pressure, temperature, sound, vibration, pollution, levels of oxygen or carbon dioxide, traffic intensity and patterns, among many others, at different sites. In general, these devices are tiny and low-cost so they can be manufactured and deployed in large quantities. One major difference between traditional MANETs (Mobile Ad hoc NETworks) and WSNs is that WSNs often have strictly limited resources in terms of power, memory, computational power, and bandwidth. The sensor node is a self-contained unit equipped with a radio transceiver, a tiny microcontroller, and a power source that is usually a battery. The nodes dynamically self-organize their configuration based on different network circumstances. Owing to the limited life of batteries, nodes are built with power saving in mind and generally spend large amounts of time in the "sleep" mode or in handing out the sensor data. Hence, each sensor is equipped with wireless communication capability, and signal processing and networking abilities. The main functions of any WSN are sensing, communication, and computing [1–15]. One scheme to categorize wireless sensor networks is based on whether the nodes are separately addressable, and another is based on whether the data in the network are aggregated. For instance, the sensor node in a parking-lot network should be individually addressable, so that one can find out the spots of all free spaces. However, if a person wants to find out the temperature or pressure in a specific corner area of a room, then addressability may not be so important. The capability of the WSN to combine the gathered data can significantly decrease the number of messages needed to be sent through the network. In some situations, it is vital to send the signal by the sensor in a timely manner such as when it is needed to send a data alert signal to the police indicating that an intruder is trying to enter someone's house or office.

1.5   Challenges of wireless sensor networks

There are several challenges that face the progress of WSNs. Among these are [1–15] the following.

(1)   Scalability. Most nodes in intelligent sensor networks are stationary. Networks of huge numbers of nodes on the order of 10 000 or more are expected. This means that scalability is a crucial issue in designing or launching any new WSN because we like to see proportional improvement in performance as the size of the network is increased. The algorithms and protocols designed for WSNs should consider communication cost with respect to network size.

(2)   Power limitation. Since WSNs are often installed in remote areas such as deserts, forests, or military zones, their nodes are usually powered by batteries with limited life. Recharging such batteries may not be feasible. Given this constraint, the lifetime of any node is decided by the life of the battery powering it. As a result, the reduction of consumed power is vital. There are protocols and schemes that have been proposed to control power consumption by WSNs. These schemes are based on energy efficient MAC protocols, data aggregation, topology management, data compression, or intelligent use of batteries. Of course, using electronic devices and chips that consume less power is also a key design issue.

(3)   Self-organization. Given the fact that WSNs may be installed in hostile environments, it is essential that they are designed to be self-organized. Nodes may fail due to harsh environment or depletion of the batteries; therefore, the network must be able to periodically re-configure itself so that it can continue to function and new nodes can be added, if possible. Individual nodes may be disconnected from the network, but the major portion of the network must continue to function.

(4)   End objective. The ultimate objective of a WSN is not only communication; it has to detect and estimate certain events of interest. In order to enhance the detection and estimation capabilities, it is helpful to merge data from multiple sensors. Such a data fusion necessitates the transmission of data and control messages, which may put a limitation on the network design and structure. Furthermore, it is vital to distinguish between false data gathered and data reflecting a real emergency. For example, a high temperature in factory may indicate a real fire or may be due to sensing or processing errors.

(5)   Querying capability. In WSN environments, a user may need to make an inquiry of an individual node or a selected cluster of nodes, for information gathering in the area. Based on the degree of data fusion performed, it may not be practicable to send a huge volume of data over the network. As an alternative, different neighboring sink nodes can gather data from a given area and generate summary messages. An inquiry may be sent to the sink node closest to the preferred location.

(6)   Interoperability. With the impressive progress in sensing and communications technology, we start to see inexpensive, short-range radios, along with wireless networking devices and links. Of course, it is expected that WSNs will be widely deployed for all sorts of applications. Each node in the network may be equipped

with different sensors including seismic, acoustic, video camera, and infrared light, among others. Nodes may be configured in groups and they can synchronize with each other in a way that makes locally transpiring events be identified by the majority, if not all, of the nodes of the cluster. Such nodes will collaborate in order to make local decisions based on the data gathered by each node in the cluster. In such an arrangement, one node may act as the master node and the rest may act as the slaves.

(7)   Cost. An important issue in the cost of wireless microcontrollers is the size of memory needed. Designers of wireless sensor networks will expect to have access to a range of chips or wireless microcontrollers with optimized memory size to meet the needs of a variety of applications. Likewise, the need for larger applications development such as gateway devices, and third party network layer development, show that there is a need for a much larger memory size, greater than 250 kB in some cases.

(8)   Transmission time. One issue that is sometimes neglected is the amount of time needed to send the packets. Transmission time affects performance, quality of service, power consumption, and interference. It is necessary to have reliable data transmission and extended battery life in wireless sensor networks. We can improve the reliability of data transmission by using a small practical packet size since this gives the highest probability of a packet being delivered to the destination in the presence of interference. Extended battery life is obtained by minimizing the on time of the radio device, where most power is consumed. In general, a small packet size and occasional transmission can help to reach this goal in saving power.

(9)   Compression of data. Compressing sensor data before transmission can offer a key decrease in transmission time. In sensor nodes like gas level, temperature, pressure, and light level sensors, the transmission of data on transition or exception, instead of normal planned transmissions, is an efficient way to minimize network traffic. Moreover, having the ability to perform digital filtering or data compression at the sensor node is a valuable approach to minimize the data size as well as the rate of recurrence of transmissions.

(10)   Interference and environment. In general, interference from other nearby wireless networks such as Bluetooth or wireless LANs, should be addressed. Usually, this only presents a transitory state of interference to the WSN. For example, the capability of an IEEE 802.15.4 or ZigBee-based network to carry out automatic repeat will probably overwhelm any effect of interference from Bluetooth. Similarly, for WSNs employing occasional transmissions and for Bluetooth with frequency hopping, the probability of a frame collision is small. By utilizing collision avoidance schemes, wireless LANs (WLANs) can listen for a clear radio-frequency (RF) channel before they send data. However, under heavy traffic conditions in WLANs, we may get limited availability of the RF channel to the WSN due to the continuous state of interference. In such a situation, it is recommended for the WSN to be set on a different channel. Surrounding building structures also affect the RF environment. Steel reinforced concrete floors, stone

walls and analogous construction resources bring in high levels of attenuation as well as multipath fading. Similarly, the movement of persons or equipment considerably affects the signal level at any specific position. In general, effects of complex building structures can be alleviated by using additional router nodes in a mesh network that are installed to get around such obstacles.

(11)   Security. Owing to the characteristics of the wireless communication medium, there are various security challenges that face WSNs including eavesdropping, man-in-the-middle attack, spoofing, and distributed denial of service (DDoS). The worry for security in WSNs can be even larger than that in a traditional ad-hoc wireless network as, in many cases, the computational and energy-consumption limitations create barriers in the implementation of powerful and effective security solutions in WSNs. Therefore, advances in the design of security mechanisms in WSNs for protecting the confidentiality, availability and integrity, are essential for the proper operation of such systems [3–15].

The acceptance of the Advanced Encryption Standard (AES) with a 128-bit key length guarantees data integrity and resistance to hacking. An AES security scheme can be implemented in software, while a dedicated hardware encryption processor offers a better solution since this reduces software overheads and permits faster encryption/decryption operation. Clearly, this is essential for sensor nodes, which must spend the least time possible awake, as staying awake consumes a lot of the power of the node's battery. Furthermore making the AES encryption chip accessible to the application software facilitates a higher level of security [3].

1.6   Comparison between wireless sensor networks and wireless mesh networks

The major applications of WSNs are logistics, environmental monitoring, industrial supervision, intelligent buildings, and military applications. WSNs have limited data capabilities and power saving requirements when compared to mesh networks. Typically, WSNs necessitate only a few bits per second per day, on average; they are not used when latency is crucial. Researchers have tried VoIP over IEEE 802.15.4, which specifies the physical layer and media access control for low-rate wireless personal area networks, with limited success. The main source of power saving of WSNs is their short duty cycle, which can be lower than 1%.

The key general differences between WSNs and mesh networks are as follows.

(1)   The WSN traffic is less complex, not real, at low data rate.

(2)   The WSN traffic is, in general, application specific; hence node design is driven by application. This means that the nodes are not flexible.

(3)   Since the nodes in WSNs are inexpensive, they are less reliable than in a mesh network.

(4)   The WSNs have more nodes at a high density, and the radio range is shorter.

Even though there are differences between WSNs and wireless mesh networks, there are many similarities [1–24].

(1)   Basically, both do not require infrastructure; however, it is worth mentioning that mesh networks make use of the access points to enhance QoS and WSNs benefit from routers to enhance power consumption for edge nodes.
(2)   The two technologies are self-organizing networks.
(3)   Both have security and privacy weaknesses.
(4)   The two networks bring in a reliance of the network on node behavior.
(5)   Both benefit from clustering to get around scalability concerns.

It is interesting to point out that both WSNs and mesh networks are cooperative networks, which means that the MAC layer is decentralized and there is an aspect of fair contention for resources related to each node [3, 4, 9–14]. After showing the difference between WSNs and mesh networks, it is worth showing the difference of WSNs from radio-frequency identification (RFID) systems. The term RFID refers to the use of an object (usually called an RFID tag) that is applied to, or integrated into, a product, animal, or person for the intention of identification and tracking using radio waves. Various tags can be read from somewhat a few meters away and more than the line of sight of the reader. The overwhelming majority of RFID tags contain as a minimum two elements. One is an integrated circuit (IC) for saving and processing the data, modulating and demodulating a radio-frequency signal as well as other necessary operations. The second element is an antenna for getting and sending the signal. Table 1.2 shows the main differences between WSNs nodes, mesh networks nodes and RFID [1–12].

1.7   Summary

In this chapter, we have shed some light on the basic foundations of wireless sensor networks including components, structure, classifications and taxonomy, and character-istics. Fundamental related background has been given along with examples. We have

Table 1.2 Comparison between WSN nodes, mesh networks nodes, and RFID

Criteria	Wireless sensor node	Mesh network node	RFID
Size	Small	It can be handheld or larger	Embeddable
Cost	Low cost	Usually expensive	Trivial price for simplest RFID
Energy resources	Limited energy resources	High-quality batteries that are rechargeable	Typically, no power source is needed
Mobility	None or nomadic	Full mobility	None or nomadic
Radio range	Small	It can be large	Usually small to medium; powered RFID has a medium range
Processing power	Limited small processing	It cannot run large protocols such as TCP/IP	Very limited

elaborated on the challenges in operating and building wireless sensor networks that include: scalability, power limitation, organization, querying, capability, interoperability with other systems, cost, transmission time and speed, compression of data, interface with working environment, and security. Finally, we compared wireless sensor networks with wireless mesh networks and RFID systems.

References

[1] S. Dhurandher, S. Misra, M. S. Obaidat and S. Khairwal, "UWSim: a simulator for under-water wireless sensor networks," *Simulation: Transactions of the Society for Modeling and Simulation International, SCS*, Vol. **84**, No. 7, pp. 327–338, July 2008.

[2] S. Misra, K. Abraham, M. S. Obaidat and P. Krishna, "LAID: a learning automata based scheme for intrusion detection in wireless sensor networks," *Security and Communications Networks, Wiley*, Vol. **2**, No. 2, pp. 105–115, March/April 2009.

[3] M. S. Obaidat, P. Nicopolitidis and J.-S. Li, "Security in wireless sensor networks," *Security and Communications Networks, Wiley*, Vol. **2**, No. 2, pp. 101–103, March/April 2009.

[4] S. Misra, M. S. Obaidat, S. Sanchita and D. Mohanta, "An energy-efficient, and secured routing protocol for wireless sensor networks," in *Proceedings of the 2009 SCS/IEEE International Symposium on Performance Evaluation of Computer and Telecommunication Systems, SPECTS 2009*, pp. 185–192, Istanbul, Turkey, July 2009.

[5] S. Misra, K. I. Abraham, M. S. Obaidat and P. V. Krishna, "Intrusion detection in wireless sensor networks: the S-model learning automata approach," in *Proceedings of the 4th IEEE International Conference on Wireless and Mobile Computing, Networking and Communications: The First International Workshop in Wireless and Mobile Computing, Networking and Communications (IEEE SecPriWiMob'08)*, pp. 603–607, Avignon, France, October 12–14, 2008.

[6] S. K. Dhurandher, S. Misra, M. S. Obaidat and N. Gupta, "QDV: a quality-based distance vector routing for wireless sensor networks using ant colony optimization," in *Proceedings of the 4th IEEE International Conference on Wireless and Mobile Computing, Networking and Communications: The First International Workshop in Wireless and Mobile Computing, Networking and Communications (IEEE SecPriWiMob'08)*, pp. 598–602, Avignon, France, October 12–14, 2008.

[7] S. Misra, V. Tiwari and M. S. Obaidat, "Adaptive learning solution for congestion avoidance in wireless sensor networks," in *Proceedings of the IEEE/ACS International Conference on Computer Systems and Applications, AICCSA 2009*, pp. 478–484, Rabat, Morocco, May 2009.

[8] A. Swami, Q. Zhao, Y.-W. Hong and L. Tong (Eds.), *Wireless Sensor Networks: Signal Processing and Communication Perspectives*, John Wiley & Sons, 2007.

[9] Y.-C. Tseng, M.-S. Pan and Y.-Y. Tsai, "Wireless sensor networks for emergency navigation," *IEEE Computer*, Vol. **39**, No. 7, pp. 55–62, 2006.

[10] C.-Y. Chong and S. P. Kumar, "Sensor networks: evolution, opportunities, and challenges," *Proceedings of IEEE*, Vol. **91**, No. 8, pp. 1247–1256, 2006.

[11] S. Cheekiralla and D. W. Engels, "A functional taxonomy of wireless sensor network devices," *Proceedings of the 2005 International Conference on Broadband Networks Conference, BroadNets 2005*, Vol. **2**, pp. 949–956, 2005.

[12] R. MacRuairi, M. T. Keane and G. Coleman, "A wireless sensor network application requirements taxonomy," in *2008 IEEE International Conference on Sensor Technologies and Applications, IEEE Computer Society*, pp. 209–216, 2008.

[13] S. Methley, *Essentials of Wireless Mesh Networking*, Cambridge University Press, 2009.

[14] J. Zheng and A. Jamalipour, *Wireless Sensor Networks: A Networking Perspective*, John Wiley & Sons, 2009.

[15] P. Nicopolitidis, M. S. Obaidat, G. I. Papadimitriou and A. S. Pomportsis, *Wireless Networks*, John Wiley & Sons, 2003.

[16] M. S. Obaidat and J. W. Ekis, "An automated system for characterizing ultrasonic transducers using pattern recognition," *IEEE Transactions on Instrumentation and Measurement*, Vol. **40**, No. 5, pp. 847–850, October 1991.

[17] M. S. Obaidat and D. S. Abu-Saymeh, "Methodologies for characterizing ultrasonic transducers using neural network paradigms," *IEEE Transactions on Industrial Electronics*, Vol. **39**, No. 6, pp. 529–536, Dec. 1992.

[18] M. S. Obaidat, "On the characterization of ultrasonic transducers using pattern recognition techniques," *IEEE Transactions on Systems, Man, and Cybernetics*, Vol. **23**, No. 5, pp. 1443–1450, Sep./Oct. 1993.

[19] M. S. Obaidat, H. Khalid and B. Sadoun, "Ultrasonic transducers characterization by neural networks," *Information Sciences Journal, Elsevier*, Vol. **107**, No. 1–4, pp. 195–215, June 1998.

[20] M. S. Obaidat and H. Khalid, "Performance evaluation of neural network paradigms for ultrasonic transducers characterization," in *Proceedings of the IEEE International Conference on Electronics, Circuits and Systems*, pp. 370–376, Dec. 1995.

[21] M. S. Obaidat and D. S. Abu-Saymeh, "Performance comparison of neural networks and pattern recognition techniques for classifying ultrasonic transducers," in *Proceedings of the 1992 ACM Symposium on Applied Computing*, pp. 1234–1242, Kansas City, MO, March 1992.

[22] M. S. Obaidat and D. S. Abu-Saymeh, "Neural network and pattern recognition techniques for characterizing ultrasonic transducers," in *Proceedings of the 1992 IEEE Phoenix Conference on Computers and Communications*, pp. 729–735, April 1992.

[23] W. Dargie and C. Poellabauer, *Fundamentals of Wireless Sensor Networks: Theory and Practice*, John Wiley & Sons, 2010.

[24] K. Sohraby, D. Minoli and T. Znati, *Wireless Sensor Networks: Technology, Protocols, and Applications*, John Wiley & Sons, 2007.

2 Inside a wireless sensor node: structure and operations

As indicated in Chapter 1, a wireless sensor network (WSN) is a wireless computer network that consists of many dispersed independent sensor nodes to coordinate physical or environmental conditions, such as vibration, pressure, temperature, sound, motion, and pollutants, at different settings. WSNs are being used in a wide variety of industrial and civilian applications, which may include industrial process monitoring and control, smart and digital homes, smart cities, environment and habitat monitoring, health care applications, and traffic control, among many others.

In terms of structure, WSNs are a blend of small sensors and the actuators along with some general-purpose small processors to perform some limited computational processing [1–12]. The WSN consists of numerous low-power, low-cost and self-organizing sensor nodes. Gathering data or physical information from the physical surroundings is one of the main aims of any WSN [1–20].

This chapter is aimed at investigating the structure of a wireless sensor node including its architecture and operation mechanism.

2.1 Limitations in wireless sensor networks

The unfriendly and remote environments at which wireless sensor nodes are usually positioned, and the restricted computational power and energy, as well as limited storage space in the nodes, are the major parameters that dictate what protocols should be employed. For example, because of such limited resources, the schemes and protocols usually used to secure WSNs are light-weight solutions, and those used for routing, are more energy-aware and should need as little execution time as possible.

Currently the processor of a typical sensor node has a speed of about 8–16 MHz with 4–8 kB of RAM, and 128–256 kB flash memory. In general, the sensor nodes are of a heterogeneous nature. Owing to their different deployment types, the nodes are installed in different environments where they are heavily susceptible to physical damage. In addition to the node restrictions, the limitations are brought together by the sensor networks where a mobile network is inferior in physical infrastructure when compared to a wired network.

A WSN may carry two sorts of information.

- Traffic. This concerns all the user-to-user information. It can be of data, voice, or video type.
- Signaling. The WSN is required to carry other information for its own operation. Its purpose can be for maintenance, security, or traffic routing control, among others. This information is usually not noticeable from the user's point of view. It is, in general, meant for management and control. Several signaling categories may exist in a WSN.
  - (1) Per-trunk signaling (PTS). Signaling and voice elements are sent on the same facility; PTS necessitates the voice component to be fully built; even if the call cannot be completed.
  - (2) Common channel signaling (CCS). There are two split paths used for information movement; one for traffic, and the other for all related signaling information. Therefore, CCS allows the voice component to be assembled individually, which lets resources be saved. For instance, there will be no voice resources assigned to the call if the called number is busy [1–21].

In general, WSNs are often employed in critical tasks including military, industrial plan monitoring, forest fire monitoring, smart homes, and health care applications. Such applications have challenging security constraints that should be dealt with at the early stages of design, in order to focus on a security strategy that will take care of as many issues as possible. There are several security issues in WSNs that need to be addressed and analyzed in detail in order to design suitable security schemes and resolve such problems that come up in the sensor surroundings. Nevertheless, designing new security procedures is limited by the resources of the sensor nodes. Among the major limitations of WSNs are the following.

(1) Arbitrary topology. Most of the time, WSNs are deployed in hostile environments in a random manner, in the form of an airplane, boat, vehicle, horse, etc. Clearly, it is difficult to know the topology of the WSN ahead of time. This means it is not easy to store encryption keys on the nodes' memory in order to create encryption among a set of nearby nodes as the neighborhood cannot be recognized ahead of time. The issue here is to design key agreement schemes that do not necessitate precise nodes to be close to other nodes. Moreover, the schemes should not necessitate encryption keys to be saved on a sensor's memory before disposition. Proper key distribution techniques have to be devised along a flexible WSN design in order to deliver robust encryption keys [1, 10–21].

(2) Energy limitations. The power limitations of wireless sensor nodes are vital issues due to their tiny physical size and absence of wires. Typically, wireless sensor nodes are driven by batteries that have limited life. Because these sensor nodes are deployed in hundreds or thousands in unfriendly or remote locations, it is hard to change or recharge their batteries. It is important to note that power limitations deeply affect security mechanisms since encryption schemes bring in a communication overhead between the nodes due to the need to exchange many messages for key management functions [9–14].

(3)   Storage limitations. Since wireless sensor nodes have small memory, this can affect the storage of cryptographic keys. Based on the encryption scheme used, each sensor node will need to know a number of keys for each other node in the WSN in order to secure communication, and hence store the keys in the node's storage device. Nevertheless, since a WSN may have hundreds or thousands of nodes, we will end up needing a lot of memory, which may not be feasible. Clearly, the issue of storage restriction is essential when designing security protocols for WSNs so that a limited number of encryption keys can be used to offer reasonable defense from hackers [2, 7].

(4)   Limited computational power. A wireless sensor node is small in size with limited computational capabilities. Given the fact that these sensor nodes have limited power, therefore, their computational power is limited as well. Keep in mind that most of the power in WSNs is spent in communication rather than in computation. Hence, the power for computation is still more limited than the total amount of power. The rustication of the computational power limits the use of some powerful routing protocols and some of the robust cryptographic schemes such as the RSA public key technique. As an alternative, routing schemes that are energy-aware and symmetric encryption techniques are employed in WSNs as they do not require a lot of computational power compared to RSA or other asymmetric encryption algorithms. Nevertheless, asymmetric encryption algorithms do not support features such as digital signatures. This means that we have here another challenge for researchers, which is to design suitable schemes to verify trust amongst the nodes taking part in the communication [1, 17–25].

(5)   Unfriendly environment. Wireless sensor networks are usually installed in inaccessible or intimidating environments such as jungles and fighting zones. In such scenarios, the wireless sensor nodes cannot be guarded from physical attacks as any person may reach the site where they are installed. An opponent may well capture/control a sensor node or even launch his own malicious nodes within the WSN. The adversary may compromise the available critical data/information. Such compromised sensitive information collected by the opponent may be used for illegitimate purpose. Clearly, it is essential that the designers of such networks pay a lot of attention to this issue, because if a node is compromised, then the opponent can get critical information stored or communicated to the node [3–6, 9–14].

There are distinctive features for WSNs that make them unique. The realization of a protocol for them must take into consideration the properties of ad hoc networks, in addition to the following.

(1)   Lifetime restrictions due to the limited energy provisions of the nodes in the network.

(2)   The communication is unreliable due to the nature of the wireless transmission medium.

(3)   The need to be small and with or without a human intervention.

(4)   The need for self-configuration and fault tolerance.

2.2   Design challenges

There are several challenges in the design of wireless sensor networks. Among these are the following.

(1)   Flexibility and redundancy. The WSN should be designed so that if a node breaks down or loses its power then the remainder of the WSN should continue its operation without interruption. The desirable protocols should be adaptable to such failure or link congestion and alternative links must be created without much extra effort. Moreover, suitable schemes should be considered so as to bring up-to-date topology information right away after the setting changes in order to lower power expenditure.

(2)   Scalable and adaptable structural design. A WSN must be flexible enough to allow expansion. The system protocols should be designed so that adding more nodes will not affect routing and clustering operations. The network protocols should be adapted to the new topology and act as anticipated, and this addition should provide proportional improvement in performance. In addition, adding more nodes to the wireless sensor network implies that extra communication messages may need to be exchanged. Of course, when adding new nodes to the WSN, we should make sure that the optimum minimum number of communication messages is needed between the nodes, which will end up in a reduction in the computational power needed, reduction in congestion and also saving in the life of the nodes' batteries.

(3)   Unreliability of the wireless transmission medium. Because WSNs can be deployed in various environments, the requirements of each application are different from each other. It is important to note that the wireless transmission medium is unreliable and can be significantly influenced by atmospheric noise, interference, and other impairments such as scattering, reflection, diffraction, hence the signal attenuates and bit error rate increases. It is worth noting that a hacker can purposely hamper and create noise to influence the communication operation in the WSN.

(4)   Real time. Since almost all operations and tasks in WSNs deal largely with real-world environments, actions and data must be sent in real-time. The overwhelming majority of WSN protocols do not take into concern the real time nature of the operations. There are many functions that need to be performed in real time like data fusion, transmission of the data, event and target detection, and many others. Keep in mind that it not only important to develop the protocols for a real-time WSN operation; it is more important to make sure that such protocols are efficient and address the real-time nature of such operations.

(5)   Security and privacy. Wireless sensor nodes have restricted power and storage space and this may have an impact on their communication capabilities and security aspects. Since sensor networks are deployed in remote and hostile environments, this make them susceptible to physical attacks. In order to have a secure WSN system, each component must be integrated with security in mind. One of the key issues is to protect the WSN links from tampering and eavesdropping. Owing to the limited resources in WSNs, security schemes used should not be complex

otherwise they will not work efficiently due to the restricted capabilities especially limited power in the nodes. That is why light-weight security and routing schemes are usually adopted for WSNs, but they have to have acceptable performance aspects.

2.3   Hardware architecture

Each node in a WSN is usually equipped with a radio transceiver, a small microcontroller, a storage space, and power source, which is typically a battery. The size of a sensor node may vary in size: it can be as small as a peanut and as big as a soda can. As for the cost of the node, it also varies from tens of cents to hundreds of dollars, depending on the required operations in the node. Of course, size and cost constraints on sensor nodes result in corresponding constraints on node's resources including memory, I/O, speed, and power.

The major elements of a wireless sensor node are: (1) transceiver, (2) microcontroller, (3) memory device, (4) power source, and (5) sensing element(s). A brief description of each of these components is given below.

(1)   Transceiver. In general, sensor nodes utilize the industrial, scientific and medical (ISM) frequency band, which is available free of charge and does not require a license. There are different choices of wireless transmission media including: infrared, radio-frequency, and optical fiber. Infrared medium is inexpensive; however, it requires line of sight and cannot penetrate opaque objects and walls. Moreover, it is sensitive to atmospheric situations. Infrared does not require antennas, so it has limited broadcasting capacity. Communication based on radio-frequency (RF) is the most-often used means of communication in WSNs. Typically, WSNs use the communication frequencies between 433 and 2.4 GHz. The term transceiver signifies the combined functions of transmitter and receiver. The modes of operation of a transceiver are: transmitter, receiver, and idle/sleep modes. It is worth noting that radios operating in the idle mode of operation consume power roughly equal to that used in receiver mode. Hence, it is recommended to turn down the radios instead of running them in the idle mode when not sending or receiving data/information. Moreover, a lot of power is spent switching from the sleep mode to the transmit mode in order to send a packet [1–5, 20, 23].

(2)   Microcontroller. The microcontroller performs important tasks needed to have proper operation of the entire WSN. These tasks include processing of data and controlling of the operations of other elements in the sensor node. In general, any process/CPU with some I/O and storage capabilities can be used as a controller. Other possibilities include microcontrollers, general-purpose microprocessors, desktop computers, laptop computers, digital signal processors, field programmable arrays, and application-specific integrated circuits. It is found that microcontrollers are the preeminent alternatives for embedded systems. Owing to their flexibility to

hook up to other gadgets, power expenditure is reduced as such devices can be set to the sleep mode while part of controller may stay active [1–23].

Although general-purpose microprocessors may be used in WSNs, most designers do not utilize them in their designs because their power consumption is greater than the microcontrollers'. Clearly, microcontrollers are often used in WSNs. Digital signal processors (DSPs) may be used; however, they are more expensive and complex for WSNs environments. Field programmable gate arrays (FPGAs) may be used and reconfigured according to the WSN requirements; however, this process takes time and power. This means FPGAs are usually not used for WSNs. The application-specific integrated circuits (ASICs) provide needed functions using hardware; however, microcontrollers offer these functions using software means, which makes it less expensive and more flexible.

(3) Memory device. A wireless sensor node should have sufficient memory space to perform the needed tasks. The kinds of memory include on-chip memory of the microcontroller and an external flash memory. Flash memories are employed due to their cost-effectiveness characteristics; they offer high capacity at low cost. It is worth noting that memory requirements are application dependent. The required memory type can be divided into two major classes: (a) data memory, and (b) program memory.

(4) Power source. In WSNs, when a sensor node collects data/signal, it consumes power. Moreover, power is consumed even more for communication and data-processing operations. Power is typically stored in batteries or capacitors. There are two major types of batteries that are used in sensor nodes: chargeable and non-rechargeable batteries. Moreover, batteries can be divided into classes depending on the electrochemical material used for electrode, including nickel–cadmium (NiCd), nickel–zinc (NiZn), nickel metal hydride (Nimh) and lithium-ion. Currently, most sensors are designed so they can recharge their battery from the Sun, heat or movement). It is noted that most WSNs employ power-saving policies in order to extend the life of the batteries in their nodes. Among such schemes that are used are the dynamic power management and dynamic voltage scaling schemes. The first technique shuts down parts of the sensor node that are not presently active. The second scheme changes power levels based on the unpredictable workload. This is performed by changing voltage besides the frequency, which allows achieving quadratic decrease in the power consumption [12–13].

(5) Sensors. Sensors are deemed to be as a vital part of any WSN. They are basically hardware devices that generate a signal proportional to the event or condition being monitored or measured. Examples of such events or conditions include: temperature, pressure, movement, and levels of CO_2 and other pollutants, among others. The sensed signal is typically converted to digital form using analog-to-digital (A/D) converters as the microcontrollers can only process digital data. Desired characteristics of a sensor node include small size, low power consumption, being able to work in high volumetric densities, being adaptive to the environment, and being independent and able to work unattended. Sensors are

usually small, and can be equipped with very limited power source. In this context, we can identify the following major categories of sensors.

- Active sensors. This category of sensors actively surveys the surroundings, for instance, a radar sensor or a seismic sensor that produces shock waves.
- Passive and omnidirectional sensors. These sensors monitor the data without really affecting the environment by active probing. Normally, they are self-powered. Power is required only to magnify their collected analog data. Further, these sensors pick up the sensed data from all directions; they are not directional.
- Passive and narrow-beam sensors. Sensors of this category are passive; however, they have distinct view of direction of measurement. Usually devices such as cameras are used. In addition to these units, a wireless sensor node may be equipped with extra units such as a global positioning system (GPS) device, and a motor to move the sensor node in specific directions, among others. Of course, such components must be built into a small module with low power expenditure and low fabrication cost.

In any WSN, the processing unit consists of a microcontroller/microprocessor with the required simple peripherals including a memory. Examples include: Strong ARM 1100, Atmega128L, Intel PXA255 (scalable from 100 to 400 MHz), MIPS SH-4 processor, Motorola HCS08, and Texas Instruments MSP430 microcontroller [25].

2.4   Operating systems and environments

Operating systems used in wireless sensor nodes are usually much less complex than these used in desktop, laptop or workstation computers, or even general-purpose microprocessor systems. This is because WSNs are meant for special applications that are typically not interactive. Owing to this characteristic, operating systems for WSNs do not need to have features for user interfaces. Moreover, resources in WSNs in terms of memory size and memory-mapping hardware support are limited, which makes the implementation of schemes such as virtual memory almost impossible to accomplish.

Since WSNs are relatively similar to typical embedded systems, therefore, it is possible to use embedded operating systems such as uC/OS for WSNs. Nevertheless, these operating systems are usually designed with real-time characteristics. Keep in mind that WSNs usually do not have real-time support whereas embedded systems do.

The first operating system that was specifically designed for WSNs is the TinyOS. It was designed as an event-driven programming model; it is not based on the multithreading model. The programs in TinyOS are arranged into *event handlers* and *tasks* through run-to-completion semantics. After an outer event such as an arrival of a packet or arrival of sensed data, the TinyOS calls the suitable event service routine to run the event, which usually can be a read or write task. Service-handling subroutines can schedule tasks that are planned by the TinyOS kernel some time later. The programming language used to write system and application programs under TinyOS is called nesC, which is a sort of C programming language [20–22].

Other operating systems that permit programming in C include SOS, Contiki, BTnut, Nano-RK, and LiteOS. SOS is basically an event-driven-based operating system that supports loadable modules. It also concentrates on supporting dynamic memory management. The Contiki kernel is also event-driven, similar to TinyOS, however, it supports multi-threading and includes proto-threads, which offer a thread-like programming abstraction except that they require small memory overhead. The BTnut is founded on cooperative multi-threading base and ordinary C code. As for Nano-RK, it is a real-time resource kernel that allows fine-grained control. Finally, LiteOS offers UNIX-like construct and supports C [20–24].

In general, middleware is the software that links various system software modules, modules or application programs. Enormous effort has been exerted by different researchers and companies in order to design efficient middleware for WSNs. The main schemes that can be employed to design middleware for WSNs are based on a distributed database, mobile agents, and event-based [5].

In the distributed database approach there are different flavors that treat the WSN as a distributed database where clients can transmit SQL-like queries in order to request the network to do a specific sensing operation. An example of this category is the TinyDB [1–2, 5, 20], which supports only "virtual" database table sensors, where each column corresponds to a specific type of sensor such as for pressure, movement, temperature, pollutants, or other types of input information such as remaining battery power information, and a node's identification. TinyDB employs a decentralized technique in which each node possesses its own inquiry processor that collects a sensor's data while moving to the user. Running an inquiry is done by following the steps explained next. A spanning tree of the network rooted at the user's device is built while the network topology varies, employing a controlled flooding scheme. The messages from floods are utilized to approximately synchronize time between the nodes. Then, a query is transmitted to all the WSN nodes by transferring it up the tree from the root to the leaves. Throughout this procedure, a time list is created in a way that a parent and its offspring concur on a time period when the parent may pay attention to the data coming from its children. At the beginning of each period, the leaf nodes get a new table row by sensing out their neighboring sensors. After that, they select the criteria for this row. If these criteria are satisfied, then a partial state trace, which has required data, is created. Then the partial state trace or record is transmitted to the parent through the planned time period. Any parent can listen to any partial state traces/records from its children through the scheduled period. After this, the parent continues by reading out its sensors, selecting the criteria, and then producing a partial state trace/record if desired. Afterwards, the parent gathers its partial state trace and the marks received from its children, which generate a new incomplete state trace. The latest partial state trace/record is then transmitted to the parent's parent during the planned period. Such a procedure continues until the root of the tree, is reached. At the root, the eventual partial state trace/record is assessed in order to get the query's outcome. It is worth noting that the entire process is repeated for each period [5].

The mobile agent approach has been motivated by mobile code and mobile agents [5–7]. In this case, the WSN is supposed to inject a program into the network that can gather local sensor data, or duplicate itself to additional sensor nodes, and then converse with these distant copies. As an example of mobile agents, let us consider SensorWare [26]. In this example, programs are described using the Tool command language (Tcl). This is a scripting language in which all operations are considered commands including language structures, and is commonly employed for rapid prototyping scripted applications, graphical user interfaces (GUIs) as well as testing applications. It is employed on embedded systems environments both in its full form and in several other small-footprinted forms. The functions that are particular to SensorWare are realized as a set of extra procedures in the Tcl interpreter. Among the famous extensions are the query, wait, and send commands. A query acquires a sensor name and a command as factors.

The address of each node consists of the node ID, a script name, and extra identifiers in order to differentiate copies of the similar script. The replicate command acquires one or more sensor node addresses as factors and generates copies of them carried on script on the described remote nodes. In SensorWare, the occurrence of an asynchronous action such as reception of information is signified by a particular event. The wait command anticipates a set of such event names as factors and postpones the running of the script pending occurrence of the specified events. Because of its recursive replication of the script to all nodes in the WSN, the user will ultimately wind up with a message that has the most volume among all nodes of the WSN [20–26].

The third approach is the events scheme, which is based on the concept of events. Here, the application specifies interest in certain state changes of the real world, and as soon as such an event is detected, a sensor node transmits an event notice to concerned applications, which may indicate specific patterns of events. To specify a complex event, we need information such as a collection of nodes interested in this complex event, event identifier, a discovery period describing the time duration of concern, a time frame, a set of basic events, and a detection range detailing the geographical area of concern, among others. After the complex event is recognized, it is sent to the concerned sensor nodes. Moreover, the WSN system in this case, contains different real-time features such as cut-off time for sending info of events as well as event validity periods.

In addition to the three approaches mentioned, there are other techniques to programming WSNs. Among these, we mention the MagnetOS [2, 20–21], which is basically a distributed virtual machine. In MagnetOS, an object-oriented byte-code program is repeatedly divided and assigned to the nodes of a WSN including communication overhead. Another example is the Enviro Track [1–2, 20–22], which is middleware mainly meant for observing portable physical devices in the surroundings.

2.5   Examples of sensor nodes

Among the early WSNs is the work at the University of California–Berkeley and the Intel Berkeley Research Laboratory, which showed a self-organizing WSN composed of over 800 small low-power sensor nodes. In this project, a 1×1.5 inch wireless sensor node was demonstrated. The nodes used were tiny in size. The core was a 4 MHz low-power microcontroller (ATMEGA 163) with 16 kB of flash instruction memory, 512 bytes of SRAM, A/D convertors, and basic I/O interfaces. A 256 kB EEPROM was used as secondary storage, and sensors, actuators, and a radio network were used as the I/O subsystem. The network employed a low-power radio (RF Monolithic T1000) running at 10 kbps. Any node in the system had four sensors: light, temperature, battery level, and radio signal strength sensors and it could trigger two LEDs, direct the signal power of the radio, send and receive signals. Through adjusting the signal level, the radio cell size could be changed from a fraction of a meter to tens of meters based on physical environment. The whole system consumed about 5 mA in the active mode. The radio and the microcontroller spent about as much power as only one LED consumes. While in the passive mode, they consumed only a few microamperes; this is whilst still looking for radio or sensor stimuli, which should instigate them to get up. The system used a new operating system, designed to operate in resource-limited environments where data and control need to be passed quickly between different sensors, actuators, and the network. As previously described, the TinyOS is an element-based, event-driven operating system framework that begins at a few hundred bytes for the scheduler and expands to a few kilobytes.

2.6   Effect of infrastructure on the performance evaluation of WSNs

The structure of a WSN can be viewed as a three-layer system.

(1)   Infrastructure. This refers to the physical sensors' properties and abilities, number of sensors, and their launching approach.
(2)   Networking protocol. This is in charge of dissemination of the sensed data by creating and maintaining paths between the sensors and the observer(s).
(3)   Application. This is responsible for explaining the observer interests into explicit network-level actions. Optimizations across the three levels are feasible in order to enhance the overall performance of the network.

In general, the infrastructure of a WSN refers to sensors' abilities, the number of sensors used, and the scheme used to launch them. We will discuss each of these in turn.

(1)   Sensor abilities. A WSN sensor usually contains five elements: sensing hardware, embedded processor, transceiver, memory, and battery. Such elements influence the overall performance of the sensor and eventually the WSN's performance. For instance, the size of the memory influences the buffering space at the sensors and the capability of the WSN to deal with transient bursts of traffic. Moreover, the battery type and size dictate the quantity of energy existing at the sensor and influence the life of the WSN. The embedded processor capacity dictates the level of optimization that is feasible at the sensors without bringing in extra loss of power or unbearable amount of delay. Furthermore, the properties of the transceiver dictate the transmission range of the WSN as well as the bandwidth of the

transmission link. There is no doubt that enhancing the properties of whichever of these elements will result in growth in the overall cost of the sensor. This means that within the on-hand budget for the WSN, the developer should make a decision on whether to spend money by using a huge number of affordable sensors, or a lesser number of costly, high-performance sensors [2–6].

(2)   Number of sensors. It makes sense to say that increasing the number of sensors installed in the field must produce enhancement in the WSN performance. For example, the precision of sensing will be enhanced as more sensors are deployed. More energy in the WSN will increase as more nodes are included. Moreover, the new added sensors to the WSN will provide more paths and links in the system and this will enhance its reliability. Nevertheless, adding more sensors to the network will create more traffic and more messages going from node to node. If this extra traffic exceeds the capacity of the WSN, then congestion and collision may occur, which will degrade the overall performance of the WSN.

(3)   Schemes of deployment. This refers to the schemes used to distribute sensors in the field. Examples of these techniques include: (a) a planned scheme, where the sensor deployment process is planned in order to offer high sensor density in regions where the occurrence is focused; (b) random deployment, where sensors are scattered in the field using the uniform distribution within the field; and (c) the regular scheme, where sensors are deployed using some regular geometric topology in the sensor field such as a grid [2, 10–15].

The major performance evaluation metrics of interest to WSNs are given below.

(1)   Delay. This refers to the time needed to get the samples to the analyst. There are several reasons for this latency, including WSN congestion, and the duty cycle of the sensors. In the case of real-time traffic such as real-time voice or video, delays in reporting the status may affect the correctness of the data and accuracy of the operation. The exact semantics of latency are application dependent.

(2)   Correctness. This metric refers to the accuracy of measurement made at the sensor. It can also be particular to a physical transducer and environment. It is expected that multiple samples will be gathered from various sensors and such samples should be merged properly in order to produce a further accurate approximation of the place of the occurrence. In this regard, it is feasible to give some sensors more say in the estimation due to their proximity to the site of occurrence. In general, there is a compromise between accuracy, latency, and energy efficiency. The specified infrastructure must be adaptive in such a way that the application gets the required accuracy and delay with the least energy being spent. The application can either request more recurrent data distribution from the same nodes or it may direct data distribution from other nodes using the same rate.

(3)   Goodput. This metric is a measure of the ratio of the total number of packets obtained by the observer to the total number of packets transmitted by all the sensors.

(4)   Scalability. This metric refers to whether increasing the number of nodes will provide proportional improvement in the overall performance of the system.

For large WSNs, it is expected that concentrating interactions throughout the hierarchy will be significant for guaranteeing scalability.

(5)   Energy effectiveness. The energy efficiency of the network can be measured by using various methods such as the consumed energy within the network. Since sensor nodes are run using batteries, the required protocols should be energy efficient in order to extend their lives, which means extending the life of the WSN. The WSN lifetime can be quantified by broad metrics like the time till 50% of the nodes pass away. Another way of measuring the lifetime is by using application-directed metrics like the network discontinuing offering the application with the needed data about the events or phenomena [1–5].

(6)   Fault tolerance. This refers to the degree at which the network can perform properly even when it loses some nodes for one reason or another, such as physical damage or running out of power. Replacing sensors is very difficult; therefore, the WSN should be fault tolerant so that non-disastrous breakdowns are veiled from the application. We can accomplish fault tolerance in WSNs by data replication [1–2, 23–25]. Nevertheless, data duplication itself necessitates energy and may congest the WSN. There is a trade-off between data duplication and energy efficiency. The data that have greater priority according to the application may be duplicated for fault tolerance, while the other data may not have to be duplicated.

2.7   MEMS technology

A micro-electro-mechanical system (MEMS) is important technology for making little, inexpensive, and low-power sensor nodes. While the electronics are manufactured using integrated circuit (IC) such as CMOS, Bipolar, or BICMOS processes, the microme-chanical elements are made using compatible micromachining processes that selectively engrave away parts of the silicon wafer or add new structural layers to produce the mechanical and electromechanical devices [11, 27].

MEMS-based sensor devices provide a periphery that can sense, treat, and/or direct the contiguous environment. In general, MEMS-based sensors are a category of devices that are constructed using very small electrical and mechanical elements on a single chip. Such sensor is a vital component in wireless devices, computer peripherals, hard-disk drives, and smart portable electronic devices such as cell phones. The key benefits of MEMS are low cost, low power requirements, small size, integration, and high perform-ance [1, 11–14].

It is worth mentioning that microelectronic integrated circuits (ICs) are considered the core of such systems and MEMS supplements this decision-making mechanism to allow microsystems to sense and manage the environment. Sensors collect data from the environment through gauging electrical, mechanical, thermal, biological, chemical, optical, and magnetic events. The electronics then take care of the information collected from the sensors and, through some decision-making means, command the actuators to respond by locating, adapting, and filtering, hence, managing the environment for some desired effect. Since MEMS devices are built by using batch manufacturing schemes like those used for integrated circuits, unparalleled degrees of functionality, dependability, and density can be set on a small IC chip at a lower cost [9–11].

The MEMS-based sensors are based on micromachining schemes that have been developed to make microscale mechanical elements, which are managed electrically. The techniques that are usually used to manufacture these devices include planner micromachining, bulk micromachining, and surface micromachining. The majority of micromachining CPUs begin with a substrate around 100 μm thick, often made of silicon, other microcrystalline semiconductors or quartz, on which several subsequent procedures are followed. These may include thin-film deposition, etching, oxidation, photolithography, machining, electroplating, and wafer bonding. By combining diverse elements together into a particular process, the volume of the sensor node can be considerably reduced. Different methods can be used to carry out post-process micromachining on foundry CMOS. With the MEMS technology, several elements of a sensor node can be reduced in size, for example, sensors, power supply units, and communication blocks, which may lead to a considerable decrease in the cost due to batch fabrication, as well as in power expenditure [1, 9].

2.8 Hardware platforms

By using the MEMS technology, the volume and price tag of sensor nodes have decreased considerably. In order to attain low power expenditure, it is essential to include awareness and energy optimization in the hardware structure for sensor networks. Advances in low-power circuit design have produced efficient ultra-low-power microcontrollers and microprocessors that are being used as sensor nodes for WSNs. Moreover, power consumption can be further minimized by using dynamic power-management techniques. In the active mode of operation, we can also have additional power savings by using dynamic voltage-scaling schemes. This latter technique has been shown to be more efficient than shut-down-based power-management techniques [1–25].

Platforms for sensor node hardware can be categorized into three key classes: (a) system-on-chip (SoC) sensor nodes, (b) augmented general-purpose personal computers, and (c) dedicated sensor nodes [2–14]. A brief overview of all of these categories is given next.

2.8.1 System-on-chip (SoC) sensor nodes

This category contains BWRC PicNode and Smart Dust, which are mainly based on MEMS, CMOS, and RF technologies. It aims at having very low power and tiny footprint with specific sensing, communication, and computation means.

2.8.2 Augmented general-purpose personal computers (PCs)

In this category, there are a variety of low-power embedded PCs and personal digital assistants (PDAs), which usually execute off-the-shelf commercial operating systems such as Win CE, Linux, or real-time operating systems, and employ standard wireless schemes such as ZigBee, Wi-Fi IEEE 802.11b and IEEE 802.11g, and Bluetooth. This platform offers more processing power than the dedicated sensor and system-on-the-chip sensor nodes, and hence can integrate a wealthier set of protocols, popular programming languages, middleware, application programming interfaces (APIs), and off-the-shelf commercial software systems. Nevertheless, they demand more power [1–20, 23, 27].

2.8.3　Dedicated sensor nodes

This group contains the MIT µAMP, Berkeley mote family, and the University of California–Los Angeles (UCLA) Medusa family. The Berkeley motes are very popular due mainly to their open source software development, and their commercial availability. This category usually uses commercial off-the-shelf chips. Nodes in this category are characterized by tiny low-power processing and communication as well as a straightforward sensor interface [4, 7].

2.9　Software platforms

In this context, we can describe a software platform as the operating system, which gives a set of applications such as file organization, memory allotment, task scheduling, and networking. Moreover, it may be the language platform which offers a library of elements to programmers. There are many software platforms for WSNs such as Mote, TinyOS, and nesC, among others [20–24].

2.10　Summary

This chapter dealt with the hardware and software aspects of the node of any wireless sensor network. We began by addressing limitations in WSNs such as arbitrary topology, energy limitation, storage limitation, limited computational power, and an unfriendly environment. We then investigated the design issues, which include flexibility and redundancy, scalability and adaptability of structural design, unreliability of wireless transmission medium, issues of real-time requirement, security, and privacy. Then we reviewed hardware architecture components which include transceivers, microcontrollers, memory devices, power sources, and sensors. After this we discussed operating systems of wireless nodes and gave some examples of these. We followed this by presenting examples of sensor nodes as well as the effect of infrastructure on the performance of WSNs. Finally, MEMS technology was briefly investigated as an important technology used to manufacture sensors, and its role in building little, inexpensive, and low-power sensor nodes.

References

[1] J. Zheng and A. Jamalipour, *Wireless Sensor Networks – A Networking Perceptive*, Hoboken, NJ: Wiley, 2009.

[2] S. Tilak, N. B. Abu-Ghazaleh and W. Heinzelman, "A taxonomy of sensor network communication models," *Mobile Computing and Communication Review*, Vol. **6**, No. 2, pp. 28–36, April 2002.

[3] F. Zhao and L. Guibas, *Wireless Sensor Networks: An Information Processing Approach*, San Francisco, CA: Morgan Kaufmann, 2004.

[4] G. J. Pottie and W. J. Kaiser, "Wireless integrated network sensors," *Communications of the ACM*, Vol. **43**, No. 5, pp. 51–58, May 2000.

[5] S. Hadim and N. Mohamed, "Middleware challenges and approaches for wireless sensor networks," *IEEE Distributed Systems Online*, Vol. **7**, No. 3, pp. 1–13, March 2006.

[6] R. Barr, J. C. Bicket, D. S. Dantas *et al.*, "On the need for system-level support for ad hoc and sensor networks," *Operating System Review*, Vol. **36**, No. 2, pp. 1–5, 2002.

[7] K. Römer and F. Mattern, "The design space of wireless sensor networks," *IEEE Wireless Communications*, Vol. **11**, No. 6, pp. 54–61, Dec. 2004.

[8] A. Perrig, J. Stankovic and D. Wagner, "Security in wireless sensor networks," *Communications of the ACM*, Vol. **47**, No. 6, pp. 53–57, 2004.

[9] R. F. Pierret, *Introduction to Microelectronics Fabrication*, Menlo Park, CA: Addison-Wesley, 1990.

[10] S. D. Senturia, *Microsystems Design*, Norwell, MA: Springer, 2001.

[11] A. E. Franke, T.-J. King and R. T. Howe, "Integrated MEMS technologies," *MRS Bulletin*, Vol. **26**, No. 4, pp. 291–295, 2001.

[12] V. Raghunathan, C. Schurgers, S. Park and M. Srivastava, "Energy-aware wireless microsensor networks," *IEEE Signal Processing Magazine*, Vol. **19**, No. 2, pp. 40–50, March 2002.

[13] L. Benini and G. DeMicheli, *Dynamic Power Management: Design Techniques and CAD Tools*, Norwell, MA: Springer, 1997.

[14] A. Chandrakasan and R. Brodersen, *Low Power CMOS Digital Design*, Norwell, MA: Springer, 1996.

[15] T. Pering, T. Burd and R. Brodersen, "The simulation and evaluation of dynamic voltage scaling algorithm," in *Proceedings of the 1998 International Symposium on Low Power Electronics and Design, ISLPED '98*, pp. 76–81, Monterey, CA, 1998.

[16] A. Savvides and M. Srivastava, "A distributed computation platform for wireless embedded sensing," in *Proceedings of the 2002 International Conference on Computer Design, ICCD02*, pp. 220–225, Sep. 2002.

[17] A. Chandrakasan, R. Min, M. Bhardawi, S.-H. Cho and A. Wang, "Power aware wireless microsensor systems," in *Proceedings of the 2002 International European Solid-State Device Research Conference, ESSDERC'02*, pp. 47–54, Florence, Italy, 2002.

[18] F. Yao, A. Demers and S. Shenker, "A scheduling model for reduced CPU energy," in *Proceedings of the 1995 Annual Symposium on Foundations of Computer Science, FOCS'95*, pp. 374–382, October 1995.

[19] J. Rabaey, J. Ammer, J. da Silva, D. Patel and S. Roundy, "Picoradio supports ad hoc ultra-low power wireless networking," *IEEE Computer Magazine*, pp. 42–48, July 2002.

[20] P. Levis and D. Culler, "Mote: a tiny virtual machine for sensor networks," in *Proceedings of the 2002 International Conference on Architectural Support for Programming Languages and Operating Systems, ASLOS'2002*, pp. 85–95, San Jose, CA, 2002.

[21] E. Cheong, J. Liebman, J. Liu and F. Zhao, "TinyGALS: a programming model for event-driven embedded systems," in *Proceedings of the 2003 Annual ACM Symposium on Applied Computing, SAC 2003*, pp. 698–704, Melbourne, FL, March 2003.

[22] http://webhosting.devshed.com/c/a/Web-Hosting-Articles/Wireless-Sensor-Networks-part2-Limitations/1/

[23] http://en.wikipedia.org/wiki/Wireless_sensor_network

[24] http://webs.cs.berkeley.edu/800demo/

[25] http://www.cse.unsw.edu.au/~sensar/hardware/hardware_survey.html

[26] http://www.sensorwaresystems.com/

[27] http://www.sensorsportal.com/HTML/DIGEST/E_24.htm

3 Wireless sensor network applications: overview and case studies

Wireless sensor networks (WSNs) have fascinated both the research and development communities. Applications of WSNs have mushroomed in both civilian and military domains. The growth of wireless sensor networks was originally motivated by military applications; however, WSNs are now used in all kinds of civilian and industrial applications [1–20]. Currently, the overwhelming majority of WSNs measure scalar tangible phenomena such as humidity, pressure, temperature, movement, and pollutants. Typically, most WSNs are built for delay-tolerant and low-bandwidth applications. Hence, most research efforts have concentrated on this latter paradigm, which is often called terrestrial sensor networks. In any WSN application the main challenge is the lifetime of the network. A WSN structure includes a gateway that offers wireless connectivity back to the wired/fixed network.

In nearly all applications of the WSNs, they are used to monitor certain physical processes. They are deployed for measuring the temperature, air pressure, radiation of the human body, chemical reactions, movement of objects, vitals of the body, and so on. By doing this we can recover some important information of the boundaries that are often called edges. To keep track of the edge of a physical development, the recognition of a boundary is imperative. The issue of discovering the boundary is considered to be the first step for resolving the edge detection. In digital processing there are many methods for identifying the edge; however, they are not easy to implement in the WSNs' environment since the sensor nodes are not equally spaced like pixels, and because of limited computational power and memory [3–6].

In this chapter, we review the major applications of wireless sensor networks (WSNs) with a description of each.

3.1 Target detection and tracking

The application of wireless sensor networks to the problem of intrusion detection and the related problems of sorting and tracking targets are of great significance to computer and telecommunication systems. In general, target detection and tracking is a well-recognized discipline. There are numerous proposed solutions in the literature that rely either on expensive and specialized sensors or on simple and inexpensive schemes and tools. The latter approaches are appealing for scalable target detection and tracking purposes.

Nevertheless, tracking with inexpensive WSN systems has its own challenges including difficult signal processing, synchronized decision making, multi-modal sensing, high-frequency sampling, and data synthesis [8].

Some researchers have relied on simple threshold-based schemes or used extra hardware on each small sensor device to carry out the related complex signal-processing operations. Threshold-based techniques have the potential of lowering valuable sensor data and hence may not be adequate for low signal-to-noise WSN environments. Of course, it is possible to add particular purpose hardware to alleviate these problems, but this will add to the cost and energy consumption of such self-powered and low-cost means.

The WSNs are designed using a huge number of sensor nodes that are intelligent in sensing the information, processing the information, and having many capabilities. Sensor nodes are employed in numerous applications for monitoring the condition of the environment, which include temperature, vibrations, pollutants, pressure, and sound; these are explained in many references such as [1–10].

Among the major applications of WSNs are intrusion detection, target detection, classification, tracking, and localizing. Latest advances in the efficiency of sensing, computing, and communications technology have made it feasible to employ a group of sensors inside a sensor network processor (node). The low cost of sensor nodes makes it realistic to set them up in large numbers over huge areas and thus, such devices have turned into capable contenders for tackling the distributed detection, categorization, and tracking problems. Many methods have been devised in the literature that can be purely centralized, or purely distributed. They can also be based on high message complexity, or be high computational, as well as being data fusion-based or decision fusion-based [10–15].

There are several works on target classification in WSNs that are based on the centralized technique. This usually entails pattern recognition/matching via time–frequency signatures generated using various kinds of targets. For example, Caruso and Withanawasam illustrated a centralized vehicle categorization system via magnetometers based on similar magnetic signatures generated by using various sorts of vehicles [21]. This work entails a high computational load on each sensor node. It also requires considerable layout and management over the environment. Moreover, the vehicle has to be driven straight over the sensor for precise categorization. On the other hand, at arbitrary directions and distances, the method can only sense the presence of the target.

Raghavendra et al. [6] illustrated a collaborative classification method based on sapping neighboring attribute vectors. The precision of this approach, nevertheless, increases just when the number of cooperating sensors enhances, which puts a burden on the WSN.

Duarte and Hu gathered a data set that is composed of 820 MByte raw time-series data, and 70 MByte of pre-processed [7]. Then they extracted spectral feature vectors, and baseline classification results via the maximum likelihood classifier. In their system, each sensor extracts feature vectors depending on its own readings and passes them to a local pattern classifier. The sensor then sends only the assessment of the local classifier and a related probability of precision to a middle node that combines all received assessments. While this approach is promising, as it only slightly loads the network, it entails huge computational power at each single node.

Zhao *et al.* formulated the complexity of distributed tracking using WSN as an information optimization issue and pioneered a number of reasonable measures of information utility. Here, the WSN can request for a knowledgeable decision about sensing and communication in an energy controlled setting [8].

Arora *et al.* [5] have shown experimentally that it is possible to distinguish between various object categories using a network of binary sensors.

They considered a surveillance application scenario called "A Line in the Sand." The aim of this setup is to recognize a violation within a region. The interfering target may be an exposed person, a soldier carrying a ferrous weapon, or a truck. The major basic user requirements of this function are: (a) target recognition, (b) categorization, and (c) tracing. The user enumerates the quality-of-service (QoS) metrics that influence the way in which the system can detect, classify, and trace the targets. Moreover, the user needs to define the perimeter to be sheltered. The detection process has to distinguish between a target's presence and nonexistence. Winning detection necessitates a node to accurately assess a target's existence while avoiding fictitious recognitions in which no objects are in existence. The major performance measures for detection are: (a) probability of accurate detection, or probability of false detection, and (b) acceptable latency between a target's existence and its final recognition [5].

In classification, the target has to be identified as belonging to one of several categories such as vehicles, soldiers, trees, and animals. Furthermore, classification relies on approximation where the pertinent parameters of the sensed signal such as its power density, period, duration, amplitude, phase, and bandwidth are estimated. The main performance measures of the classification process are: (a) the probability of correctly classifying the ith class, and (b) the probability of error in classifying the ith class as the jth class.

Tracking entails keeping on the target's position as the target moves in the area covered by the WSN's area of coverage. In order to have a successful tracking process, the system estimates of the target's primary point of entry and recent position are within the acceptable detection latency.

There are several design considerations that need to be considered when designing a WSN for target detection. These include: (a) energy, (b) dependability, and (c) complexity. Such design considerations play an important role in choosing the schemes used to solve the problem under study. A brief description of each of these considerations is given below.

3.1.1   Energy

Nodes in a wireless sensor network should employ either stored energy-based batteries or harvested-based energy solar cells. The speed at which energy is spent is limited either by the node's required lifetime for stored energy or by the mean rate of energy gathered using harvesting. There are four major operational modes in which nodes use energy: sensing the environment, computing, saving data, and communicating to peripherals, other nodes and interfaces.

Every one of these operations uses a specific amount of energy that depends on the effective work that it does. Actually, the sending and receiving operations engaged in communication themselves have different energy expenditures.

3.1.2   Dependability

The instability of WSNs has a great impact on the system design for classification and tracking, especially while choosing the feature that provides the basis of classification. We can use one of two main techniques while working on feature selection: centralized and distributed. The first method entails performing a time–frequency series investigation followed by a signature-matching scheme. Nonetheless, since the sensor nodes usually have restricted computational power, carrying out these difficult tasks will overload the individual node. Moreover, the event to be sensed and classified, namely, the target moving throughout the area, is itself distributed in both space and time domains. However, selecting the correct distributed attribute is difficult as it entails a number of design compromises such as the limited resources of a sole node that impose restrictions on the characteristics that can be obtained. In addition, the undependable character of the WSNs and the disappointing performance of the WSNs under heavy load conditions oblige the users to minimize the quantity of data to be transmitted over the network in order to avoid congestion. These limitations on local sensor node and network load may lead users to try to find distributed attributes whose projection on a single node could be computed without overloading the WSN and can still give the required precision for the classification and tracking processes [6–8].

3.1.3   Complexity

The schemes that carry out the detection, estimation, classification, tracking, synchronization, and routing are the ones that will basically perform the sensing, computing, storage, and communications operations. This means that the user should concentrate on optimizing the time, space, and message complexity of the schemes. As an example, the user may be concerned with the tracking methodology as a function of the number of messages. Moreover, consideration of the choice and method of collecting characteristics for classification needs to be investigated. A classifier based on centralized data synthesis would have a high message complication as high-dimensional data must be relayed. On the other hand, a distributed classifier may have a small message complexity, sending a signal only when the target is sensed, but a great time or space complexity owing to the classification algorithm's computational and/or memory need [6–20].

Choosing the right sensor is an important task in the design of WSNs, as selecting the correct set of sensors for the task at hand can considerably enhance the system's performance, reduce their cost, and increase their lifetimes. Nonetheless, there is a basic anxiety between the wealth of a sensor's output and the resources needed to handle the signals it produces. For instance, most of today's small cameras have tens of thousands of pixels that offer huge volumes of data; however, the image processing schemes have to manipulate this enormous amount of data that usually have high time, space, or message complexity and this necessitates a lot of computational power and resources.

The sensing modes suitable for detecting the target categories which could be, for example, an unarmed person, an armed soldier, and a vehicle based on the variations they produce. The system must first recognize the target perturbations to the environment that prospective targets are expected to produce. Next, a set of sensors that can distinguish these must be identified.

3.1.4   Recognition of the target perturbations to the environment (phenomenology)

Phenomenology is basically the study of the basic nature of things. In this context, our aim is to find a set of critical characteristics whose importance is very similar for objects in the same classes and extremely different for objects in dissimilar classes. Features are mainly identified in the six main energy domains: electrical, mechanical, magnetic, optical, thermal, and chemical. The designers may need to consider all six domains as there is significant ongoing research and development (R&D) in MEMS sensor technology in all of these areas. Keep in mind that a variety of sensors may possibly detect various facets of the same energy realm.

For instance, an unarmed individual is expected to interrupt the environment acoustically, electrically, thermally, optically, seismically, and chemically. The person's body heat is produced as infrared energy omnidirectionally from the source. The individual's footsteps are impulsive signals that generate ringing at the normal frequencies of the ground. Resonant fluctuations are weakened and spread in the ground.

A human body can be regarded as a dielectric that produces a change in the surrounding electric field. It is known that human beings release a chemical trace that dogs can easily sense. Dedicated sensors can identify particular chemical discharges. A human being reflects and takes in light beams and can be identified by using a camera. A human also reflects and disperses acoustic, optical, electromagnetic, and ultrasonic signals [5–10].

An armed warrior is likely to have a mark that is a superset of an unarmed person's signature. It is expected that a soldier (warrior) holds a gun and other tools that contain metal. Consequently, it is expected that the soldier will have a magnetic signature that nearly all unarmed persons would not have. The soldier's magnetic signature is mainly due to the disruption in the surrounding magnetic field produced by the availability of ferro-magnetic substances [6–15].

An automobile may disturb the surroundings electrically, magnetically, thermally, optically, seismically, acoustically, and chemically. In general, automobiles have a thermal signature that consists of hotspots like the engine area. In addition, automobiles have evident seismic and acoustic signatures. Automobiles generate carbon monoxide and carbon dioxide as a side effect of burning fuel. They also absorb, reflect, diffract, and scatter electromagnetic, optical, ultrasonic, and acoustic signals [5–10].

3.1.5   Sensing selection

Deciding on the correct set of sensors for the system to be designed is an essential task as it can provide a cost-effective design as well as extend the lifetime of the overall system.

Although there are a huge number of choices of sensors, no primitive sensors exist that detect people, automobiles, animals, or any object of interest to the system. Typically, sensors are employed to detect a variety of attributes such as electrical or optical signature. The output of a sensor is really an estimation, which is not 100% accurate. Furthermore, noise affects almost all signals picked up by any data-acquisition system, and this may limit the system's efficiency. We can distinguish between two main types of sensors: active sensors and passive sensors. Active sensors, such as ultrasonic transducers, can gauge a target's existence or velocity by how the target changes, or reflects a signal sent by the sensor. On the other hand, passive sensors identify and gauge various analogs of a target such as its acoustic, magnetic, or thermal mark. Major types of sensors are: electric, thermal, optical, ultrasonic, magnetic, radar, acoustic, seismic, and chemical sensors. Each type has its own strengths and weaknesses. For example, the strengths of the electric-based sensors are: no line-of-sight constraint, cold, silent, fragrance-free, steady, operate with high-speed or slow-moving environments. Their weaknesses include a requirement for electrode positioning, active nature, and interference from nearby electric circuits and devices [9–14].

The specification of the target detection and tracking system, its design concerns, target phenomenology, and sensing mode and domain dictate the selection process of the sensor used. The guidelines to be considered in such a process include the following [18–20].

(1)   No need for a special hardware.
(2)   The sensor can operate regardless of its orientation.
(3)   The signal processing schemes needed for related signal detection and parameter estimation are realistic given the limitations of the situation.
(4)   The sensors are properly picked out and can be found by using off-the-shelf commercial sources.
(5)   Co-location of sensors will not create interference or crosstalk.

3.2   Contour and edge detection

There is no doubt that advances in WSNs have prompted new applications in recent years. Examples of such applications include weather monitoring, smart homes and cities, health care applications, military applications, and infrastructure protection. The key issue behind these applications is data aggregation. Because of the limited bandwidth, the high communication cost, and the restricted processing power in WSNs, data aggregation from sensors can be a challenge if data processing and fusion are not performed correctly, preferably in a distributed manner.

There are two main schemes that are typically used to thwart the problem of data aggregation and lengthen the lifetime of WSNs. The first scheme relies on distributed data compression, which lessens the size of data to be sent by utilizing redundancy in the data itself. The second scheme is distributed data fusion, which pre-handles the gathered data in order to obtain synthesis results for broadcasting. Although distributed data synthesis may noticeably reduce the size of data to be sent, the particular algorithm greatly relies on the application domain and the sensors' surroundings [10–15].

Applications such as edge detection, which is used for environmental monitoring, present only a binary result of 1 or 0 based on running the algorithm on usually noisy data. For some real-world applications, the fundamental physical events are characterized by a choice of values, not binary values. Examples of these include pressure, pollutants, temperature, and velocity. When applying edge detection, we digitize measured values while some aspects of the monitored event vanish because of the quantization process involved. The contour line is delineated as a two-dimensional curve on which the worth of a function of relevance is a constant. This means that contour lines can offer additional information on the event being observed in terms of the spatial dispersal and a processed set of values of the collected data. Contour line removal has applications in the environment and in weather forecasting, among others. Some researchers have proposed using a distributed algorithm in order to determine contour lines accurately [10–15].

Liao *et al.* devised a distributed algorithm for contour line extraction with wireless sensor networks [10]. The scheme consists of three main steps.

(1)   Consecutive extremum search. The gradient here is approximated and used to find the extremum (either the maximum or the minimum) point of the uni-modal field. This crucial point is referred to as the reference point of the sensor field.
(2)   Sensor clustering and contour point finding. Depending on finding the location of the extremum, sensors are grouped into a number of regions and clusters according to their relative location in regard to the reference point. Next, contour points using collected data samples in each cluster are determined.
(3)   Contour line creation. The information of contour points is transmitted to class leaders (heads) and then to the base station for the contour line creation.

3.2.1   Consecutive extremum search

There are two main concerns to deal with in extremum point search over a sensor area. The first is to employ data trials for gradient assessment at each sensor locally. The second one is to utilize the predictable gradient to find the extremum point. Simic and Sastry [22] devised a scheme to predict the gradient in noise-free surroundings that congregates to the correct gradient in probabilistic logic. Nevertheless, the noisy measurements used in this effort make the issue harder. The sparsely gathered data trials and the noise consequence make the inaccuracy of the estimated gradient greater.

Liao *et al.* [10] devised a scheme for the extremum search in noisy surroundings. It consists of the following steps.

(1) Initialization phase. The seek process can be initialized at any sensor.
(2) Consecutive ascending (or descending) process. Here the gradient of the present sensor is computed by utilizing data samples from its nearby sensors. Next, the scheme computes the inner product of the predicted gradient and the normalized directional vector linking its neighbor to itself. Then, the user has to pick the nearby sensor, which has the greatest (or the smallest) inner product value as the subsequent sensor.
(3) Ending the process. The iterations stated above are reiterated for all sensors. The sensor between its first and second calls makes a closed path. The centroid of these sensors provides the extremum point, which is also often referred to as the reference point.

3.2.2 Sensor grouping and contour point finding

After the maximal point is located, the observation beams have to be identified and the sensors need to be grouped into directional regions. In order to find out contour lines, we identify virtual points alongside each observation beam with an equal spacing that is found by the system. It is likely that a virtual point does not run over with a sensor. In order to distinguish the contour point along the inspection beam, data samples are gathered from sensors that are within the data-collection limit of the target virtual point. Such sensors make one sub-cluster. The sensor that has the shortest distance from the target virtual point may be elected as the neighboring synthesis center. Given the fact that sensors are arranged arbitrarily, it is probable that there might not be sufficient sensors for a number of observation beams to identify contour points. Hence, the precision in that specific beam might become less.

The key job here is to find out whether a target virtual point alongside every observation beam is a contour point or not, depending on gathered data specimens. Because, in this context, a stationary environment is tested, data gathering and filtering over space rather than time are dealt with. The key question is whether the signal may possibly fluctuate from one spot to another. Hence, a scheme has to be found so that it alleviates the noise influence while at the same time maintains the spatial deviation of the signal [10–12].

3.2.3 Contour line creation

The different types of data that are sent from sensors all the way through cluster heads to the base station for contour lines creation are: (a) the coordinates of the extremum, (b) the predicted signal level at the extremum, (c) the coordinates of contour points, and (d) the signal level at every contour point. Depending on these data kinds, the base station can chart contour lines and identify the extremum point. An easy interpolation scheme can be employed to link contour points to build the contour lines. One possible interpolation method that is described by Liao $et\ al.$ [10] relies on converting the Cartesian coordinates of the two contour points (x_1, y_1), and (x_2, y_2) to the equivalent polar coordinates, (r_1, θ_1) and (r_2, θ_2), which are centered at the estimated extremum. Next, the polar coordinates (r^*, θ^*) of interpolated points can be estimated using:

$$\theta^* = \theta_1 + \kappa(\theta_2 - \theta_1),\ 0 \leq \kappa \leq 1,\ \text{and}\ r^* = [(r_1 - r_2)/(\theta_1 - \theta_2)](\theta^* - \theta_2) + r_2.$$

By using these interpolated values, we will be able to obtain soft contour lines connecting calculated contour points. The authors in [10] have used simulation analysis to show that their proposed algorithm works well in noisy environments.

3.3   Types of applications

Wireless sensor networks (WSNs) can be used for a variety of applications that range from civilian applications to military applications. They are meant to identify or observe various physical parameters or conditions including pressure, pollutants, temperature, humidity, air quality, water quality, soil structure, and characteristics of an object in terms of its weight, height, size, position, speed or direction.

Wireless sensor networks have an advantage over fixed sensor networks as they can be deployed in hostile territories, battlefields, outer space or under seas, rivers, and oceans. A brief description of the popular applications is given below [1–22].

3.3.1   Environmental applications

Wireless sensor networks can be used effectively to monitor the environment. This class of applications is considered one of the earliest applications of WSNs. In this context, sensors are deployed to monitor different environmental factors and conditions. Under this application category, we can identify the following cases.

(1)   Sensors can be deployed to observe wild animals or plants in forests and wild habitats as well as to monitor environmental factors of these habitats.

(2)   Water or air quality control: WSNs can be used to monitor water or air quality by deploying sensors on the ground or under water. Air quality monitoring can be used for air pollution control while water quality monitoring can be used in hydrochemistry areas.

(3)   Disaster monitoring: In this case, sensors can be deployed in forests to detect fires or in rivers to detect floods. In addition, seismic sensors can be installed in a building to find out the path and degree of earthquakes and offer an estimation of the safety of the building.

3.3.2   Health care applications

In this application, WSNs can be employed to observe and trace patients and senior citizens for health care reasons. This has a great potential to reduce the overall cost of health care, especially these days as costs have skyrocketed.

Furthermore, sensors may be installed in patients' homes in order to monitor their behavior or movement; this is especially useful for elderly patients. Such systems can alert the responsible nurse or physician if a patient falls or needs immediate attention.

Vital signs of patients along with GPS coordinates can be monitored so that, in case of emergency, such noninvasive and ambulatory monitoring of patients and elderly persons can alert nurses and physicians of the need of an immediate attention. This also can provide real-time health status and updates as well as location of the patients so that they can be reached or rescued, if needed. This is achieved by having wearable sensors that can be incorporated into wireless body area networks (WBANs) in order to monitor vital signs in real time.

3.3.3   Manufacturing process control

We can employ WSNs to monitor the conditions of manufacturing machines and equipment as well as for observing the manufacturing processes in assembly lines of factories. For example, small sensors can be set into the regions of specific machines and devices that are unattainable to humans in order to monitor the condition of a machine and send warning signal in case of any failure. In these ways, WSNs can help decrease the cost of maintenance in factories, as well as enhance the life of equipment and even save human lives.

3.3.4   Intelligent and smart home

Wireless sensor networks may be used to provide more comfortable and smart living environments for their inhabitants. For example, sensors can be set into homes and be networked to make an independent home network. An intelligent refrigerator linked up to a smart microwave oven or stove can arrange a menu based on the stock of the refrigerator and send related cooking factors to the smart stove or microwave, which in turn can set the required temperature for cooking. Furthermore, the contents and timetables of TV, DVD, and CD players can be managed remotely in order to meet the needs of the family.

Utility companies have started to use WSNs to read gas, water, and electricity meters remotely using wireless technology and without the need to go inside homes to read the meters, if they are installed inside the homes. This can save such companies money and time, and will not disturb the inhabitants.

3.3.5   Homeland security

Wireless sensor networks are good tools for many security and surveillance applications. Sensors can be embedded in public buildings and plants such as airports, bridges, subways, bus stations, and other critical infrastructures such as nuclear power plants, telecommunications centers, and petroleum refineries in order to recognize and follow intruders and offer sensible alerts and safeguard from possible terrorist attacks.

3.3.6   Underwater applications

Wireless sensors can be installed in the bottom of oceans, seas, rivers, and lakes in order to monitor events and report on such events in a wireless and timely manner. Ocean-bottom sensor nodes are considered excellent means for facilitating applications for oceanographic data gathering, contamination observation, offshore investigation, calamity avoidance, and assisted route-finding applications. Numerous unattended or independent underwater vehicles (UVs) furnished with underwater sensors can find applications in the discovery of natural undersea resources such as oil and minerals, and in the gathering of data in shared monitoring undertakings. In order to make these applications feasible, we have to enable underwater communications between underwater apparatuses. In such scenarios, underwater sensor nodes and means should be able to be self-configurable so that they can manage their actions by switch over configurations, locations, and movement information, and be able to forward monitored data to a ground post [11–12].

For such applications, we need to use wireless underwater acoustic sensor networks. These consist of a number of sensors and vehicles that are set up to act in a collaborative manner in order to monitor related events over a given area. To this end, sensors and vehicles self-organize in an independent network that can adjust to the attributes of the ocean surroundings. In underwater networks, acoustic communications are the classic physical layer technology. Actually, radio waves spread at long distances throughout conductive sea water only at lower frequencies in the range 30–300 Hz, which necessitates huge antennas and great communication power. Although optical waves do not experience such attenuation, they suffer from dispersion. In addition, transmission of optical signals needs high accuracy in directing the narrow optical beams. Consequently, links in underwater communications typically relied on acoustic wireless systems.

Rapidly growing fuel requirements [23–24] motivate us to find new resources of energy including oil and natural gas. It is well known that about 70% of the Earth is covered with water, which is a huge home for many resources that are essential in meeting the needs of our daily lives. Therefore, searching for oil under water in an effective way is an essential concern to oil explorers as well as nations [23–24]. The oil exists in the sea floor and leaks out from the cracks. If there are sensors embedded in such regions, then oil can be discovered by them.

Conducting underwater search processes is challenging owing to the specific characteristics of water that make it very difficult to employ the existing search methods [24–27] for terrestrial use. These attributes include: varying levels of salinity or measure of dissolved salts in sea water, temperature, and pressure [24, 28] under the surface of water. Such factors influence underwater wave propagation [28]. The traditional scheme for ocean-bottom or ocean-column monitoring is to install underwater sensors that record data during the monitoring mission, and then pick up the instruments. This tactic has the following drawbacks [11–12].

(1)   Real-time monitoring is not feasible. This is significant particularly in surveillance or in environmental monitoring applications such as seismic monitoring.

(2)   The amount of data that can be recorded throughout the monitoring mission by each sensor is restricted by the capability of the on-board memory devices.

(3)   No communication is possible between onshore control systems and the monitoring devices. This issue can hinder any adaptive tuning of the tools, nor is it possible to reconfigure the system after certain events happen.

(4)   If a *malfunction* happens, it may not be possible to notice it before the instruments
       are recovered.

Hence, there is a need to install underwater networks that can enable real-time monitoring
of chosen sea/ocean areas, remote configuration, and communication with onshore
individual operators. This can be achieved by linking underwater devices by using
wireless linkages based on acoustic communication technology [11–12].
    An underwater acoustic network (UAN) is basically a small-scale network that is
deployed under the water and is able to gather data using acoustic modes. A UWSN has
characteristics unlike those of general UANs. The key differences between a UWSN and
a UAN are in respect to scalability, self-organization, and localization [24]. Major
research challenges and design issues in the field of UWSN, as discussed in [18], are
as follows: (1) traffic congestion control; (2) security, resilience and robustness; (3)
dependable data transfer; (4) effective multi-hop acoustic routing; (5) distributed localization and time synchronization; (6) resourceful multiple access; and (7) acoustic
physical layer.
    Several factors may negatively affect the communication quality between nodes in
underwater environments, compared to the communication that takes place on ground-
based networks. The major factors that influence underwater communication are briefly
reviewed below [11–12].

(1)   Underwater currents. The changeable flow or current speed influences the relative
       position of sensors in the network. They also affect the communication quality
       because of the noise produced by them.
(2)   Salinity. Increased level of salt in the water increases the density of water. This
       produces delay in the signal path.
(3)   Multipath delay spread. Because of the various reflections of transmitted signals,
       the signal is received at diverse times, which results in inter-symbol interference
       [25–27].
(4)   Surrounding noise. Such noise is generated by underwater currents, marine life
       and shipping in the harbor.
(5)   Temperature instability. Experiments have revealed that tidally driven temperature
       fluctuations have consequences on acoustic communication [24].
(6)   Multipath fading. Whenever we have waves of multipaths that are out of phase, the
       signal strength at the receiver is decreased. This is usually called Rayleigh fading.
(7)   Pressure. It has been found that high pressure at the bottom of the ocean/sea can
       affect the signal communication.

Next we review the main limiting factors in UWSN [12, 27–29].

(1)   Restricted memory. Since sensors have limited memory, there is a requirement to
       reduce the data saving and communication. The on-demand routing schemes
       experience increased delay, which could make them inadequate for use in these
       application realms. Table-driven schemes may be thought of as an option that would
       take additional memory. Hence, we can also think of an answer to the number of
       needed retransmissions. Since retransmissions necessitate store-and-forward means

for data packets, and if these retransmissions are reduced, then the issue of small memory will be lightened.

(2)   Propagation delay. It is found that delay in underwater networks is relatively high compared to the delay in air, decreasing the speed of propagation to about 1.5×10^3 [24–29]. Hence, all the propagation-delay-connected calculations that need to be performed whilst doing the implementation should be done by using speed of sound in water.

(3)   Low frequency. Communication between nodes in UWSNs takes place in very low (acoustic) frequencies and not radio frequencies. Absorption attenuates the signal. The latter is greater for higher frequencies [28]. Therefore, an acoustic channel with frequency in the range 20 Hz–20 kHz is favored. Hence, slighter attenuation can be obtained up to a certain level at the expense of the propagation time.

(4)   Large transmission power. Owing to the resistance provided by water to wave propagation, attenuation rises more [28]. Hence, we have to send signals at greater source power in order to receive a good signal.

(5)   Modest bandwidth. The obtainable bandwidth for underwater acoustic channel is restricted, and mainly relies on both frequency and transmission distance. This band is restricted because of the underwater absorption with most acoustic systems operating below 30 kHz [12, 27]. Hence, the packets that sensor nodes send to each other should be as short as possible so that they do not incur high delay.

An easy way to appreciate the communication among different sensors in a UWSN is given in Figure 3.1, which illustrates a small network of only seven sensors.

Figure 3.1 shows sensor 1 is communicating with sensor 2, sensors 4 and 7 are communicating with each other via sensors 5 and 6 and sensor 3 is lonely since none of the sensors is in its transmission range.

In reference [12], the authors proposed an architecture for underwater sensor networks. Figure 3.2 depicts the proposed architecture from the viewpoint of the horizontal view at

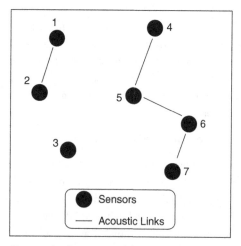

Figure 3.1   Communicating sensors in an underwater wireless sensor network (UWSN).

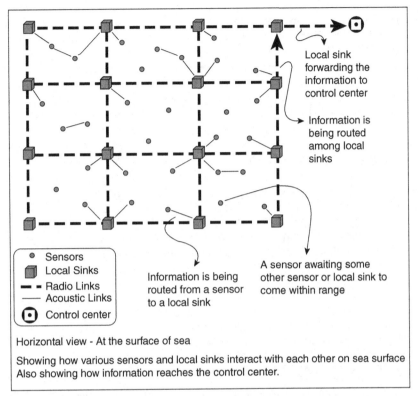

Local sink forwarding the information to control center

Information is being routed among local sinks

A sensor awaiting some other sensor or local sink to come within range

Information is being routed from a sensor to a local sink

- Sensors
- Local Sinks
- Radio Links
- Acoustic Links
- Control center

Horizontal view - At the surface of sea

Showing how various sensors and local sinks interact with each other on sea surface
Also showing how information reaches the control center.

Figure 3.2   Horizontal view of the architecture proposed in [12].

the sea plane. The figure illustrates the location of local sinks and sensors on the surface of the sea. The local sinks are inherently static in the shape of a grid pattern. Such local sinks are capable of communicating with each other via both acoustic and radio-frequency forms. They utilize acoustic method to communicate with sensor nodes and radio-frequency allows them to communicate with the other local sinks and the control center. Moreover, such local sinks are in charge of the implementation of different launching and controlling procedures. The control center is linked to the local sinks using multiple hops. The arbitrarily installed sensors talk to each other using the acoustic mode. Clearly, these sensors are the essential nodes since all the information in the system is produced only by these sensors, and they are in charge of sending such information to the local sinks on the surface.

Figure 3.3 shows our proposed architecture reported in reference [12], from the angle of the vertical view at the sea surface and the relative place of the sensors in the water column under the sea surface. This may be regarded as showing the side-view of the system. Here, the vertical levels are preset; however, the sensors are moveable at such levels. It should be noted that the term "fixed," in this context, means that the net altitude of the sensors at that level is unchanging; however, it is still an insignificant vertical movement relative to that level that may be present because of water currents. This must not be perplexed with the tied-up sensors. At the beginning, sensors are installed at this

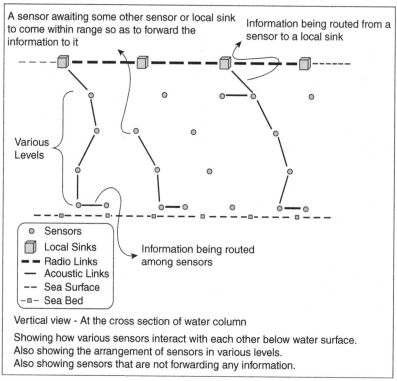

Figure 3.3 Vertical view of the architecture proposed in [12] – at the cross-section of the water column.

leveled structure and throughout the operation they may go up or down by one level at a time. Such sensors are linked to their neighbors by using an acoustic channel and they transmit the data to the local sinks via the sensor-to-sensor hopping (SSH) scheme.

3.3.7 Agriculture

Wireless sensor networks are being used more and more in the agricultural industry. Gravity-fed water systems can be observed via pressure transmitters in order to check water tank levels. Pumps can be managed by using wireless I/O devices, and water use may be estimated wirelessly. Irrigation automation can help in better water consumption and so reduce wastage. In addition, WSNs can be used to control the temperature and humidity levels inside greenhouses so that, when they fall below threshold levels, the greenhouse guard should be notified by using a cell-phone text message or e-mail message. In such a case, host systems can trigger a broad spectrum of responses in order to fix the problem [1–29].

3.3.8 Military applications

Wireless sensor networks have many applications in the military domain. They may be used for military situation awareness, detection of an enemy unit's movements on land or sea, chemical or biological threats, offering logistics in urban warfare, battlefield surveillance, command, control, communications, reconnaissance, computing, intelligence, surveillance, and targeting systems, among others [12–16].

In military command, control, communication, and intelligence, WSNs have become an essential technology. Wireless sensors can be deployed rapidly in any battlefield or hostile territory without the need for any infrastructure. Here is a brief review of major applications of WSNs to the military domain [12–17].

3.3.8.1   Intelligent assistance

Wireless sensors can be installed on unmanned robotic vans, tanks, fighter planes, missiles, or submarines to direct them near barriers to their goals as well as to enable them to coordinate and collaborate with each other effectively.

3.3.8.2   Object protection

Wireless sensors can be mounted around sensitive objects such as tactical bridges, telecommunication centers, electrical power generation stations, oil pipelines, water purification plants, and military nerve centers.

3.3.8.3   Battlefield observation

Wireless sensors can be mounted in battlefields to monitor and track vehicles, and force movement, all of which permit close surveillance of enemy forces.

3.3.8.4   Remote sensing

Here, sensors may be installed for remote sensing of biological, nuclear, and chemical weapons, as well as for uncovering possible terrorist attacks.

3.4   Summary

This chapter has addressed wireless sensor networks applications. We reviewed the major applications of WSNs along with some detailed case studies. Typical applications range from weather monitoring and smart homes to battlefield surveillance. We reviewed target detection and tracking, and contour and edge detection as related to WSNs. We also investigated applications of WSNs to several areas including environmental monitoring, health care, intelligent home, homeland security, underwater applications, agriculture and greenhouse monitoring, and military applications.

References

[1] J. Zheng and A. Jamalipour (Eds.), *Wireless Sensor Networks: A Networking Perspective*, Wiley, 2009.
[2] M. Ilyas and I. Mahgoub, *Handbook of Sensor Networks: Compact Wireless and Wired Sensing Systems*, Boca Raton, FL: CRC Press, 2004.

[3] N. Ahmed, Y. Dong, T. Bokareva *et al.*, "Detection and tracking using wireless sensor networks," in *Proceedings of the 2007 International Conference on Embedded Networked Sensor Systems*, pp. 425–426, Sydney, Australia, November 2007.

[4] T. Bokareva, W. Hu, S. Kanhere *et al.*, "Wireless sensor networks for battlefield surveillance," in *Proceedings of Land Warfare Conference*, Brisbane, October 2006.

[5] A. Arora, P. Dutta, S. Bapat *et al.*, "A Line in the Sand: a wireless sensor network for target detection, classification, and tracking," *Computer Networks*, Vol. **46**, No. 5, pp. 605–634, December 2004.

[6] C. S. Raghavendra, C. Meesookho and S. Narayanan, "Collaborative classification applications in sensor networks," in *Proceedings of 2002 IEEE Sensor Array and Multichannel Signal Processing Workshop*, pp. 370–374, August 2002.

[7] M. Duarte and Y. Hu, "Vehicle classification in distributed sensor networks," *Journal of Parallel and Distributed Computing*, Vol. **64**, No. 7, pp. 826–838, July 2004.

[8] F. Zhao, J. Shin and J. Reich, "Information-driven dynamic sensor collaboration for tracking applications," *IEEE Signal Processing Magazine*, Vol. **19**, No. 2, pp. 61–72, March 2002.

[9] P. K. Dutta and A. K. Arora, "Sensing Civilians, Soldiers, and Cars," The Ohio State University Department of Computer and Information Science Technical Report OSU-CISRC-12/03-TR66, 2003.

[10] P.-K. Liao, M.-K. Chang and C. Kuo, "Contour line extraction with wireless sensor networks," *Proceedings of the 2005 IEEE International Conference on Communications, ICC 2005*, Vol. **5**, pp. 3202–3206, May 2005.

[11] S. Dhurandher, S. Khairwal, M. S. Obaidat and S. Misra, "Efficient data acquisition in underwater wireless sensor ad hoc networks," *IEEE Wireless Communications*, Vol. **16**, No. 6, pp. 70–78, December 2009.

[12] S. K. Dhurandher, S. Misra, M. S. Obaidat and S. Khairwal, "UWSim: a simulator for underwater wireless sensor networks," *Simulation: Transactions of the Society for Modeling and Simulation International, SCS*, Vol. **84**, No. 7, pp. 327–338, July 2008.

[13] I. F. Akyildiz, D. Pompili and T. Melodia, "Underwater acoustic sensor networks: research challenges," *Journal of Ad Hoc Networks, Elsevier*, Vol. **3**, No. 3, pp. 257–279, March 2005.

[14] C.-C. Shen, C. Srisathapornphat and C. Jaikaeo, "Sensor information networking architecture and applications," *IEEE Personal Communications*, pp. 52–59, August 2001.

[15] E. Shi and A. Perrig, "Designing secure sensor networks," *IEEE Wireless Communications*, pp. 38–43, December 2004.

[16] I. F. Akyildiz, W. Su, Y. Sankarasubramaniam and E. Cayirci, "A survey on sensor networks," *IEEE Communications Magazine*, pp. 102–114, August 2002.

[17] C. F. García-Hernández, P. H. Ibargüengoytia-González, J. García-Hernández and J. A. Pérez-Díaz, "Wireless sensor networks and applications: a survey," *International Journal of Computer Science and Network Security*, Vol. **7**, No. 3, pp. 264–273, March 2007.

[18] J.-H. Cui, J. Kong, M. Gerla and S. Zhou, "Challenges: building scalable mobile underwater wireless sensor networks for aquatic applications," University of Connecticut, USA, UCONN CSE technical report UbiNET-TR05-02; last update, September 2005.

[19] A. Mainwaring, J. Polastre, R. Szewczyk, D. Culler, and J. Anderson, "Wireless sensor networks for habitat monitoring," in *Proceedings of the 2nd Annual International Conference on Mobile and Ubiquitous Systems: Networking and Services, MobiQuitous '05*, pp. 479–481, July 2005.

[20] C.-Y. Chong and S. P. Kumar, "Sensor networks: evolution, opportunities, and challenges," *Proceedings of the IEEE*, Vol. **91**, No. 8, pp. 1247–1256, August 2003.

[21] M. J. Caruso and L. S. Withanawasam, "Vehicle detection and compass applications using AMR magnetic sensors, AMR sensor documentation," http://www.magneticsensors.com/datasheets/amr.pdf

[22] S. N. Simic and S. Sastry, "Distributed environmental monitoring using random sensor networks," in *Proceedings of Information Processing in Sensor Networks, IPSN 2003*, pp. 582–592, 2003.

[23] M. Stojanovic, "Underwater acoustic communication," *Encyclopedia of Electrical and Electronics Engineering*, J. G. Webster, Ed., Wiley, 1999, Vol. **22**, pp. 688–98.

[24] J. Michel, "Assessment and recovery of submerged oil: current state analysis," Research & Development Center, U.S. Coast Guard, 2006; http://www.crrc.unh.edu/workshops/submerged_oil/submerged_oil_workshop_report.pdf (accessed May 10, 2008).

[25] J.-H. Cui *et al.*, "Challenges: building scalable mobile underwater wireless sensor networks for aquatic applications," UCONN CSE technical report UbiNET-TR05–02; last update, Sept. 2005.

[26] J. Kong *et al.*, "Building underwater ad-hoc networks and sensor networks for large scale real-time aquatic application," *Proceedings IEEE MILCOM*, 2005.

[27] M. Stojanovic, "Recent advances in high speed under water acoustic communication," *IEEE Journal of Oceanic Engineering*, Vol. **21**, No. 2, pp. 125–36, 1996.

[28] J. Heidemann *et al.*, "Underwater sensor networking: research challenges and potential applications," USC/ISI technical report ISI-TR-2005–603.

[29] N. M. Carbone and W. S. Hodgkiss, "Effects of tidally driven temperature fluctuations on shallow-water acoustic communications at 18 kHz," *IEEE Journal of Oceanic Engineering*, Vol. **25**, No. 1, pp. 84–94, 2000.

4 Medium access in wireless sensor networks

Wireless sensor networks (WSNs) use the same medium (which is the air) for wireless transmission that the nodes in a wireless local area network use. In order for nodes in a local area network to communicate properly, standard access protocols like IEEE 802.11, IEEE 802.15.4, and ZigBee, are available. However, these and other protocols cannot be directly applied to the wireless sensor area networks. The major difference is that, unlike the devices participating in local area networks, sensors are equipped with a small source of power (usually a battery), which drains very quickly. In addition, sensor nodes have limited resources of memory and computational power. Hence, there is a need to design new protocols for medium access control (MAC). This chapter investigates some of the major MAC protocols available [1–7].

4.1 Medium access control in wireless networks

There exists a very good set of standard protocols for wired and wireless area networks, which are proven to operate efficiently. There are many reasons for not using the wired and traditional wireless MAC protocols for WSNs. The standard protocols in wired local area networks use the well-known carrier sense multiple access with collision detection (CSMA/CD) scheme for individual stations to access the medium. This protocol, the IEEE 802.3, is known as Ethernet [1–46]. In this scheme, each station senses a medium for a random amount of time. If no activity is detected it starts its own transmission. If it detects any activity in the medium it defers its transmission until the activity ceases. If it senses a collision of packets from different nodes in the networks it backs off a random exponential amount of time and then starts contending for trans-mission using the backoff exponential algorithm. Slotted ALOHA networks tradition-ally use a time division multiple access (TDMA) scheme [9–10]. In this scheme, the time is divided into an equal number of slots such that a node is allowed to transmit only in its allotted slot. Here, the disadvantage is that a node is allowed to transmit if and only if it owns the slot; otherwise it has to listen to the medium for the data intended for it. Therefore, if the node has no data to send the bandwidth is simply wasted because there will be no activity for the amount of slot time allotted to a node with no data. On the other hand, since there is the division of time in this scheme there will be no collisions.

The Ethernet that implements the CSMA/CD scheme performs very well when the density of the nodes in the network is low. When the number of nodes increases, the number of collisions tends to increase and hence the efficiency decreases.

Wireless sensor networks cannot directly use the scheme described above since saving energy is a prime goal for them. The saving of energy comes from identifying the main factors consuming energy and how to get around them, if possible. As the energy needs to be saved, the nodes have to use as small an amount of energy as possible in communication. This indirectly means that sensor nodes are to be equipped with transceiver devices with very restricted range of transmission.

Clearly, traditional MAC approaches are not enough to deal with the characteristics of MAC required in WSNs. Hence, WSNs have specific requirements that have to be met. These are briefly described below [1–7, 10–15].

Energy efficiency

Unlike the wired or wireless ad hoc networks, one of the major requirements that needs to be specified, not just by the MAC protocol but by any sort of protocol (routing protocol, or training protocol), is the conservation of energy as much as possible. The major sources of energy consumption identified during the communication are as follows [7–15].

(a) Transmission of overhead. Transmission of overhead in each packet is identified as one of the major sources of energy depletion. The WSNs typically exchange data once in a while and those data too have a very small packet sizes. The overhead for each packet needs to be as small as possible [10–22].

(b) Idle listening. In traditional wired or wireless networks, the node keeps on listening to the medium for a possible reception of data destined to itself in the idle node because the nodes there do not need to worry about how much power they consume since in most of the cases they are directly plugged on to the alternative current (AC) supply [10–29]. For applications of WSNs in which nodes are deployed in hostile environments like those in enemy territories, nodes have to save their nonrenewable source of energy power as long as possible. Therefore, schemes to turn off the battery during the idle mode for a significant amount of time should be implemented in order to save as much as possible of the battery power and life.

(c) Localization. Localization in general refers to finding the placement of the coordinates of all the nodes in a particular network. This problem has been a very challenging issue for researchers since it also involves taking into account the movement of the nodes when designing the protocol [11–20].

(d) Autoconfigurable networks. The nodes in WSNs are prone to failure due to the attacks on the nodes or when the battery of one of the nodes gets exhausted. Such a network should indeed be aware of this type of occurrence so that it can get rid of the faulty node, or one that is no longer working, and reorganize itself to carry out the task it is supposed to do. In general, the term "autoconfigurable networks" is used for such networks. In other words, they should be self-organizing.

(e) Collaborative function. Another important characteristic, which distinguishes WSNs from traditional networks, is that the main functions of these networks are to collect and detect events and map this useful information. It is often quite useful to aggregate or fuse data coming from multiple nodes to increase the efficiency of data collection.

This chapter will cover some of the MAC protocols in detail. The following are among the major MAC protocols for WSNs.

(1)   S-MAC protocol: S-MAC stands for sensor MAC protocol. This is one of the early protocols developed for WSNs and laid the foundation of basic designing principles.
(2)   L-MAC: A lightweight medium access protocol based on the TDMA scheme.
(3)   D-MAC: The dynamic scheduling MAC protocol.
(4)   Energy-efficient QoS-aware medium access protocol.
(5)   Energy-efficient application-aware medium access protocol.
(6)   A location-aware access control protocol.
(7)   An energy-efficient MAC approach for mobile WSNs.
(8)   O-MAC: A receiver-centric power management protocol.
(9)   PMAC: An adaptive energy-efficient MAC protocol for WSNs.
(10)   T-MAC protocol: Timeout-MAC protocol, which is an adaptive energy-efficient MAC protocol for WSNs that minimizes idle listening, while considering wireless sensor communication patterns and hardware limitations.
(11)   BMAC protocol: Berkeley media access control for low-power wireless sensor networks.

The detailed discussion of these protocols is given in the next sections.

4.1.1   S-MAC: An energy-efficient protocol

The primary goal of this protocol is to increase energy conservation and support self-configuration. These features come at the expense of latency and fairness on a per node basis. This protocol identifies the following main energy consuming mechanisms [12–14].

Collision
Whenever a transmitted packet undergoes collision it turns out to be corrupted. The packet has to be retransmitted to avoid data loss. The retransmission procedure is a highly energy consuming process. In addition, collisions lead to increase in the network's latency.

Control packet overhead
Each MAC protocol has some type of control information, which is transmitted along with the actual data. Usually, sending and resending of these control packets also consume energy.

Overhearing
The other source of high energy consumption is overhearing, in which a node keeps listening to the data packets that are not intended for itself.

Idle listening

Idle listening implies listening to the channel to expect useful data at any time. Experiments have showed that, in protocols like IEEE 802.11, the station constantly tries to listen in order to receive the useful data, which leads to almost 50%–100% energy utilization of the actual receipt of the data.

Therefore, S-MAC tries to reduce the consumption of energy from all the above sources, which in turn leads to a reduction in fairness and latency. However, it does not result in low end-to-end latency or fairness. The fairness or latency is not an important issue here as sensor nodes are intended for a distinct application. In these kinds of situations the nodes usually have a very large amount of data to send for which they use a message-passing scheme. In this technique, each message is divided into various bursts. The outcome of this is that the node that has more data gets a longer time interval in which to send than the other nodes [12–24].

Latency can be essential or not, depending on the application type. Throughout the time there is no sensing event, there is little data flow activity. Hence, S-MAC lets nodes periodically enter into the sleep cycle or idle mode, which leads to savings in energy consumption. This, however, leads to an increase in latency as the transmitting node has to wait for the receiver to wake up to send useful data.

The PAMAS (power aware multiple access protocol) [13–20] uses the same method to avoid idle listening, which leads to the reduction of energy consumption. However, there is a fundamental difference between PAMAS and S-MAC. PAMAS employs two independent channels, one for control signaling and the other for data signaling, whereas the S-MAC uses an in-band signaling channel [11–24].

4.1.1.1   Design of S-MAC protocol

As described above, the S-MAC protocol is designed to be an energy-efficient protocol that reduces energy consumption from the above-mentioned sources of collision: overhearing, idle listening and control overhead. The design of this protocol makes some assumptions, which are as follows [12–20].

(1)   Network and applications assumptions. WSNs are in many ways different from traditional IP networks. Certain assumptions need to be made regarding their overall operation. A WSN consists of many tiny devices that are separated by a very small distance from each other and from the target. Because of the very small distance between them, the nodes can use a straightforward signaling mechanism rather than a very complex one. Also the communication is mainly hop-based rather than the direct communication. Since these nodes are designed for a specific application or a set of collaborative applications, instead of providing a per-hop fairness, it is imperative to offer the fairness at the overall application level [11–22].

An assumption is also made about the application having a greater period of sleep time. Typically, a regular surveillance application requires a longer network lifetime, but it can tolerate higher latency.

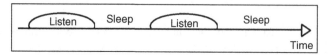

Figure 4.1   The division of time into listen and sleep states in the S-MAC protocol.

Figure 4.2   Illustration of two nodes, A and B, having different schedules; thus they synchronize with nodes C and D respectively.

Periodic listening and sleep

For this protocol, nodes are assumed to have a sleep cycle period and a periodic listening period as shown in Figure 4.1. In this case we take a situation such that, in a cycle of one second, a node listens to the medium for half a second and sleeps for the other half; this will lead to a 50% reduction in the duty cycle. This implies 50% saving in the energy. Normally, the sleep and listening time depend on the application; however, the values are same for all nodes at any instance [14–20].

Basic scheme

The node goes into periodic sleep cycles if there is nothing to listen to on the channel. It then wakes up to see if any other node wants to talk to it. This method requires periodic synchronization with nearby nodes to treat their clock drift. There are two schemes used for offering strength against the synchronization errors. In the first one, the timestamps are not exchanged with respect to the actual or absolute time clock, but they are relative. In the second method, the listening period is longer than the clock drift. For instance, a typical listening period of 0.5 is 10^5 times larger than the clock period. In a TDMA scheme, the time slot is smaller than the listening period, hence it requires a tighter synchronization process compared to this protocol. All nodes are permitted to pick their own schedule; nevertheless, in order to reduce control, the neighboring nodes are required to synchronize with each other in a periodic manner [13–16].

Nodes exchange their schedules by broadcasting them to their neighbors; see Figure 4.2. This ensures that even though all nodes have different schedules they can listen to each other. For example, if node A wants to send its schedule information to node B, it waits until B starts listening. If multiple nodes want to talk to a single node, they need to contend for the medium by using the IEEE 802.11 mechanism. This involves sending request-to-send (RTS) and clear-to-send (CTS) packets. The node that first sends the RTS packet to a given receiving node gets control of the medium [14–17].

One of the downsides of this approach is that, because of the recurring sleeping period of each node, the latency tends to increase. Moreover this latency tends to accumulate on each and every hop. Hence, it places a fundamental limit on the latency requirement of the application.

Selecting and maintaining schedules

Before each node goes into a sleep and wake up schedule, it first sends its needed information to all of its neighbors. Every node maintains a table called the scheduling table, which includes the sleep–wake up cycle information about all its neighboring nodes. The steps through which the scheduling is done are summarized below [12–29].

(1)   A node first listens to the medium for the schedule information from all the other nodes. If it does not hear anything for a specific amount of time, it broadcasts its own schedule in a short message called "SYNC" and straightaway goes into the sleep mode. Such a node is often called a synchronizer node as it selects its own schedule individually and other nodes will get synchronized to it. The SYNC message contains the information that the node will go to sleep after t seconds.

(2)   If a node receives a neighboring schedule before it goes to sleep, it follows what it has received. This type of node is usually called a "FOLLOWER" node. The follower node rebroadcasts the message indicating that it will go to sleep mode after time $(t - t_d)$. The random delay of t_d is needed for collision avoidance, to avoid having multiple followers rebroadcasting the schedules from colliding with each other.

(3)   If a node gets a different schedule after it starts acting upon its own schedule, then it follows both schedules.

Maintaining synchronization

It is important periodically to update the schedules to bring synchronization between the elements [12–26]. This is done by periodically sending the SYNC packet that includes the address of the node and the next time at which the sending node will go to sleep. The next-time sleep is relative to the time when the transmitter terminates its data and reaches the receiver.

In order for a node to receive the SYNC packet and the data packets, its listen period is divided into two parts. The first part is kept for receiving the SYNC packet and the second part is reserved for receiving a request-to-send (RTS) packet. The timing relationship is shown in Figure 4.3. The listen portion is divided up into several time slots. The sender begins sensing for the carrier, during which time the receiver is listening for any event. The sender then arbitrarily picks off any slot and begins sending the SYNC packet. If the sender does not detect any action by the end of the time slot, it will suppose that it has won the medium. A similar process is followed when sending data packets.

Figure 4.3 shows three sorts of configurations that the sender can be in. A brief description of each is as given next [14–18].

(1)   Sender 1. In this configuration, the sender begins sensing the medium when the receiver also starts listening. Next, the sender arbitrarily selects a time slot in which it transmits the SYNC message. If there is no activity till the next time slot then the sender has won the medium, otherwise it backs off.

(2)   Sender 2. In this case, after sensing for some time, if no activity is found on the medium, the sender sends an RTS (request-to-send) packet. If a CTS (clear-to-send)

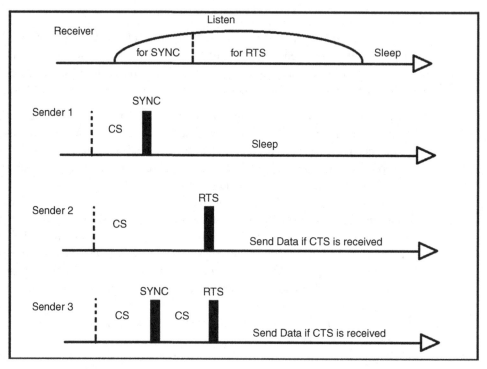

Figure 4.3   Illustration of different sender configurations [14–18].

packet is received within the listening period of the receiver, then data will be transmitted.

(3)   Sender 3. In this case, the sender sends SYNC, RTS and data in a single listening cycle of the receiver.

Collision and overhearing avoidance

Avoiding collisions is one of the major duties of the S-MAC protocol. Since this scheme utilizes a contention-based scheme, it is better to avoid the overhearing, as a message sent by each node can be heard by all nodes.

To avoid collisions as much as possible, the protocol uses the RTS/CTS mechanism of the IEEE 802.11 protocol, which also tackles the hidden terminal (node) issue [41]. Moreover, whenever a packet is sent by any node, there is a field that specifies for how long the transmission is going to be alive so that when the receiver is not the intended one, it knows for how long it can sleep before it can wake up again. Every receiver registers this value in a variable called the NAV (network allocation vector) and rests a timer for it. The sender also finishes in order to bring virtual sensing into action by looking at the NAV timer value. If the NAV timer has a nonzero value, it can be logically thought that the medium is busy until the value of the timer goes to zero, which is usually called virtual sensing.

Performance characteristics of S-MAC

S-MAC uses mechanisms such as collision avoidance, overhearing prevention, and less idle listening, which makes it a widespread scheme for wireless sensor networks compared to the standard IEEE 802.11 from the point of view of energy consumption.

4.1.2   L-MAC: a light-weight medium access protocol

This is a very light-weight energy-efficient scheme, which employs time-division multiple access for collision-free transmission of the data. The main objective of this protocol is to minimize the overhead at the physical layer. It attempts to decrease the states of the transceiver switches and hence tends to minimize energy wasted in the preamble transmission [15–20].

Usually, wireless sensor nodes sleep and wake up a couple of times during their full life-cycle in order to save energy. Hence, the core hardware or the transceivers have to be trained each time a node wakes up from sleep. Such training is often accomplished by using a predefined sample of the data both at the transmitter and the receiver, which is known as the preamble. The re-waking of the crystal oscillator of the receiver every time the sensor node wakes up lessens the battery power. The L-MAC protocol learns to save power by making the sleep interval of the node adjust to the volume of traffic. This mechanism reduces the complexity at the physical layer.

The S-MAC protocol [12–22] usually works with two stages for the transceiver operation: a sleep period and a listen period. In the sleep period, the nodes completely switch off their transceivers and discontinue any activity. Throughout the listen period, which comes after the sleep period, the nodes turn on their transceiver either for listening to the channel or for sending their own data. This means that the sleep and listen periods are locally synchronized between a cluster of nodes or the entire network in some situations. Local synchronization also means that nodes can be taken out of their sleep period in order to perform useful tasks of transmission. In order to produce synchronization, nodes tend to transmit a short message called the "SYNC" message. The listen period on a per node basis is typically divided into two sorts of slots: one part is reserved for the "SYNC" messages and the other one is meant for the RTS messages [1–10].

The design of the L-MAC protocol assumes the following main characteristics at the physical layer.

(1)   The transceiver is assumed to have a single channel, which has one of the three states of operation: receive, transmit and standby.
(2)   A transmission mode uses more energy than the receive mode. Typically, the standby operation consumes far less energy than the other two modes of operations.

This protocol is based on a time division multiple access (TDMA) approach, which means that there is no contention. Every node is allocated a single time slot in a frame time. In this time slot the node may send data to other nodes [15–29]. Unlike the traditional TDMA scheme, the time slots are not assigned by any central manager;

Table 4. 1 Different fields in the control field of the L-MAC protocol

Field type	Size in bytes
Identification field size	2
Present slot number	1
Taken slots	4
Distance to gateway	1
Collision in slot	1
Destination ID size	2
Data size	1
Total size	12

however, they are decided using a distributed mechanism. During its time slot, a node will send a message that is composed of two parts: the control part and the data part. The control field is again divided into different parts as shown in Table 4.1 [17–24].

A brief description of each of the control fields is given next.

Identification field. This field indicates the identification number for the time-slot controller. It is like a MAC address that is unique.

Present slot number. This specifies the slot position in a frame for the protocol.

Occupied slots. This gives the number of slots occupied.

Distance to gateway. This gives the number of hops the source node is away from the gateway in the network.

Destination ID. This signifies the identification number of the node for which the data are intended.

Data size. This field shows the size of the data. Since one byte is reserved for this field, it can hold up to 255 units of the data.

The control frame is used for maintaining the synchronization between the nodes. If the nodes are not addressed in the control word, they switch off their radio until the next time slot in order to save energy. A node can send only one message per frame.

Network setup

All the nodes at the beginning of the network setup [18–20] are unsynchronized. At the beginning of a network setup the gateways take the initiative and start controlling slots. The immediate one-hop neighbors will start synchronizing their own clock with the gateways. Next, the recently synchronized nodes will pick up random time slots, which are already occupied. All nodes are required to maintain a lookup table of the entire neighborhood so that a node can come to know which slot is occupied. This guarantees that no single slot is occupied by two slots in order to avoid collision.

Nodes will maintain their time slots until their battery runs out or the node finds out that there is a collision on its time slot. In case of collision, the nodes back off, this backoff time being dependent on the identification number of the node. Each node also maintains the information about the hop distance from the gateway and broadcasts this in its control message.

Performance characteristics

In terms of network lifetime the MAC protocol performs reasonably well against the S-MAC and the very old EMAC protocols due to its light-weight nature. This light-weight nature comes from the time being divided into equal slots so that this protocol is a TDMA type protocol. Since there are no contentions, the nodes do not waste their energy in retransmitting lost or damaged messages [18–30].

4.1.3   Dynamic scheduling MAC protocol

This sort of protocol, which is a neighbor-aware medium access protocol, is designed by keeping in mind the requirements of very-large-scale sensor networks. At a high level this protocol adds a dynamic scheduling scheme to the modified distributed mediation device (MDMD) protocol to enhance the performance in dense environments [20–24].

The large-scale networks have very distinctive characteristics, which require a newly designed medium access protocol. Firstly, the large-scale sensor nodes reduce the overall amount of computing power at each node and enhance the fault probability per node. The data rates are also very low. Generally, before designing any new MAC protocol for WSNs, the following points need to be taken into consideration.

(1)   Energy efficiency. This is the necessity for any type of wireless sensor network. Prolonging the life of a network is an essential goal and task.

(2)   Scalability. The MAC protocol must adapt itself to the changing size of the network. In the case of the TDMA protocol, which is one of the medium access protocols, if the number of nodes in the network is small and the nodes do not have always some data to send, the bandwidth is usually wasted for most of the slots. If the network size is too large, the time available for a slot decreases, which in turn makes a node that has a large amount of data to send wait although the other nodes do not need to transmit. Thus we see that TDMA is not a good technique of MAC protocol for WSNs. The same applies to code division multiple access (CDMA) [19–20] and the standard wireless 802.11 MAC protocols.

(3)   Adaptivity to changes in node density and topology. In many applications, sensor nodes are mobile and change their relative positions continuously. Moreover, in some cases the nodes fail due to batteries being discharged or due to enemy attacks, among other possible reasons. In such cases the MAC protocol should adapt to such situations.

There are other attributes that are not given main importance but should be taken into consideration when designing a new MAC protocol. They are bandwidth utilization, fairness, latency, throughput, and latency.

One of the other characteristics of a sensor network is that nodes are usually used for the transmission of the control information and not the data; hence the packet size is small and normally the nodes only need a fraction of the time for transmission. This means that the nodes are inactive throughout the remaining time.

As previously discussed, one of the best-known sensor network protocols is S-MAC, which employs a scheme of synchronized periodic listening and regular sleep where each node in the beginning listens to the medium for a random amount of time, and if no transmission is sensed it assumes its own schedule and conveys the listening–sleep cycle information to other nodes in the form of a SYNC message. We may say that S-MAC utilizes a virtual clustering method in which neighbors create a virtual cluster. The neighboring nodes of two or more clusters are able to pursue more than one listening–sleep cycle so they use more energy. The major benefit of this category of protocols is its small latency. However, each node can listen to only one of its bordering nodes. As the network density grows, so the contention grows. This scheme does not perform well in terms of successful transmission performance [14–20].

The dynamic scheduling MAC [19–28] protocol is founded on the device mediation device (DMD) protocol in which the listen–sleep intervals are not synchronized. The DMD protocol brings in a new kind of node called the mediation device (MD) node. Such a node acts as an arbitrator/mediator between the other wireless sensor nodes; see Figure 4.4 [19–24].

The MD node saves and then sends the control message between the source and destination nodes. All nodes go into the mode called the "MD" state where they listen to the inquiries from other nodes. As soon as any given node enters the ON state it sends inquiries to the nearby MD node, which notes down the query. This way the MD node loads its database with the schedule information of all the nearby nodes. The MD node is in charge of synchronization between the source and the destination nodes. The drawback of this scheme over the S-MAC scheme is that the delay in the network is enhanced since its transmitting source has to wait for a duration of time until the destination is synchronized.

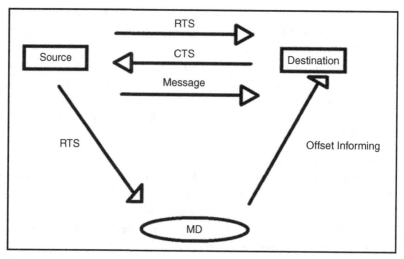

Figure 4.4   Illustration of RTS–CTS exchange in a dynamic scheduling MAC.

4.1.3.1   Modified distributed mediation device (MDMD)

This is a modified version of the distributed mediation device protocol, which minimizes the delay introduced by the distributed mediation device protocol. In this scheme, each node has got its ON state followed by its OFF state. Like the S-MAC, each node sends a query packet at the end of its ON state, which includes the information about its schedule. All nodes make and fill their database in the listening period. Every entry in the database includes the information about the ID and its timetable information.

Every time there is any fresh node joining the network, it has to listen to the full cycle of the sleep–listen period in order to fill up its database with the information about the nearby nodes schedules. Every node also updates its database periodically to get rid of offset error or to become familiar with the new nodes entry.

Every time a node needs to transmit data to another node there may be a third node that can interfere with its transmission. First of all, the source node notifies the interfering node about its wish to send the data to the sink node. After receiving the RTS packet, the interfering node will switch off its radio transmission for the next cycle during which the actual data transmission takes place. The transmitting node in its subsequent ON cycle transmits the RTS packet for which the sink node replies by sending a CTS packet. After receiving the latter packet, the sending node can initiate the data transfer at its next ON state when the receiving packet is in its OFF state. Next, the sink node replies with an ACK packet, which signifies the start of the data transfer.

4.1.3.2   Neighbor-aware dynamic (NAD) scheduling

The neighbor-aware dynamic (NAD) scheduling scheme is a variation of the MDMD protocol described earlier, with the key difference that NAD introduces a dynamic schedule scheme to resolve the hidden and exposed terminal/node problems [15–19].

The exposed node/terminal problem happens when there are two groups of neighbor clusters such that the receivers are not in direct range of each other but the transmitter nodes are in direct range of each other so that, when a transmitting node is sending in one of the groups, the transmitter in the other group is averted from transmitting even though that transmission will not result in collision. Figure 4.5 shows an instance of this type of problem [41].

The hidden terminal (node) problem occurs when we have more than one transmitting node that wants to send data to the same receiver, such that the receiver is in the range of both transmitters, but the two transmitters are not in the direct range of each other. In such cases if just a DATA-ACK sending mechanism is used the data transmission will result in a collision at the receiver. In order to avoid such situations a common mechanism called RTS–CTS (request-to-send–clear-to-send) communication is used [41].

Here, the transmitters first send the receiver the RTS packet. After receiving the RTS packet the receiver sends back a CTS packet only to the transmitter from whom it received the first RTS packet. Then, the receiver does not respond to any other RTS packets from other transmitters until the start of the data transfer. Figure 4.6 illustrates the concept [12, 41].

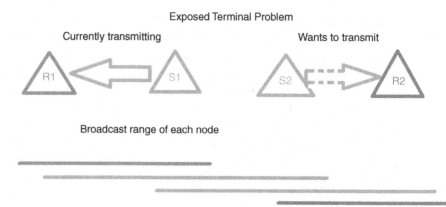

Figure 4.5   Illustration of the exposed node terminal problem between nodes S1–R1 and S2–R2.

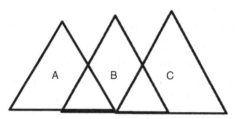

Figure 4.6   Illustration of the hidden node/terminal problem.

In order to resolve the exposed node dilemma, it is essential that the two neighboring nodes do not have the same schedule, as is the case with the S-MAC protocol, since if the schedules of the two nodes coincide then they will send and receive data at the same time, which will lead to collision. To avoid this, when the nodes go into the listening period they have to dynamically modify their schedules if they find any overlap.

As seen in Figure 4.7, the schedules of nodes A, B, C, and D overlap even though they are not direct neighbors. In these situations, each node along with its own schedule information transmits its scheduling information as well as a query packet to its neighbors. While in the listening mode the sender's neighbor receives this query packet and checks to see if its own schedule overlaps with either its neighbor's or neighbor's neighbor's schedule. It dynamically modifies its own schedule if there is an overlap, and transmits this information to its next query packet [17–20].

Performance evaluation studies such as the one in reference [19] have indicated that due to the dynamic scheduling features of the protocol, the normalized energy spending is low when compared to S-MAC. Furthermore, the average energy consumption is also a little less in DMD compared to the S-MAC protocol.

4.1.4   Energy-efficient QoS-aware medium access (Q-MAC) protocol

The QoS-aware MAC (Q-MAC) protocol is a protocol that pays more attention to the QoS aspects of the network along with the other aspects that are essential, such

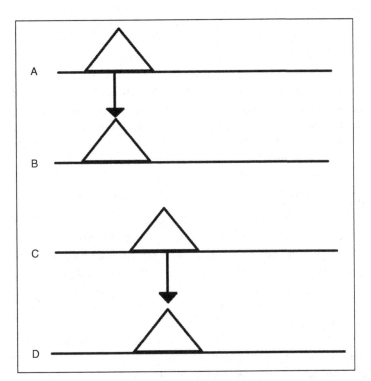

Figure 4.7   The hidden node problem in wireless sensor networks.

as energy efficiency and transmission dependability. This scheme may be called the priority-based random access protocol because data packets with top priority are given the top importance and forwarded before any other packets minimizing their latency [15].

The Q-MAC scheme consists of both intra-node scheduling and inter-node scheduling. The intra-node scheduling utilizes a multi-queuing architecture to organize data packets based on their application and MAC layer abstractions. The methods employed for determining what packet goes next are based on the MAX-MIN fairness algorithm and the packetized GPS scheme. The inter-node scheduling algorithm deals with giving access to the channel competing with various neighboring nodes. The techniques used in this case are the power-conserving MACAW and the loose priority random access scheme.

4.1.4.1   Intra-node scheduling

The intra-node scheduling procedure employs multiple First In First Out (FIFO) order queues. The received packets are ordered and stored in different queues based on the criticality. The service rank in the network is decided by the number of queues for each node. The issues here are the number of queues to keep, the size of these queues, and finding a trade-off between the node resources and the QoS needed for the application.

The priority of an incoming packet is decided by the application and MAC layer abstractions. The application layer allocates priority to the packets based on significance of the contents of the packet. Normally, this protocol adds five additional bits of information to every packet. Of the five bits, two are employed for identification of the application and three bits are utilized for the sensing data [11]. The MAC layer abstraction seeks to offer fairness in terms of network resources among the self-produced packets and the relayed packets, and also among the various kinds of relayed packets in the multi-hop networks. Overall, packets which have traveled the maximum number of hops are given highest priority [17].

4.1.4.2   Inter-node scheduling

One of the important functions of the MAC layer in any protocol is to manage and schedule the data transmission between various nodes using the same channel. Because of the high cost connected with retransmissions, inactive listening, collision, communication overhead and overhearing, Q-MAC employed basic and distributed protocols such as the power conservation–multiple access with collision avoidance for wireless (PC-MACAW), and loosely prioritized random access (LPRA) protocol [22].

In a PC-MACAW protocol, each frame is described as the set of RTS–CTS–DATA–ACK cycle. Each frame period is associated with a small space known as frame space in which no action takes place; see Figure 4.8. The frame period consists of a conflict (contention) period (CP) and a transmission period (TP). There is also some little time space between a CP and a TP, which is known as a short space (SS). A contention period is identified as a time that begins at a point when the node transmits an RTS frame and waits for a CTS frame [12, 21, 41].

There are two kinds of collision recovery techniques that are employed in Q-MAC protocol: (a) doubling the contention/conflict window size and (b) setting of the packet dropping limit/threshold.

4.1.5   Energy-efficient application aware medium access protocol

Flow aware medium access (FLAMA) is an energy-efficient medium access protocol, which achieves energy efficiency by avoiding idle listening, data collision, and transmissions to a node that is not prepared to get data. It experiences less delay when compared with the conventional TDMA-based schemes like TRAMA [19] and fewer collisions when compared with the contention-based techniques such as the S-MAC.

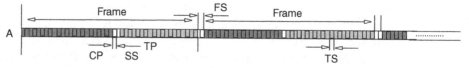

Figure 4.8   The frame format in PC-MACAW protocol.

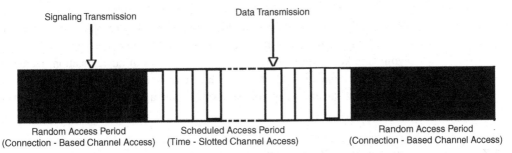

Figure 4.9   Illustration of the time slot division into different components in a QoS-aware MAC [27–30].

To achieve energy-efficient channel access, this scheme utilizes a simple traffic adaptive distributed election method. In order to perform the election process, we will need the two-hop neighborhood and flow information. Scalability is attained via the two-hop information. As shown in Figure 4.9, time is arranged into the period of random and scheduled access slots. Typically, a single channel is used for sending the control and data information. Clearly, the scheme employs the in-band signaling [27–30].

Getting into channel is a contention-based process in the random-access period and it is time slotted for the planned access period. Throughout the random-access period tasks such as the neighborhood discovery, time synchronization, and topological changes recognition are carried out. The data transmission is achieved only in the scheduled access period [21–24]. The scheduling data about the time slotting are exchanged in the random-access phase only.

4.1.6   Location-aware access control protocol

The location-aware access control protocol is centered on roughly defined areas of several access points which overlie in space. The location key is obtained from the beacon information of the direct access which can be employed to track the moving mobile gadgets [22].

Note that WLAN technology usually uses the open-air medium to transport information along nodes taking part in the transfer process. Nevertheless, this leads to wireless access of information which makes the information an easy target to the intruders. In order to safeguard data transmission in an open-air medium, numerous techniques can be used. Typically, all of these methods need some kind of password, ticket, biometric or access control list (ACL) [23] to verify an entity. Among these methods are the ones listed below.

(1)   Access control list (ACL). Access control list keeps a list that contains info on what each user is entitled to access in a specific setting. This method is used in today's state-of-the-art operating systems.

(2)   Secure ID. This is a token-based authentication method employed in various big companies in which the person who seeks to have access is required to enter a password along with a random number token.

(3)   Kerberos. This is a ticket-based authentication technique in which each user is assigned a ticket for a session. It builds on symmetric key cryptography and needs a trusted third party, and optionally may use public-key cryptography by using asymmetric-key cryptography during certain phases of authentication. Kerberos has been developed by MIT and is available as free software for users. Its main application is in providing robust client server applications by cryptographic keying techniques [42].

4.1.7   An energy-efficient MAC approach for mobile wireless sensor networks

Almost all of the protocols that we discussed above have good performance; however, they do not support mobility of wireless sensor nodes. Since in wireless networks the nodes are on the move, this creates a problem called the Doppler effect/shift, which leads to the losses [41]. The technique described here reduces the Doppler effect by effectively changing the frame length of the data during transmission. A filter identified as the extended Kalman filter [25] is used to forecast frame size for every transmission, which also improves the efficiency of the system from the point of view of energy. This scheme deals with how to increase the energy efficiency for a network that includes mobile nodes instead of static nodes.

In a setting with mobile sending and receiving ends, the sent frame always experiences a Doppler-shift effect, which causes frequency distortion in the signal that causes bit errors. The noise in the wireless channels also leads to a low signal-to-noise ratio, which gives a high bit error rate. The high bit error rate leads to losing frames at the receiver. The retransmission of such dropped frames can lead to the loss of a large amount of the energy [41]. This scheme suggested an algorithm that may lead to the reduction of the retransmission energy losses, which will result in significantly lower energy consumption. Furthermore, it will preserve the fairness at the MAC layer.

Among the techniques that can be employed in these types of situations is always sending small-sized frames, which has the following two benefits [26, 41].

(1)   Smaller-sized frames will always require less energy to retransmit when compared to large-sized frames.
(2)   Smaller-sized frames have less probability of having errors compared to large-sized frames. This will enhance the reliability and improve the life of the battery as fewer frames will need to be retransmitted.

Nevertheless, if we send a smaller-sized frame each time, this will reduce the bandwidth, which is unwanted. Therefore, there needs to be an answer which has to be in the middle of smaller- and larger-sized frame transmission that utilizes varying-sized frames based on the signal attributes.

Extended Kalman filters can be used to find out the best frame size for transmission based on the length of the frame in the previous transmission, the characteristic of the channel at the preceding transmission, and the present channel characteristics. An estimate approach can be employed to address the timely changeable nature of the wireless sensor channel. When the medium's quality is bad, a small frame size can be selected [26, 41]. If it is improved, a large frame size can be selected. If all nodes of the wireless sensor network are not moving, the frame can be selected to have the maximum length as there would be less likelihood for retransmissions in such situations.

4.1.8   O-MAC: a receiver-centric power management protocol

It is broadly accepted that the leading power-spending candidate for WSNs is the radio communication among the nodes [28]. The rise in the transmitter power can be caused by the improvement in the data rates. The growth in the receiver power can be accredited to the sophistication of operations linked to processing of the data. It is because the receiver has to take care of error-correction schemes and contention-resolution mechanisms, among other tasks. This increase in the energy consumption has led researchers to think of developing protocols that are receiver centric. The majority of the MAC protocols are basically sender centric. In such protocols, a sender wakes up and sends data, and the receiver has to synchronize with the transmitter. Alternatively, the receiver-centric protocols are developed in such a way that the transmitter has to synchronize with the receiver schedule to recognize when it will be in a state to get the data [29].

Among the schemes that are receiver centric is the O-MAC protocol. The authors in [28] designed a new MAC protocol, which they called O-MAC. This is based on Pseudorandom Staggered On; O-MAC can achieve near optimal energy efficiency. The authors also devised two variations of the O-MAC protocol – with local broadcast channel and preamble-sized slots.

Solving issues such as time synchronization, and management of neighbor tables is essential for the development of the O-MAC protocol.

Time synchronization. The Staggered On and Pseudorandom Staggered On processes need time synchronization. Typically, the period of time synchronization is 2–10 min. If the usual slot length is considered to be 5 ms, then the duty cycle for time synchronization will be lower than 0.004%, which is insignificant. Sometimes, synchronization protocols can be mixed with O-MAC. Moreover, it is not necessary to send timestamps since the time difference can be computed by the receiver based on the anticipated receiving time and actual receiving time.

Asynchronous neighbor discovery. O-MAC contains an inhabitant discovery mechanism based on load-balanced beaconing. We have two types of nodes: synchronized and unsynchronized. An unsynchronized node is the virtual node, which has a schedule related to it, which it listens to. The task of the networked nodes is to wake up and beacon when the virtual node is awake. Such beaconing occurs at the start of the frame and the networked nodes must arbitrarily listen to the beacon in order to preserve the neighborhood. Such a virtual node wakes up for a single frame length in order to locate the network and afterwards connect to the network using the beacon channel [28].

4.1.9   PMAC: an adaptive energy-efficient MAC protocol for wireless sensor networks

Schemes for WSNs such as S-MAC are put to sleep at times in order to extend the life of the battery. Since these kinds of protocols have a preset duty cycle, the throughput of the network is reduced under heavy load situations whereas power is simply wasted under light load situations. In a new type of protocol, named pattern MAC (PMAC) [31], instead of having preset sleep–wake up cycles, the duty cycles of the nodes are found out dynamically. When compared with S-MAC [4], this scheme can offer power savings throughout low-load situations and enhancement of the throughput throughout high-load situations.

There are several techniques that can be used to design MAC protocols for WSNs. Essentially, power saving is the key factor in devising any MAC protocol for WSNs as nodes are usually provided with a limited source of power. In addition, the throughput and latency must be given priority when devising wireless sensor MAC protocols.

The pattern MAC scheme considers all the major factors for its design; namely power efficiency, latency, and throughput. Here, a node adaptively decides its own duty cycle based on its own traffic conditions and that of its neighbors. Figure 4.10 illustrates the point that, under a no-traffic situation, the S-MAC scheme with a fixed duty cycle and the T-MAC scheme with a variable duty cycle run into some waste in the power, whereas the PMAC scheme, which adjusts itself based on the traffic status, will inform the nodes

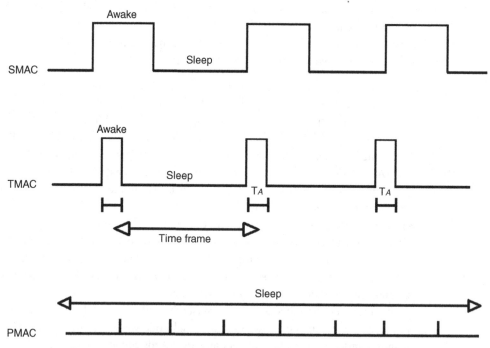

Figure 4.10   An illustration that compares the lengths of the idle listening periods of S-MAC, T-MAC and LMAC with the no-traffic condition.

to sleep for the duration of the time slot and thus it will end having a significant reduction in the expenditure of the power.

Note that PMAC is considered a time-slotted protocol like S-MAC. In S-MAC, time is partitioned into slots and in every slot the node is in a sleep state and wake up state for a specific duration of time. In PMAC, throughout a time slot the node may only be in either the sleep or wake up state during a time slot [31].

The PMAC scheme relies on the traffic pattern of its own node and nearby nodes in order to adaptively find out the duty cycle of a specific node. Such a pattern is basically a sequence of bits signifying what nodes are expected to do (sleep or wake up) in the time slots. For instance, a binary sequence pattern of 1010 illustrates what the node may do in every time slot. The string indicates that: (a) during time slot 1, the node wakes up, (b) during time slot 2, the node sleeps, (c) during time slot 3, the node again wakes up, and (d) during time slot 4, the node again sleeps. Hence the sleep–wake up timetable may then be obtained from the sleep–wake up patterns. This tactic of exponentially growing the sleep time leads to considerable saving of power. Basically, a node's pattern is an estimate of the node's sleep–wake up schedule. The real sleep–wake up timetable is obtained from the pattern of the node and its nearby nodes. Therefore, in order for every node to obtain a schedule, it has to exchange patterns. In order to achieve this exchange, time is partitioned into what is termed a super time frame (STF) [31]. Every super time frame is in turn divided into two smaller frames that are often called the pattern repeat time frame (PRTF) and the pattern exchange time frame (PETF).

The pattern repeat time frame has N time slots, i.e. N is the number of slots. Throughout a slot, one of the N nodes begins sending its present pattern of transmission. At the ending of N time slots, there is an additional time slot in each node in the WSN that stays awake. This is because, when the upstream nodes have some data to transmit to the downstream nodes, the latter nodes could be in the sleep state sensing that there are no data meant to be sent to them where, due to this, the upstream node may hold up their transmission as they wait for the next time the downstream nodes become available. The additional slot can also be utilized for broadcasting purposes. In the PRTF period of STF, every node sends its pattern to its nearby nodes. The slot time in PRTF is signified by T(R). If a node doesn't get any pattern from its neighbor due to a collision, it continues to replicate its specific pattern; see Figure 4.11.

The number of 0s in the pattern increases exponentially during situations where traffic is low or absent. Therefore, the node stays in the sleep mode for a very long time, which directs it to saving power; nevertheless, if the presence of data is sensed during the PRTF time period more 1s will be produced. Obviously, PMAC [32] is also adaptive in higher

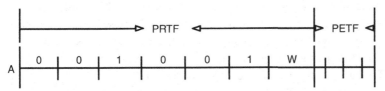

Figure 4.11   Illustration of pattern exchange.

traffic-loads situations. This is because PMAC utilizes the selective wake up methods by permitting a subset of nodes that are engaged in data transfer to continue being awake. This approach provides reduced idle listening, which leads to saving power. Because time is slotted in the order of hundreds of milliseconds, the sequence time synchronization is not required as in the situation of S-MAC because of clock drifts.

If we compare the PMAC protocol with the S-MAC [5] protocol, we see that PMAC gives power savings under low-load situations, while it leads to greater throughput under higher-load situations. This leads us to conclude that the pattern exchange framework is a good approach for WSNs.

4.1.10   T-MAC

The timeout-MAC protocol (T-MAC) is one of the contention-based MAC protocols for wireless LANs. To save more power in wireless sensor area networks, this scheme introduces the idea of a variable duty cycle compared with the fixed duty cycle of S-MAC and the no-duty cycle of CSMA. In situations where there is a changeable load, T-MAC demonstrates an enormous amount of performance increase (as large as five times) when compared to the S-MAC protocol.

It is essential to make the design choices of the MAC protocols based on the traffic type, as this is what they are expected to deal with. In WSNs, we have two major communication patterns [33].

(1)   Local broadcast. Here, the communication occurs when the node is required to transmit data only to its direct neighbors instead of sending it to all its neighbors.
(2)   Nodes to sink reporting. After processing the local event, nodes may need to transmit the collected data to a sink node or a group of sink nodes. Here, the communication pattern is the unidirectional flow of data from the nodes in the network to the path in which sink nodes are in attendance.

Frequently, power is wasted in snooping on the transmission medium even when there are no data sent by other nodes; this is often called idle listening. For instance, in a sample WSN in which nodes are needed to send messages at a speed of around 1 message per second, messages are quite short; they take only 5 ms to be sent. Normally, it takes 5 ms for a transmitting node to send and then 5 ms for a receiver to receive the message. This indicates that the node is set to idle for the remaining 990 ms; a big waste.

The S-MAC protocol is a contention-based scheme that includes a concept known as duty cycle. The active part in a total frame time of 1 s is only about 200 ms, which is low. Given that the active part is only a portion of the entire time frame, the power consumed because of idle listening is minimized to a large value. This will reduce the throughput and raise the latency. Even though energy waste because of idle listening is minimized by the S-MAC scheme, the concept of a fixed duty cycle is not really optimal [33]. The T-MAC protocol brings in the idea of a changeable duty cycle.

The basis for the changeable duty cycle is that, typically, the rate of the message will not stay the same in WSNs. If the message rate is much lower than the fixed active time, this will lead to the waste of power because of idle listening. If the message rate is especially high, then the latency will increase due to buffering. In addition, this will decrease the throughput as nodes would only be able to send/receive in the preset active time phase.

In T-MAC, each node regularly awakens in a specified time frame, begins receiving and transmitting data if there are any, then returns to sleep. The activation episode here indicates the following: (a) the event of the notice of the recurring timer, (b) the event of the delivery of data via the channel, (c) the event of perceiving the communication, and (d) the event that shows data swap of the neighbor has finished.

In general, a node will at all times return to sleep after it has not felt any event activity for a period of Ta. Hence, the upper limit of idle listening is decided by the Ta. A node must not exit to sleep as long as one of its neighbors is still in communication. Because a node might not be in reach so that it may hear the nearby node information conversation, Ta must be long enough so that the node stays active till the delivery of a CTS packet for a neighbor. So, typically: $Ta > C + R + T$, where C, R, and T are the contention period, the length of the RTS packet, and the turnaround time, respectively. Typically, $Ta = 1.5 (C + R + T)$.

4.1.11   BMAC protocol

The Berkeley medium access control (BMAC) protocol is a configurable protocol that is simple in its design and implementation, but is efficient. This scheme provides a good interface for WSNs consuming low power. It is very efficient in collision avoidance and has high channel utilization. To accomplish the low power consumption trait, the BMAC scheme provides a preamble sampling mechanism to minimize the duty cycle and decrease idle listening.

The designer of BMAC considered the major goals of an efficient MAC protocol that meets WSNs should include [35]: (a) small power operation, (b) implementation must be as simple as possible, (c) channel utilization should be as optimal as possible at both high and low data rates, (d) it must have an effective collision avoidance technique, and (e) it should be scalable with the increasing network.

In general, MAC protocols are based on traditional schemes such as the RTS/CTS exchange mechanism to avoid the hidden terminal problem, synchronizing with the neighbor before initiating data transfer, and fragmentation of the message for very large length. In such protocols, however, the application at the higher level is totally uninformed about the network. In other words, the application is absolutely opaque to the lower layer functionalities.

On the other hand, the BMAC technique is a very small core medium access protocol. It does not offer any type of network layer organization or synchronization strategies. At its lower layer, BMAC uses a method called clear channel assessment and packet back-offs for channel arbitration, low-power listening (LPL) for lower power consumption, and link layer acknowledgements for reliability.

Note that BMAC does not offer any multi-packet methodology such as RTS/CTS exchange for getting around the hidden terminal problem, nor does it offer the packet fragmentation scheme, but it provides a group of interfaces to the higher-level application to fine adjust its performance based on the sort of data to be sent/received in the network, which in turn is application dependent. By revealing this interface to the applications at the higher level, the BMAC protocol permits the higher levels to optimize power expenditure, delay, throughput, reliability, and fairness.

To afford an effective collision avoidance tool, the protocol must be able to make simple assessment on whether the channel is clear or not: BMAC employs software that automatically forecasts the noise level in the channel so that the data transfer will not take place when noise level is high. The software tests the signal intensity at well-defined time intervals when the channel is presumed to be free; these are basically the time intervals after the transmission of the packets or when the node is not getting any valid data. The mean of these samples is then added to an exponentially weighted mean with a constant decay value. The median value is employed as a low pass filter to insert robustness. As stated earlier, the BMAC [35] protocol offers different types of interfaces. If the backoff interval is not organized via the interface, a random backoff is selected. As soon as the random backoff ends, the channel is tested for activity. If any activity is found, then an event is scheduled for a congestion backoff time.

Furthermore, BMAC often offers the link layer acknowledgement support. If this mechanism is triggered, the receiver, after the delivery of a packet, successfully sends the acknowledgement back to the transmitter. The BMAC also utilizes a scheme of low-power listening. Throughout a given period of time, if a node senses any activity, it receives the packet and then returns to sleep. If the node does not sense any activity, it stays awake for a random period of time and then returns back to sleep. There is a requirement that must be satisfied: accurate channel assessment (CCA). In order to get data reliably, the length of the preamble should match the interval for which the activity on the channel is tested.

The authors in [36] have found out that the performance of the basic BMAC from the viewpoint of throughput is higher than that for S-MAC under same network environments. We can attribute this mainly to the fact that BMAC utilizes a collision avoidance methodology. Even in terms of power consumption and latency, BMAC has better performance than S-MAC.

4.2   MAC issues in wireless sensor networks

Designing an effective medium access layer protocol for WSNs is vital to the proper operation of the network. A WSN should employ a MAC protocol to decide access to the common medium to circumvent collision from various nodes. Of course, this also requires reasonable sharing of the resources' bandwidth among the sensor nodes in the network. Clearly, a MAC protocol has an essential role to play in facilitating proper network operation efficiently and with good performance.

Traditional wireless network MAC protocols, such as those for cellular systems and mobile ad hoc networks, cannot be used directly for WSNs as they do not consider the power, computation, and memory limitations. In addition, WSNs have many applications, which mean WSN protocols should not only consider the limitations of resources; they also must consider the requirements of various applications.

We must address the unique characteristics of WSNs when their MAC protocols are designed. The main ones are as follows [1–24, 43–46].

(1)   A WSN consists of a much larger number of nodes than traditional wireless networks.
(2)   The WSN topology changes more frequently due to high probability of node failure and mobility.
(3)   Nodes of a WSN have limited computational power and storage.
(4)   Nodes in a WSN are typically powered by batteries, hence they have limited life as it is difficult, if not impossible, in most cases to replace the batteries or recharge them.
(5)   Nodes in a WSN are usually set up in an ad-hoc manner, thus they should arrange themselves into a communication network.

Hence, due to these unique features, MAC protocols used for traditional wireless networks cannot be used as they are for WSNs; they have to be modified if they are to be used for WSNs. While there are different MAC protocols for sensor networks, there is no single protocol that has been accepted as a standard. One key reason behind this is that the MAC protocol choice will, in general, depend on the application, which means that there will not be *one* standard MAC scheme [12].

Time division multiple access (TDMA) has the advantage of collision-free medium access; nevertheless, it suffers from clock drift problems and reduced throughput at low load conditions due to idle slots. The inconvenience with TDMA systems is the synchronization of the nodes and adjustment to topology variations that are caused by the addition of new nodes, collapse of battery capacities, damaged links because of interference, sleep timetables of relay nodes, and scheduling due to clustering algorithms. Hence, the period allocations should be performed by considering such possibilities. Nevertheless, it is not simple to modify the slot assignment within a distributed environment for classical TDMA system as all nodes should agree on the slot assignments.

The carrier sense multiple access (CSMA) scheme has a smaller delay and potential throughput at low load conditions, which is normally the situation in WSNs. Nevertheless, more collision avoidance or collision detection schemes must be used to deal with collision, especially at high load conditions.

The frequency division multiple access (FDMA) method provides a collision-free medium; however, it requires more hardware in order to dynamically communicate with different radio channels. This will raise the cost of sensor nodes.

Code division multiple access (CDMA) provides a collision-free medium. However, it requires high computation, which is a major issue when we deal with WSNs as energy conservation is a major objective. In order to achieve low computational cost for wireless CDMA sensor networks, there has been some effort into examining source and modulation methods, especially signature waveforms, and devising simple receiver versions. In some situations, the high computational complexity of CDMA can be justified as an access scheme for WSNs given its excellent feature of collision avoidance [12].

4.3   Summary

In this chapter we have reviewed the major medium access control (MAC) protocols for wireless sensor networks. A MAC protocol offers somewhat different functionality depending on the network's topology, device ability, and upper layer constraints; however, many functions exist in the majority of MAC protocols including framing, reliability, MAC access, error control, and flow control [1–48].

Medium access protocols offer the greatest influence over communication means and provide major direct control over the exploitation of the transceiver, the key energy consumer in almost all WSNs.

The overwhelming majority of the work on WSN MAC protocols has concentrated on medium access techniques, as the transceiver expends a large amount of energy and the MAC protocol has major direct control over its use. Constrained energy resources offer the main limitation for WSN protocol design, thus almost all proposed MAC protocols emphasize minimizing energy losses associated with the transmission medium. Moreover, there are other design limitations such as fairness, throughput, and latency that are important for some applications.

Conventional wireless MAC protocols endeavor to offer high throughput, low delay, fairness, and mobility management; however, they usually pay no attention to energy saving. On the other hand, WSN MAC protocols should offer the highest performance without consuming a lot of energy due to the limited energy resources on hand. The WSN MAC protocols usually swap performance characteristics, such as latency, throughput, and even security, for a reduction in energy expenditure in order to extend the sensor node's lifetime. The most widely used schemes to minimize energy consumption entail rotating the sensor node hardware between high-power active states and low-power sleep states. Wireless sensor nodes cannot operate in the network during the sleep mode; nevertheless, setting the node in the sleep mode when not needed may noticeably enhance a sensor node's lifetime. More energy saving can be obtained by operating the WSN in a multi-hop manner where sensor nodes advance messages to the destination for other nodes.

Infrastructure MAC protocols can consume a lot of energy for WSNs deployed over large geographical areas because the transmit power needed to properly receive a message increases geometrically with distance; usually the factor is between 2 and 4, with 2 most often used in models. The wide variety of proposed applications for WSNs offers a challenge for protocol designers as each application may generate traffic with particular characteristics and need considerably different QoS metrics. Typically, messages in WSNs applications have smaller sizes compared to these for conventional wireless networks. The smaller message sizes of WSNs mean the MAC protocols do not need to keep long time periods for the transmission of usual messages.

In this chapter, we reviewed the main MAC layer protocols that are often used for wireless sensor networks and gave case studies and examples. We have also shed some light on the challenges in the design and operation of these protocols.

References

[1] C. Ramachandran, M. S. Obaidat, S. Misra and F. Pena-Mora, "A secure, and energy-efficient scheme for group-based routing in heterogeneous ad-hoc sensor networks and its simulation analysis," *Simulation: Transactions of the Society for Modeling and Simulation International*, Vol. **84**, No. 2–3, pp. 131–146, Feb./March 2008.

[2] S. Misra, V. Tiwari and M. S. Obaidat, "LACAS: learning automata-based congestion avoidance scheme for healthcare wireless sensor networks," *IEEE Journal on Selected Area on Communications (JSAC)*, Vol. **27**, No. 4, pp. 466–479, May 2009.

[3] A. Marco, R. Casas, J. L. Sevillano *et al.*, "Synchronization of multi-hop wireless sensor networks at the application layer," *IEEE Wireless Communications*, Vol. **18**, No. 1, pp. 82–88, Feb. 2011.

[4] S. Misra, M. P. Kumar and M. S. Obaidat, "Connectivity preserving localized coverage algorithm for area monitoring using wireless sensor networks," *Computer Communication Journal, Elsevier*, Vol. **34**, No. 12, pp. 1484–1496, 2011.

[5] A. Chalak, S. Misra and M. S. Obaidat, "A cluster-head selection algorithm for wireless sensor networks," in *Proceedings of the 2010 IEEE International Conference on Electronics, Circuits and Systems, ICECS 2010*, pp. 130–133, Athens, Greece, Dec. 2010.

[6] S. Dhurandher, M. S. Obaidat, G. Jain, I. Mani Ganesh and V. Shashidhar, "An efficient and secure routing protocol for wireless sensor networks using multicasting," in *Proceedings of the IEEE/ACM International Conference on Green Computing and Communications, Green Com 2010*, pp. 374–379, Hangzhou, China, December 2010.

[7] S. K. Dhurandher, M. S. Obaidat, D. Gupta, N. Gupta and A. Asthana, "Energy efficient routing for wireless sensor networks in urban environments," in *Proceedings of the 2010 Workshop on Web and Pervasive Security (WPS)-GlobCom 2010*, Miami, FL, 2010.

[8] J. Hill, R. Szewczyk, A. Woo *et al.*, "System architecture directions for networked sensors," in *Proceedings of the 9th ACM International Conference on Architectural Support for Programming Languages and Operating Systems*, pp. 93–104, Cambridge, MA, USA, Nov. 2000.

[9] M. Stemm and R. H. Katz, "Measuring and reducing energy consumption of network interfaces in hand-held devices," *IEICE Transactions on Communications*, Vol. **E80**-B, No. 8, pp. 1125–1131, Aug. 1997.

[10] G. J. Pottie and W. J. Kaiser, "Embedding the internet: wireless integrated network sensors," *Communications of the ACM*, Vol. **43**, No. 5, pp. 51–58, May 2000.

[11] K. Sohrabi and G. Pottie, "Performance of a novel self organization protocol for wireless ad hoc sensor networks," in *Proceedings of the IEEE 50th Vehicular Technology Conference*, pp. 1222–1226, 1999.

[12] I. Demirkol, C. Ersoy and F. Alagöz, "MAC protocols for wireless sensor networks: a survey," *IEEE Communications Magazine*, Vol. **44**, No. 4, pp. 115–121, April 2006.

[13] A. Woo and D. Culler, "A transmission control scheme for media access in sensor networks," in *Proceedings of the 7th Annual International Conference on Mobile Computing and Networking*, Vol. **6**, No. 5, pp. 221–235, Rome, Italy, 2001.

[14] W. Heinzelman, A. Chandrakasan and H. Balakrishnan, "Energy-efficient communication protocols for wireless microsensor networks," in *Proceedings of the Hawaii International Conference on Systems Sciences*, Vol. **8**, Jan. 2000.

[15] L. van Hoesel and P. Havinga, "A lightweight medium access protocol (LMAC) for wireless sensor networks: reducing preamble transmissions and transceiver state switch," in

Proceedings of the 1st International Workshop on Networked Sensing Systems, pp. 205–208, Tokyo, Japan, 2004.

[16] T. Nieberg, S. Dulman, P. Havinga, L. van Hoesel and J. Wu, "Collaborative algorithms for communication in wireless sensor networks," in *Workshop on European Research on Middleware and Architectures for Complex and Embedded Systems*, Pisa, Italy, April 2003.

[17] W. Ye, J. Heidemann and D. Estrin, "An energy-efficient MAC protocol for wireless sensor networks", in *21st Annual Joint Conference of the IEEE INFOCOM*, Vol. **3**, pp. 1567–1576, June 2002.

[18] D. Johnson and D. Maltz, "Dynamic source routing in ad hoc wireless networks," in *Mobile Computing*, T. Imelinsky and H. Korth, Eds., Kluwer Academic Publishers, pp. 153–181, 1996.

[19] V. Mansouri, M. MohammadNia-Awal, Y. Ghiassi-Farrokhfal and B. Khalaj, "Dynamic scheduling MAC protocol for large scale sensor networks," in *Proceedings of the 2005 IEEE Mobile Adhoc and Sensor Systems Conference*, Nov. 2005.

[20] B. Sinopoli, C. Sharp, L. Schenato, S. Schaffert and S. Sastry, "Distributed control applications within sensor networks," *Proceedings of the IEEE*, Vol. **91**, No. 8, pp. 1235–1246, Aug. 2003.

[21] E. Callaway, Jr., *Wireless Sensor Networks: Architectures and Protocols*, Auerbach Publications, 2003.

[22] Y. Liu, I. Elhanany and H. Qi, "An energy-efficient QoS-aware media access control protocol for wireless sensor networks," in *Proceedings of IEEE MASS 2005*.

[23] W. Heinzelman, A. Chandrakasan and H. Balakrishnan, "Energy-efficient communication protocol for wireless microsensor networks," in *Proceedings of the 33rd Hawaii International Conference on System Sciences*, Vol. **8**, p. 8020, January 2000.

[24] R. Iyer and L. Kleinrock, "QoS control for sensor networks," in *Proceedings of the 2003 IEEE International Conference on Communications*, Vol. **1**, May 2003.

[25] V. Rajendran, J. Garcia-Luna-Aceves and K. Obraczka, "Energy-efficient, application-aware medium access for sensor networks," in *Proceedings of the 2005 IEEE International Conference on Mobile Ad-Hoc and Sensor Systems*, Nov. 2005.

[26] V. Rajendran, K. Obraczka and J. Garcia-Luna-Aceves, "Energy-efficient collision-free medium access control for wireless sensor networks," in *Proceedings of the 2003 International Conference on Embedded Networked Sensor Systems*, Nov. 2003.

[27] J. Hill, R. Szewczyk, A. Woo *et al.*, "System architecture directions for network sensors", in *Proceedings of the International Conference on Architectural Support for Programming Languages and Operating Systems (ASPLOS), ASPLOS 2000*, Cambridge, November 2000.

[28] Y. Cho, L. Bao and T. Michael, "LAAC: a location-aware access control protocol," in *Proceedings of the 2006 Annual International Conference on Mobile and Ubiquitous Systems: Networking & Services*, pp. 1–7, July 2006.

[29] J. Hightower and G. Borriello, "Location systems for ubiquitous computing," in *IEEE Computer*, Vol. **34**, No. 8, pp. 57–66, Aug. 2001.

[30] P. Raviraj, H. Sharif, M. Hempel and S. Ci, "An energy efficient MAC approach for mobile wireless sensor networks systems communications," in *Proceedings of the 2005 IEEE Computer Systems and Applications*, pp. 370–375, 2005.

[31] W. Ye, J. Heidemann and D. Estrin, "An energy-efficient MAC protocol for wireless sensor networks," in *Proceedings of the 2200 IEEE INFOCOM*, pp. 1567–1567, June 2002.

[32] H. Pham and S. Jha, "An adaptive mobility-aware MAC protocol for sensor networks (MS-MAC)," in *Proceedings of the 2004 IEEE International Conference on Mobile Ad-hoc and Sensor Systems*, pp. 558–560, October 2004.

[33] H. Cao, K. Parker and A. Arora, "O-MAC: a receiver centric power management protocol," in *Proceedings of the 2006 IEEE International Conference on Network Protocols*, pp. 311–332, 2006.

[34] J. Polastre, R. Szewczyk and D. Culler, "Telos: enabling ultra-low power wireless research," in *Proceedings of the 2005 International Symposium on Information Processing, PSN 2005*, pp. 364–369, April 2005.

[35] R. Zheng, J. Hou and L. Sha, "Asynchronous wakeup for ad hoc networks," in *Proceedings of the 2003 ACM International Symposium on Mobile Ad Hoc Networking and Computing*, pp. 35–45, 2003.

[36] T. Zheng, S. Radhakrishnan and V. Sarangan, "PMAC: an adaptive energy-efficient MAC protocol for wireless sensor networks," in *Proceedings of the 2005 IEEE Parallel and Distributed Processing Symposium*, April 2005.

[37] S. Weinmann, M. Kochhal and L. Schwiebert, "Power efficient topologies for wireless sensor networks," in *Proceedings of the 2001 International Conference on Parallel Processing, ICPP 2001*, pp. 156–163, Valencia, Spain.

[38] J. Hill and D. Culler, "Mica: a wireless platform for deeply embedded networks," *IEEE Micro*, Vol. **22**, No. 6, pp. 12–24, Nov.–Dec. 2002.

[39] J. Mainwaring, R. Polastre, R. Szewczyk, D. Culler and J. Anderson, "Wireless sensor networks for habitat monitoring," in *Proceedings of 2002 ACM International Workshop on Wireless Sensor Networks and Applications*, pp. 88–97, Sept. 2002.

[40] K. Kredo II and P. Mohapatra, "Medium access control in wireless sensor networks," in *Computer Networks Journal*, Vol. **51**, pp. 961–994, 2007.

[41] P. Nicopolitidis, M. S. Obaidat and G. Papadimitriou, *Wireless Networks*, Wiley, 2003.

[42] M. S. Obaidat and N. Boudriga, *Security of e-Systems and Computer Networks*, Cambridge University Press, 2007.

[43] J. Zheng and A. Jamalipour, *Wireless Sensor Networks: A Networking Perspective*, Wiley, 2009.

[44] http://www.rfm.com/products/data/tr1000.pdf

[45] http://citeseer.ist.psu.edu/article/ye02energyefficient.html

[46] www.cs.nyu.edu/~nikos/personal/pubs/sow2002.pdf

[47] http://www.consensus.tudelft.nl/documents_papers/vanDam03.pdf

[48] http://www.polastre.com/papers/sensys04-bmac.pdf

5   Routing in wireless sensor networks

A number of factors have to be considered carefully when implementing network protocols for wireless sensor networks (WSNs). Any routing protocol for WSNs should take into consideration the limited energy resources in the sensor nodes. Moreover, WSNs are distinctive from basic ad-hoc networks with respect to communication channels. Generally, the sink nodes are more concerned with an overall depiction of the environment than with precise readings from their own sensor devices. Hence, the communication in the WSN is known as data-centric rather than address-centric and the data may be collected locally rather than transmitted raw to the sink(s). All of these distinctive features of WSNs play a part in the network layer and thus data routing protocols need to be thought of carefully.

In order to access data, sensors rarely have knowledge of their own locations and these locations can be used for routing functions in the network layer. If a WSN is well connected, the services of the topology control have to be utilized in conjunction with the general routing protocols. To the end of dealing with specific network issues in the routing layer, we find ourselves dealing with a different routing layer from traditional networks.

In this chapter, we will review the major aspects of routing in WSNs, including various types of routing protocols and related advantages, and disadvantages, as well as challenges [1–58].

5.1   Fundamentals of routing and challenges in WSNs

The major design issues of routing in WSNs are node deployment, the effect of energy consumption on accuracy, data reporting, node heterogeneity, link heterogeneity, fault tolerance, scalability, network dynamics, transmission media, connectivity, coverage, data aggregation, and quality-of-service (QoS).

Wireless sensor networks may contain numerous nodes; they may even be in the thousands. These wireless sensors can have the ability to communicate among each other or may directly go to the external station, which is the base station (BS). The great number of wireless sensors can allow the sensing of the network over larger geographical areas with a greater accuracy [1–30].

Figure 5.1 shows a schematic diagram of the components of a WSN. Typically, each of the wireless sensor nodes has the ability of sensing, processing, and transmission as well as being able to be a mobilizer [1–15].

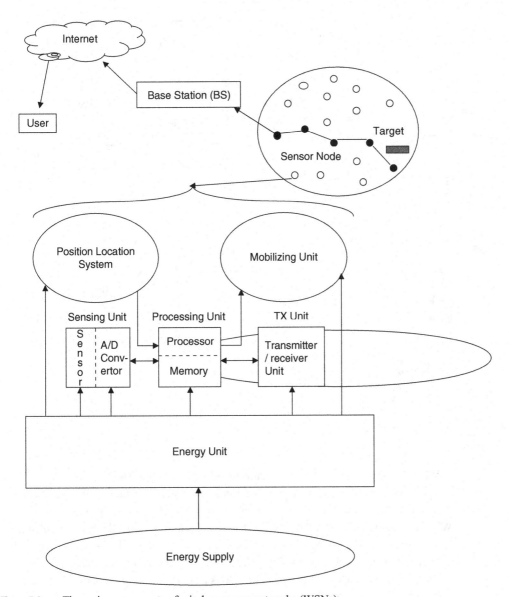

Figure 5.1   The main components of wireless sensor networks (WSNs).

The sensor nodes in WSNs are scattered in different sensor fields, and they coordinate amongst themselves in order to produce high-quality and secure information about the physical environment being monitored.

Each and every node of the scattered wireless sensor nodes has the efficient capability to collect and route the data either to different sensors or back again to the external station(s), which are at the base. A BS might be a fixed or mobile node, which is capable of connecting to the sensor network or to the communications infrastructure that already exists. It also may be the Internet [1–4, 27–30].

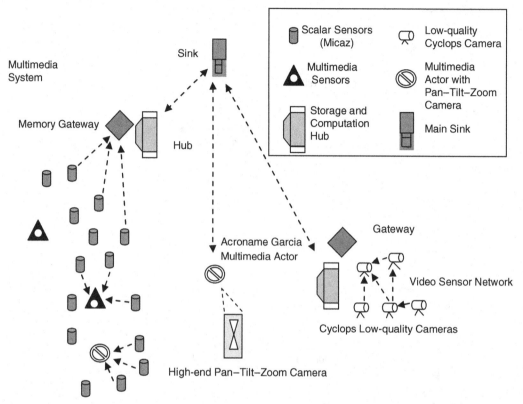

Figure 5.2   A typical wireless sensor network (WSN).

Wireless sensor networks have a huge number of applications, however, these networks have some limitations including limited computation power, limited memory, limited power supply, limited bandwidth and data rate (see Figure 5.2). When designing a WSN and its routing algorithms, we need to bear in mind essential design goals of WSNs such as moving data while trying to extend the lifetime of the network and avoid connectivity degradation by utilizing efficient energy management schemes.

The major challenges in the design of WSNs from the point of view of routing can be summarized as follows [1–30].

(1)   Power spending without affecting correctness. The sensor nodes in a WSN can use up their restricted supply of energy in order to perform computations and communication of data. Hence, doing these necessary tasks in an energy-conserving form is important. In a typical multi-hop WSN, each node has to act as both a data sender and data router. Obviously, a fault in a sensor node may cause some topological modifications, which may lead to re-routing of data.

(2)   Node setup. Node installation in WSNs is dependent on the specific application the system is intended for, which in a way influences the overall performance of the routing scheme being employed. In general, launching of nodes may be random,

where the nodes are distributed randomly or deterministic, where the nodes are placed by hand and routing is done using static (predefined) routes. In all cases, if the setup is not uniform, optimal clustering is needed in order to permit connectivity and facilitate power-efficient network operation.

(3)   Data-sending paradigm. It is known that data sensing and reporting in WSNs is dependent on whether the application is in real-time or not. Moreover, we can categorize data reporting in WSNs into four classes [1–20]: (a) time-driven, (b) event-driven, (c) query-driven, and (d) hybrid. The time-driven delivery scheme is meant for applications that need cyclic monitoring. Hence, the nodes will periodically turn on their sensors and transmitters, in order to sense the environment and send data periodically. In the case of event-driven and query-driven schemes, sensor nodes reply instantly to abrupt modifications in the sensed features because of the occurrence of a specific query, which is suited to real-time situations. A blend of the above schemes is also feasible.

(4)   Fault tolerance. In general, WSNs are considered fault tolerant due to the numerous nodes in them. However, if numerous nodes fail, then the MAC and routing protocols should be able to create new routes to the base stations. Of course, this may necessitate adjusting transmission power and signaling rates on the present links in order to minimize power consumption, or to re-route packets to sections in the WSN that have more energy. Clearly, a hierarchy of levels of redundancy may be required in efficient WSNs.

(5)   Node dissimilarity. Most practical applications of WSNs include nodes that are dissimilar, which means that the nodes do not have similar computation, communication, and power consumption characteristics/capabilities. Nevertheless, in most analytic and simulation analysis of WSNs, the analysts assume that the network has similar nodes. This means that having a heterogeneous set of sensor nodes brings several technical concerns, especially related to packet routing. Keep in mind that hierarchical protocols select a cluster-head node that is different from the ordinary sensors in that it is supposed to be more powerful in terms of bandwidth, computation power, memory size, and energy capability. Hence the task of transmission to the BS is performed by cluster heads [1–15].

(6)   Scalability. In a typical WSN, the number of nodes deployed is in the hundreds or thousands. Hence, it is essential that routing schemes work under such a large number of nodes. Moreover, the routing algorithms must have good scalability characteristics in order to accommodate new nodes; furthermore, they should be able to react to incidents in the working setting. Of course, until the proper condition is satisfied for reading or writing, the majority of sensor nodes will be in the sleep mode; only a low number of sensor nodes will be providing data [1–25].

(7)   System dynamics. The nodes in a WSN are not always static; they could be moving. The regular nodes and the base stations (BSs) may need to move in order to have proper operation. Routing under dynamic environments such as moving nodes is more difficult than under static environments as issues such as bandwidth, energy, and route stability become essential. Furthermore, the events

being monitored and sensed may be dynamic or static depending on the application of the WSN. Bear in mind that, in general, monitoring dynamic environments necessitates cyclic reporting and hence, generates a huge volume of traffic
that needs to be routed to the BSs [1–30].

(8)   Transmission media. Usually, in WSNs, the transmission medium is wireless,
which suffers from high bit error rate, hidden terminal problems, and energy
limitations. In multi-hop WSN environments, the nodes are connected together
using the wireless transmission medium, which suffers from fading along with
other unreliability issues. In addition, the required bandwidth is usually small:
around 200 kbps or even lower. As for medium access control (MAC), TDMA-
based schemes are preferred as they save energy when compared with other
schemes such as the CSMA. Furthermore, schemes like Bluetooth and ZigBee
can be used as well [1–30].

(9)   Connectivity. In WSNs, nodes are usually well connected. Given the fact that the
number of nodes in such a setup may get reduced due to sensor failure or
expiration of its battery life, the network may shrink in size. This will definitely
affect the network degree of connectivity.

(10)   Quality-of-service (QoS). Traffic in WSNs can be of any type including constant bit rate, variable bit rate (whether real-time or non-real-time), guaranteed
bit rate, available bit rate, and unspecified bit rate. Guaranteeing the required
QoS for each of these possible traffic types such as low delay, low jitter (delay
variation), or minimum data rate, may mean consuming more power from the
batteries of the mobile nodes, which may reduce the life of the entire network.
Clearly, the routing protocols to be used for WSNs have to be energy-aware
schemes.

(11)   Data aggregation. Owing to the nature of the operation, the nodes in WSNs may
produce a huge volume of data that has multiple copies. In order to avoid such
redundancy, identical packets from various nodes have to be aggregated to reduce
the traffic in the network and reduce the number of transmissions. Data aggregation is the process that combines data from various sources in the network
based on a certain aggregation function/algorithm. Signal-processing schemes
may be used for this process and, in such a situation, the method is called data
fusion [1–20].

(12)   Coverage. One of the aspects of a sensor node is its range or coverage scope. In
general, a sensor node is limited in terms of range. This feature needs to be looked
at carefully when designing a WSN for a specific application as it affects its
overall performance [1–30].

5.2   Network architecture-based routing protocols for wireless sensor networks (WSNs)

As stated earlier, routing is an important issue for the proper operation of WSNs. In this
section, we review the major network architecture-routing schemes for WSNs. Routing

in WSNs can be divided into [1–3, 10–17]: (a) flat-based routing, (b) hierarchical-based routing, and (c) location-based routing depending on organization of the WSN.

In addition, a routing protocol is called adaptive if specific parameters can be managed in order to adjust the current network conditions. Such protocols can be categorized into multipath-based, query-based, negotiation-based, QoS-based, or coherent-based routing techniques. Moreover, we can divide routing schemes into proactive, reactive, and hybrid, based on how the source discovers a route to the target. In proactive schemes, all routes are calculated ahead of the time when they are needed. In reactive schemes, routes are calculated when needed. The hybrid protocols use a blend of these two schemes. Next we review the major routing protocols for WSNs.

5.2.1   Multi-hop flat routing

In this type of protocol, each sensor node performs similar tasks to the others. The nodes cooperate together to do the task. Here, it is not realistic to give a global ID to each sensor node. This situation has led to data-centric routing, in which the base station (BS) transmits inquiries to specific areas and waits for the data from the sensors located in there. Examples of centric routing include sensor protocols for information via negotiation (SPIN) and directed diffusion [18–30]. In these schemes, energy is saved by using data bargaining and exclusion of redundant data.

5.2.1.1   Sensor protocols for information via negotiation (SPIN)
This family of protocols has been proposed by Heinzelman *et al.* [3, 7, 10–30]. It spreads the information from every sensor node to all other nodes in the WSN with the assumption that every node may be a candidate for the BS. Hence, this makes the process query any node and obtain the needed information directly and easily. Sensor protocols for information via negotiation (SPIN) makes use of the feature of clustering where nodes close to each other have similar data, and thus there is a necessity just to disseminate the data that other nodes do not have.

In SPIN, data negotiation and resource-adaptive algorithms are often used. The main concepts on which the design of SPIN schemes is based are [1–9, 20–30]: (a) nodes function well and save energy by sending some of the data, rather than all the data, as collected data have redundancy, and (b) traditional schemes such as flooding or gossiping-based routing protocols [6] dissipate energy and bandwidth when transmitting unneeded copies of data by sensors covering overlapped regions. The main disadvantages of flooding are: (a) implosion, which is due to the extra messages transmitted to the very same sensor node, (b) overlap in cases when two sensor nodes collect data from the same area, which end up sending the same data packets to the same neighbors, and (c) source loss of sight by spending a huge amount of energy without concern for the energy limitations.

Meta-data negotiation in SPIN can solve the traditional issue of flooding, hence getting high energy effectiveness. Since the nodes use three kinds of messages – ADV, REQ and DATA – to correspond, SPIN is considered a three-stage scheme. The ADV message is meant to advertise fresh data, the REQ message asks for data, and DATA is the real message. The algorithm begins when a SPIN node gets new data that it is keen to distribute. This is performed by transmitting an ADV message containing meta-data. If a nearby node is in need of the data, it transmits a REQ message for the DATA and the DATA are then transmitted to this nearby node. Then the latter repeats this procedure with its neighbors. At the end, the whole sensor region will get a copy of the data [1–5, 25–30].

There are many protocols that have been devised under the SPIN family of protocols. These include SPIN-1 and SPIN-2, which involve negotiation before sending data to guarantee that only needed information will be sent. Here, every sensor node has its own source supervisor that tracks resource consumption, and is polled by the nodes before sending the data. SPIN-2 is basically an extension to SPIN-1, which has a threshold-based resource alertness method. Of course, this is in addition to the negotiation process. When energy is abundant in the nodes, the SPIN-2 protocol corresponds to using the three-stage protocol of SPIN-1. Nevertheless, when the energy becomes very low in the sensor node, it reduces its involvement in the scheme. SPIN-1 and SPIN-2 are well-matched for situations where the sensor nodes are moving as they base their passing on decisions on neighboring information.

There are other protocols in the SPIN family. Among these are SPIN-BC, SPIN-PP, SPIN-EC, and SPIN-RL. The SPIN-BC scheme is intended for broadcast channels while SPIN-PP is meant for point-to-point communications. The SPIN-EC functions like SPIN-PP, however, with an energy heuristic added to its mechanism. SPIN-RL is designed for lossy channel environments; here modifications are added to the SPIN-PP protocol to address lossy channels.

5.2.1.2   Directed diffusion

This was devised as a data-aggregation scheme [1–15, 26]. It is a data-centric and application-aware model as all data produced by nodes are named by attribute-value pairs.

In the direct diffusion protocol, data are combined from different sources while traveling by getting rid of the redundancy, and reducing the number of transmissions, hence saving the power of nodes, and extending the overall network lifespan. This scheme finds routes from multiple sources that are intended for the same destination and makes sure that unneeded redundant data are removed.

In this protocol, sensor nodes assess events and build gradients of information in their neighborhoods. The base station asks for data by sending inquiry messages describing a specific task to be done by the WSN. As the interest (inquiry) message is broadcast all over the WSN, gradients are established to move data filling the query needs towards the interested node.

Here, the sensor node establishes a gradient toward the nodes from which it obtains that interest. The procedure continues on until gradients are established from the source nodes back to the base station. Keep in mind that a gradient identifies a characteristic and a path.

The directed diffusion protocol has two major different characteristics from the SPIN scheme [1–10, 20–30].

(1) It delivers on-demand data requests as the BS sends queries to the nodes by flooding some jobs. Nevertheless, in SPIN, sensor nodes publicize the availability of data which allows concerned nodes to seek the data.

(2) Every communication in this protocol is from neighbor-to-neighbor, with every sensor node having the potential to do data caching and aggregation. There is no necessity to keep a global network topology, unlike a SPIN scheme. Nevertheless, the directed diffusion protocol cannot be used for applications that need constant data supply to the BS.

5.2.1.3 Minimum cost forwarding algorithm (MCFA)

The MCFA scheme [18] utilizes the situation that the path of routing is at all times known, which is to the base station. Therefore, a node does not need to have a distinctive ID nor keep a routing table. Each sensor node keeps the smallest cost estimate from itself to the base station. All messages that have to be sent by the sensor node are relayed to its neighbors. If a node gets the message, it tests if the message is on the cheapest cost path connecting the source node and the base station. If it is true, then it re-sends the message to the nearby nodes. Such a procedure is replicated until the base station is reached. Each node knows the cheapest cost path approximation from itself to the base station. Such info is obtained by having the base station broadcast a message with the cost set to zero while each sensor node at the beginning puts its least cost to the base station to infinity. Then, upon the receipt of the message initiated at the base station, every node tests to find out if the approximation in the message and the link on which it arrived are less than the present estimate. If this is the case, then the estimate in the broadcast message is updated. So, if the delivered broadcast message is brought up to date, it is retransmitted. If not, it is eliminated and no other action is performed. Nevertheless, the preceding mechanism can result in having some nodes with several updates, and nodes far away from the base station will have more updates when compared to these nearer to the base station. In order to shun this issue, the protocol was modified so that it executes a backoff algorithm at the initial stage. This latter algorithm determines that a sensor node will not transmit a recent message until a specific amount of time elapses from the moment at which the message is updated. This amount of time is given by the product $(a*lc)$ where "a" is a constant and "lc" is the link cost from which the message was received [1–21].

5.2.1.4 ACQUIRE protocol

This protocol [30] is basically based on querying the WSN and it was called ACtive QUery forwarding In sensoR nEtworks (ACQUIRE). It perceives the network as a distributed database with compound queries that can be broken down into a number of sub-queries. In ACQUIRE, the BS node transmits a query that is then advanced by each node getting the query. In the meantime, each node attempts to reply to the query in part by utilizing its pre-cached information and then moves it to another node. If the pre-cached information is not current, then the nodes collect information from their neighbors in a look-ahead of hops. As soon as the query is addressed, it is sent back using the reverse or shortest-path to the BS.

5.2.1.5   Energy-aware routing protocol

The main aim of this scheme is to extend the lifetime of the WSN by saving consumed power as much as possible. This protocol differs from the directed diffusion scheme in that it keeps a set of paths instead of keeping one best possible path at high data rates. Such paths are preserved and selected using a specific probability function, which depends on how little the power spending of each pathway can be accomplished. Since paths are selected at different times, the energy of any path will not diminish rapidly; this can prolong the network's life since energy is spent more uniformly between sensor nodes. In contrast to the directed diffusion scheme, this protocol offers overall enhancement of over 21% energy saving and a 44% increase in WSN lifetime. Nevertheless, it necessitates collecting the location data and setting up the addressing method for the sensor nodes, which makes it more complex than the directed diffusion scheme. The major performance metric for this protocol is network survivability [3, 12, 30].

5.2.1.6   Rumor routing

This routing protocol is a variation of the directed diffusion scheme [3, 32]. It is used for situations where geographic routing is not possible. It employs flooding to insert the query to the whole WSN in case there is no geographic condition to diffuse. Nevertheless, there are cases where only small amounts of data are requested, hence using flooding is not necessary. Another way is to flood the events if the number of events is small and the number of queries is large. The rationale for this is to send the queries to the nodes that have experienced a specific event instead of flooding the whole WSN in order to find information on the happening events. Here, the rumor routing scheme uses special, long-lived packets, which are called agents, in order to flood events through the WSN. Hence, if a node spots an event, it includes such an event in a special table called an events table, and then produces an agent. The agents pass through the WSN to disseminate information on local events to remote nodes. If a sensor node produces a query for an event, the nodes that are familiar with the route may reply to the query by checking their tables. Clearly, there is no need to flood the entire WSN, which minimizes communication overhead. However, rumor routing keeps only a single path between source and destination as opposed to directed direct fusion, which uses multiple routes [3, 14–16].

It has been reported that rumor routing provides considerable energy savings if compared to event flooding. Nevertheless, rumor routing works well only when the number of events is low; for large events, it is impractical to maintain agents and event tables in each sensor node [3, 14–25].

5.2.1.7   Gradient-based routing

This protocol is a variant of the direct diffusion scheme [3, 15, 32]. The main notion of the gradient-based routing (GBR) protocol is to learn the number of hops when the attention is diffused throughout the entire WSN. Hence, every sensor node is able to compute a parameter termed the height of the node, which is the least number of hops to get to the BS. This approach uses some ancillary techniques such as data accumulation and traffic distribution in order to evenly distribute the traffic over the network. If various paths disseminate across a node, then this node will act as a relay and it could pool data based on a specific function. We can have three data dissemination schemes: (a) a stochastic technique, where a node chooses arbitrarily one gradient if two or more subsequent hops have similar gradient, (b) an energy-based technique, where a node enhances its altitude if its energy goes under a specific edge so that other sensor nodes are not recommended to send data to that sensor node, and (c) a stream-based technique, where new streams are not sent over the nodes that presently belong to the paths of additional streams. Performance evaluation studies using simulation have shown that the gradient-based routing (GBR) scheme is better than that for the directed diffusion from the point of view of overall communication energy consumed.

5.2.1.8 Routing protocols with random walks

This family of protocols obtains load balancing in a statistical manner by employing multipath routing in wireless sensor networks. It is meant only for giant networks with partial movement for the nodes. Here, it is understood that nodes are switched on/off randomly. Moreover, every sensor node has a distinctive identifier; however, no position information is necessary. Sensor nodes are arranged in a way that every node maps correctly on one crossing point of a grid on a plane; however, architecture may be random. In order to discover a route from a source to endpoint, the position information is acquired by calculating distances among nodes by the distributed asynchronous form of the Bellman–Ford routing protocol. An intermediary sensor node can select as the following hop a nearby node that is nearer to the endpoint based on an estimated probability. By calibrating this probability, a sort of load balancing may be achieved in the WSN [30].

5.2.1.9 Information-driven sensor querying (IDSQ) and constrained anisotropic diffusion routing (CADR)

These two schemes were proposed by Chu *et al.* [33]. In IDSQ, the probing node can decide which node can deliver the most valuable information with the extra benefit of equalizing the power bill. Nevertheless, IDSQ does not precisely describe how the request and the data are transmitted among sensor nodes and the BS. Hence, this scheme may be considered as a harmonizing optimization process. Performance evaluation results using simulation analysis have shown that this scheme is more energy-efficient than the directed diffusion scheme where requests are diffused in an isotropic manner.

Next, CADR is meant to be a general form of directed diffusion protocol. The major concept here is to probe sensor nodes and move data in the network so that the information improvement is exploited whereas latency and bandwidth are reduced. Queries are diffused by CADR through the use of a set of information standards in order to choose the sensor node that should receive the data. We can accomplish this by activating only sensor nodes that are near to a specific event and dynamically tuning the data paths. Here, CADR has an edge over the directed diffusion scheme owing to information gain and reduced communication cost [1–10, 33].

5.2.1.10   COUGAR

This protocol is considered a data-centric protocol [34], which regards the WSN as a giant distributed database structure. Basically, it employs declarative queries so as to extract query handling from the network layer tasks like selection of applicable sensors. It includes a structure for the sensor database system in which nodes choose a head node in order to achieve aggregation and send data to the base station (BS). The latter is in charge of producing a probe strategy, which states the needed information about the data movement. Moreover, the probe scheme defines the method to choose a head for the query. The structure offers in-network calculation capability, which can offer energy effectiveness in cases where the produced data are large. Although COUGAR offers network-layer-independent schemes for data query, it suffers from some drawbacks, namely: (a) the extra energy and memory needed due to the additional query layer, (b) the requirement for synchronization among nodes before sending data to the head node for effective in-network data calculation, and (c) the head nodes being dynamically sustained in order to avoid them turning into hotspots, which make them susceptible to breakdown [1–5, 34].

5.2.2   Hierarchical/cluster-based routing schemes

The hierarchical or cluster-based routing schemes were originally devised for fixed communications networks. Their key advantages are scalability and efficient communication characteristics. The concept of hierarchical routing is also utilized to accomplish energy-efficient routing in WSNs. In a hierarchical architecture, higher-energy nodes are able to process and send the information whereas low-energy nodes can carry out the sensing in the vicinity of the target.

Hence, the creation of clusters and assigning special tasks to cluster heads can greatly provide overall system scalability, lifespan, and energy efficiency. Hierarchical routing is an effective way to reduce energy spending inside a cluster and by doing data aggregation and fusion so as to reduce the number of sent messages to the base station (BS) [35].

Hierarchical routing is basically a two-layer routing scheme where one layer is employed to select cluster heads and the other one is used for routing. Nevertheless, most techniques in this class are not about routing, but are about who and when to transmit or process/accumulate the information, which may be orthogonal to the multi-hop routing function.

LEACH (low-energy adaptive clustering hierarchy) is a well-known routing protocol designed for wireless sensor networks in situations that require an end-user to remotely observe the environment. The reason we need a WSN protocol like LEACH is because a node in the WSN becomes useless when its battery dies. The data from the discrete nodes should be sent to a central base station, usually located away from the network, via which the end-user can read the data. The key desirable characteristics for protocols on these networks include: (a) the use of a large number, such as thousands, of nodes, (b) maximizing the system lifespan, (c) maximizing the network coverage, and (d) having identical battery-operated nodes [3, 35].

Traditional protocols like direct transmission, minimum transmission energy, multi-hop routing, and clustering all have downsides that do not allow them to accomplish all the anticipated features. LEACH contains distributed cluster formation, local processing in order to minimize global communication, and randomized variation of the cluster heads. These features allow LEACH to achieve the preferred properties. This protocol permits us to space out the lifespan of the nodes, which makes it perform the least amount of work in order to send out data.

LEACH is a cluster-based scheme, which includes distributed cluster formation. It arbitrarily chooses a few sensor nodes as cluster heads (CHs) and alternates this role to uniformly distribute the energy load between the sensors in the network. The cluster head (CH) nodes compress data received from nodes that belong to the relevant cluster, and then relay an aggregated packet to the BS to minimize the volume of information that needs to be sent to it.

LEACH uses a TDMA/CDMA MAC scheme in order to decrease inter-cluster and intra-cluster collisions. Nevertheless, data gathering is unified and is done regularly. Thus, this scheme is most suitable when there is a requirement for continuous monitoring by the WSN. Since the user may not need all the data straightaway, periodic data transmissions are not needed. After a specified interval of time, a randomized alternation of the role of the CH is performed to obtain uniform energy dissipation in the WSN.

LEACH operations can be divided into two main stages: the setup stage and the steady-state stage. In the first stage, the clusters are structured and CHs are chosen. In the steady-state stage, the real data transfer to the BS happens. The length of this stage is longer than the length of the setup stage, which helps in reducing the overhead. During the setup stage, a prearranged portion of nodes, p, designate themselves as CHs. A node selects a random number, r, between 0 and 1. If this random number is lower than a maximum value, $T(n)$, the node turns into a cluster head for the present round [3, 35].

5.2.2.1   Power-efficient gathering in sensor information systems (PEGASIS)

This scheme is an improvement over the LEACH scheme [3, 36–37]. The key concept in this scheme is to prolong the WSN lifetime. Nodes should correspond only with their nearest neighbors. Moreover, nodes should alternate in corresponding with the base station (BS). When the set of all nodes interacting with the BS ends, a new set will commence and so forth. Of course, this can minimize the power needed to send data per cycle since the power depletion is stretched evenly over the WSN nodes. This scheme has two main aims: (a) to increase the lifespan of every node by employing cooperative methods, which also raise the life of the WSN, and (b) to facilitate only neighboring management between nodes that are near to each other in order to reduce the bandwidth needed for communication. This is different to LEACH, as this protocol prevents cluster creation and employs only a single node in a sequence to send to the BS instead of using several nodes.

In this scheme, in order to locate the nearest-neighbor node, every node employs the signal strength to determine how far the node is to its nearby nodes and then modify the signal power in order to have only one node to be heard. The sequence here will involve the nodes that are nearest to each other and make a path to the BS. The combined structure of the data will be transmitted to the BS by any node in the group. It was found that PEGASIS can improve the lifetime of the network by double the amount of the LEACH scheme. This was due to the fact that LEACH suffers from the overhead affected by dynamic cluster establishment, and because in PEGASIS we have lower numbers of transmissions and reception as it uses data aggregation.

One important note about the PEGASIS scheme is that, despite the fact that the clustering overhead is prevented, PEGASIS necessitates dynamic topology tuning as a sensor node requires familiarity with the energy condition of its neighboring nodes to make decisions on data routing matters. Of course this kind of regulation may produce substantial overhead, particularly under heavy load conditions.

Furthermore, the PEGASIS scheme assumes that every node can communicate with the BS using a single hop; however, in reality the nodes employ multi-hop communication to correspond with the BS. Also, the PEGASIS scheme supposes that all nodes in the WSN have a similar power level, which means they will die simultaneously. It is worth noting here that PEGASIS produces disproportionate latency for faraway nodes in the chain and the sole leader may become a bottleneck node.

A modified version of PEGASIS, called hierarchical-PEGASIS, was proposed with the aim of reducing the latency experienced by packets during communication with the BS [3, 36–37]. A technique that combines signal coding and spatial transmissions can be used to prevent collision when concurrent communications of data are present.

5.2.2.2   Threshold-sensitive energy-efficient protocols

There are two known hierarchical routing algorithms that fall under this category: (a) threshold-sensitive energy-efficient sensor network protocol (TEEN) [3, 38], and (b) adaptive periodic threshold-sensitive energy-efficient sensor network protocol (APTEEN) [39]. These schemes are recommended for time-constraint applications.

In the TEEN method, the nodes sense the medium constantly, but data transmission is rarely performed. Each cluster head sensor transmits to its group a strict limit that is the ceiling value of the sensed feature and a lenient limit that is a minor modification in the value of the sensed feature. This prompts the node to turn on its transmitter and send data. Hence, the strict limit attempts to minimize the number of transmissions through letting the nodes send only whenever the sensed feature is in the radius of concern. Moreover, the lenient limit minimizes the number of transmissions that could then have taken place when there is minor or no modification in the sensed feature. A reduced value of the lenient limit offers a more-precise image of the network at the cost of higher power expenditure. Hence, the user can manage the trade-off amongst power effectiveness and information correctness.

The major disadvantage of this protocol is that, if the limits are not reached, the nodes will never correspond and the user will not be able to acquire data from the WSN.

The APTEEN scheme is a hybrid protocol, which modifies the ceiling values used in the TEEN technique based on the user demand and the nature of the use. In this protocol, the cluster heads send the following parameters: (a) the schedule, which is a TDMA timetable that assigns a time to each node; (b) the count time, which is the largest time period between two consecutive reports transmitted by a node; (c) attributes, which are a group of physical values that the user has interest in getting reports on; and (d) limits (thresholds), which are a set of strict or soft ceilings/thresholds [38–40].

Here, the node senses the setting constantly, and only the nodes that sense a data value at or outside the strict limit/threshold can send. After a node senses a value higher than the strict limit, it sends data only when the level of that feature varies by a degree equal to, or higher than, the lenient/soft limit. Now, if a node cannot transmit data for a time duration equal to the count time, it is obliged to sense and re-send the data. For this scheme, a TDMA timetable is employed and every WSN in the cluster is given a transmission time frame. Therefore, the APTEEN protocol employs an altered TDMA timetable to realize the mixture network.

The key aspects of the APTEEN protocol are: (a) it blends together proactive and reactive strategies, (b) it provides a huge flexibility as it permits the user to set the count-time period, and (c) the limit values for the power spending can be managed by modifying the count time and the limit/threshold levels.

On the negative side, the protocol suffers from the extra complexity needed to realize the limit/threshold functions and the count time. Performance evaluation studies have shown that the TEEN and APTEEN schemes have better performance than LEACH. The APTEEN performance is between LEACH and TEEN from the viewpoint of the energy dissipation and network lifetime performance metrics. The TEEN protocol has the best performance when compared to APTEEN and LEACH as it reduces the number of the needed transmissions [3, 38–49].

5.2.2.3   Small minimum energy communication network (MECN)

Here, the scheme figures out an energy-efficient subnetwork, specifically the minimum energy communication network (MECN), for a particular sensor network by using low power global positioning systems (GPS) [40]. It identifies a relay region for each sensor node in the network. The communication area comprises nodes in a nearby area. Sending data within such nodes can provide better energy efficiency when compared to sending data directly [41]. The addition of a node i is then done by acquiring the union of all communication areas that node i can touch. The key concept of MECN is to locate a sub-network that can have a smaller number of nodes and entails lower power for communication between any two nodes. Hence, the overall lowest power paths are identified without the need to consider all sensor nodes in the WSN.

This scheme is self-reconfiguring and, therefore, it can adjust to the failure of nodes as well as to the placement of new ones [42]. Another efficient scheme, which is an extension of MECN, is called small minimum energy communication network (SMECN) [3, 42]. It is assumed in the MECN scheme that each sensor node is able to send data to every other node, which may not be practical all the time. Potential barriers between any two nodes are contemplated in SMECN. Nevertheless, the network is usually presumed to be completely linked as in MECN [3, 42].

5.2.2.4   Self-organizing protocol (SOP)

In [43], the authors present a self-organizing scheme and a classification that was employed to form a structure used to sustain mixed types of sensors. In addition, such sensors can be transportable or fixed. Certain sensors inquire about the surroundings and pass information to a selected group of nodes that behave as routers. Router nodes are fixed and they establish the pillar for interaction. Gathered information is sent via routers to BSs. Every node is expected to get to a router to become a member of the WSN [3, 43].

5.2.2.5   Sensor aggregates routing

Several schemes have been devised to build and support sensor collection in [44]. The aim is to jointly observe target action in a specific setting. A wireless sensor collection consists of those nodes in a WSN that fulfill a predicate for a cooperative processing mission. These factors rely on the mission and its resource constraints. The creation of suitable sensor collections has been investigated [3, 44] from the point of view of giving resources to sensing and communication missions. Hence, sensor nodes in a field are grouped together based on their sensed signal intensity. There is a single peak for each group/cluster. Next, the local cluster heads are selected. In order to choose a leader, data interchange among immediate sensors is essential. The sensor node interchanges packets with the nearby sensor nodes and if it discovers that it is greater than all of its one-hop neighbors from the point of view of signal strength, then it announces itself as a head. The tracking scheme supposes that the head recognizes the geographical area of the cooperation. Reference [44] presents some algorithms in this regard.

(1)   The distributed aggregate management (DAM), for creating sensor aggregates for a target observation mission, consists of a determination predicate P for every sensor node to determine if it must be part of an aggregate and a message exchange process. Any node decides if it is part of the aggregate depending on the outcomes of employing the predicate to the information of the sensor node. This is in addition to data from other sensor nodes.

(2)   The energy-based activity monitoring (EBAM) scheme approximates the power intensity at every sensor node by calculating the signal influence region. This is done by fusing a weighted configuration of the sensed target power at every affected sensor node supposing that every aimed sensor node has identical or stable power intensity.

(3)   The expectation-maximization like activity monitoring (EMLAM) scheme eliminates the steady and equal aimed power intensity hypothesis. It approximates the aimed positions and signal power by employing the obtained signals and utilizing obtained estimates to envisage how signals from the targets could be assorted at every sensor node. The procedure is repeated till the approximation is satisfactory [3, 45].

5.2.2.6   Virtual grid architecture routing scheme (VGA)

This protocol was devised as an energy-efficient routing scheme [45]. It employs data accumulation and in-network handling in order to extend the network's life-cycle. Because of the sensor node's exceptionally low movement in most WSN applications, a pragmatic tactic is to place sensor nodes in a static layout [1–7, 45].

Square clusters have been utilized to get a static rectilinear virtual topology [45]. In every region, a sensor node is chosen to operate as a head of the cluster it belongs to. Information is processed at the local and global level. The group of heads of clusters, which are often called local aggregators (LAs), execute the local aggregation, whereas a subset of these aggregators worked to do global aggregation. Nevertheless, finding the best among the global aggregation points, usually called master aggregators (MAs), is considered an NP-hard problem [45, 46].

5.2.2.7   Hierarchical power-aware routing (HPAR)

This scheme is basically a hierarchical power-aware routing protocol that splits the WSN into groups of sensors [47]. Every group of sensors is dealt with as one entity. For routing, each sector decides how to route a message across the other sectors in a way that makes the batteries of the nodes live as long as possible. The message is sent over the path that experiences the greatest overall minimum of the remaining power, usually called the max–min path. The idea is that selecting the sensor nodes with the great remaining power might be expensive when compared to the path with the least power expenditure. In [47], a scheme, called the max–min zP_{min} algorithm, was devised. The scheme is based on the compromise between reducing the total power expenditure and making the most of the minimal remaining power of the WSN. It attempts to improve a max–min route by restraining its power spending as described next. Initially, the scheme discovers the path with the minimum power spending (P_{min}) by utilizing the popular Dijkstra technique. Then it discovers a link that exploits the smallest residual power in the WSN. So, the scheme attempts to improve both yardsticks [3, 47]. Zone-based routing, which depends on max–min zP_{min}, is a hierarchical technique where the region covered by the sensor network is split into smaller regions [47]. In order to send a message through the area, a path from region to region needs to be discovered. Sensors in a region independently guide neighboring routing and take part in assessing the region power intensity. Every message is moved through the regions by utilizing information about the region power approximations. Here, an overall manager for message movement is given the respon-sibility of handling the regions. This manager can be the node with the most power level. When the WSN is divided into small regions, the range for the overall routing scheme is minimized. Basically, the needed information to pass every message is reduced to the estimates of power level of every region [3, 47].

5.2.2.8   Two-tier data dissemination (TTDD)

The two-tier data dissemination (TTDD) scheme offers data provision to various mobile base stations (BSs) [48]. Here, every source of data constructs a grid that is employed to spread data to the mobile sinks by supposing that the nodes are fixed and location aware.

In this scheme, the nodes are fixed and location aware; however, the sinks may move their locations. When an event happens, nearby sensor nodes manipulate the signal and one of the nodes turns out to be the source to produce the data reports.

In order to construct the grid, a data source announces and selects itself as the initial crossing point of the grid, and conveys a message for its four adjacent crossing points by a simple greedy geographical forwarding scheme.

After the message arrives at a node that is the nearest to the crossing point, it will halt. Throughout this course, every in-between sensor node saves the source report and then sends the message to its nearby crossing points with the exception of the one from which the message originates.

This procedure continues until the message halts at the boundary of the WSN. Here, the nodes that save the source information are selected as distribution outlets. By utilizing the grid, a BS can swamp a query, which will be advanced to the nearest distribution outlets in the local cell in order to get data. Then the request is sent along other distribution outlets to the source. The needed data then move down in the opposite path to the sink. Trajectory advancing is used while the BS moves in the sensor region. Despite the fact that TTDD is an effective routing scheme, we have specific concerns on how the scheme gets location data. Moreover, the length of a forwarding path in TTDD is longer than that of the shortest path. Performance evaluation results between TTDD and directed diffusion algorithms have shown that the former can achieve extended lifetimes and data-delivery delays. Nevertheless, the overhead associated with keeping the grid as the network topology varies may be excessive. In addition, TTDD assumed that we have a very exact positioning system technology, which is not yet possible for WSN environments [48].

5.2.3   Location-based routing schemes

In this class of routing, the nodes are addressed using their locations. The distance between nearby nodes may be approximated based on the power level of the received RF signal. Comparative coordinates of adjacent sensor nodes may be acquired by exchanging these data among nearby nodes [37, 49, 50]. Otherwise, the position of sensor nodes can be found directly by cooperating with a satellite system employing GPS; this assumes that the nodes have tiny low-power GPS receivers [46].

In order to save power, various location-based techniques require nodes to nap when there is no action. Additional power reduction may be attained by putting as many nodes to sleep as possible. The key issue in scheming the nap cycles for each sensor node in a localized fashion has been investigated [3, 51, 46].

Below is a brief review of the major location (geographic)-based routing schemes.

5.2.3.1   Geographic adaptive fidelity (GAF)

This is an energy-aware location-based routing scheme that was devised mainly for mobile ad-hoc network systems; however, it can be applied to WSNs. Here, the WSN is partitioned into regions that make an implicit grid. Within each zone, the sensor nodes team up with each other to perform various tasks. The nodes in the WSN can elect one of them to remain alert for a specific period of time and then they turn to the sleep mode. This selected node is in charge of observing and sending information to the BS on behalf of the sensor nodes in the region. This scheme saves power by switching off the nodes in the WSN that are not needed without modifying the intensity of the routing reliability. Nodes use their GPS-designated location to correlate themselves to points in the implicit grid. The nodes related to the same point on the grid are deemed the same from the viewpoint of the cost of packet routing. This similarity is utilized in preserving some nodes positioned in a specific grid region in the sleep mode in order to save power. Hence, this scheme has the capability of significantly enhancing the lifespan of the WSN with the increase of the number of nodes [3, 46].

The operation of this scheme is based on three states: the discovery, active, and sleep states. To manage the mobility of the system, every sensor node has to approximate its departure time and relays this to its nearby nodes. In order to keep proper routing accuracy, the napping/sleeping nearby nodes modify their sleeping time as needed. Here, sleeping nodes awaken and one of them turns to be active immediately before the departing time of the active node expires. The GAF scheme is usually realized for both mobile and non-mobile environments.

Figure 5.3 illustrates a scenario of fixed zoning, which may be employed in WSNs [46].

The choice of the square size is based on the required sending power and the transmission path. An upright and straight communication is assured to occur if the signal moves a distance of $x = R/(5)^{0.5}$, selected in a way to make any two sensor nodes in nearby upright or straight clusters correspond quickly. In the case of transverse communication, the signal has to traverse a distance of $y = R/(2)^{1.5}$. A cluster head may request the node group to turn on and begin collecting data if it detects a target. Subsequently, the head takes charge of getting basic data from the nodes in its cluster and sends them to the

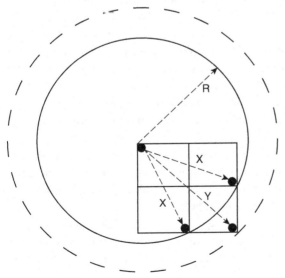

Figure 5.3 One possible scenario of fixed zoning in WSNs.

BS. This scheme endeavors to have the WSN linked by having an agent node constantly in the active state for every zone in its grid. Performance evaluation studies using simulation analysis have shown that this protocol does as well as a normal ad-hoc routing scheme from the view point of delay and packet loss, and enhances the lifespan of the WSN as it reduces power consumption [1–3, 46].

5.2.3.2   Geographic and energy-aware routing (GEAR)

The utilization of geographic information while distributing requests to applicable regions as data requests usually contain geographic attributes has been investigated [52]. Geographic and energy-aware routing (GEAR) limits the number of notices in a straight transmission; this is achieved by having a specific region instead of transferring the request to the entire WSN. Hence, this scheme can save more power than directed diffusion.

Every sensor node in GEAR maintains an approximated cost and a learned cost of reaching the target node via its neighbors. This approximated cost is a mixture of residual power and distance to target. The learned cost is an enhancement of the approximated cost that takes care of the routing near holes in the WSN. When a sensor node does not have any nearby nodes to the aimed zone, we say then that there is a hole. When there are no holes, then the approximated cost is the same as the learned cost. The latter is sent one hop back whenever a packet arrives to the target in order to amend the route setup for the next packet. The operation of this scheme can be summarized by two stages [3, 52]: (a) forwarding packets towards the aimed zone, and (b) forwarding the packets inside the zone. In the first stage, as soon as a packet is delivered, a node examines its adjacent nodes to find out if there is a neighbor that is nearer to the aimed zone than that one. If more than one node is found, then the closest neighbor to the aimed zone is chosen. However, if all nodes are away from that particular node, then this implies that there is a hole. In such a case, one of the nearby nodes is selected to move the packet forward depending on the related learned cost function. Such a selection may then be brought up to date based on the convergence of the learned cost over the distribution of packets, and moving these packets inside the zone. When the packet reaches the zone, it then can be disseminated in the zone by using whichever recursive geographic forwarding or restricted flooding approach is applicable. In general, in the densely populated WSNs, the recursive geographic flooding scheme is better in terms of energy efficiency when compared to the restricted flooding. Here, the zone/region is partitioned into four subdivisions and four copies of the packet are generated. This dividing and redirecting procedure prolongs up until the zones with only one sensor node are left. Reference [52] compares the GEAR scheme with analogous non-energy-aware routing technique, GPSR [53], which is one of the first efforts in geographic routing. The latter utilizes planar graphs in order to resolve the issue of holes.

Moreover, in a GPSR scheme, packets pursue the boundary of the planar graph in order to locate their route. As for GEAR, it not only minimizes power expenditure for the route arrangement, it also does better than GPSR in terms of packet distribution [3, 52–53].

5.2.3.3 MFR (most forward within radius), GEDIR (the geographic distance routing), and DIR (compass routing method referred to in the literature as DIR)

These three schemes are based on the concepts of basic distance, progress, and direction [29]. The main concern here is how to handle the forward direction and backward direction. The GEDIR scheme is of a greedy nature, as at all times it routes the packet to the node that is close to the present peak whose distance to the destination is reduced. The scheme falls short if the packet traverses the same edge twice in sequence. Normally, the MFR and greedy schemes use the same route to the target node. However, in the case of the DIR scheme, the best neighbor takes the nearest direction to the target. This means that the nearby node with the least angular distance from the virtual line connecting the present node and the target node is chosen. In the case of the MFR scheme, the favorite neighbor can reduce the scalar product $DA \cdot DS$, where D and S are the destination and source nodes, respectively, and SD represents the Euclidean distance between the source and destination nodes, S and D. Otherwise, the dot product $SD \cdot SA$ can be maximized. Every scheme stops routing the message at the sensor node that has the optimum choice of taking the message back to the preceding node. Both the GEDIR and MFR schemes are without loops. On the other hand, the DIR technique may have loops, except if the previous traffic is remembered or a timestamp is administered. Performance evaluation studies of these three techniques have shown [28] that all three basic protocols have pretty much similar performance outcomes from the viewpoint of the delivery rate metric. Moreover, the simulation analysis results have found out that the nodes in greedy and MFR techniques chose an identical routing neighbor almost all the time; and the entire chosen routes were the same in the overwhelming majority of the cases [3, 28].

5.2.3.4 The greedy other adaptive face routing (GOAFR)

This scheme usually selects the neighbor nearest to a sensor node to be the following node for transmitting [29]. Nevertheless, it may not find any closer node except the current node. The other face routing (OFR) scheme is an alternative of the familiar face routing (FR) algorithms. The latter scheme [28, 29] is the earliest technique that ensures realization if the sender and the receiver are linked. The poorest-case cost of FR is proportionate to the number of nodes in the WSN.

The adaptive face routing (AFR) scheme is the major technique that is able to contend with the optimum path in the poorest case.

The OFR scheme exploits the face configuration of planar graphs in a way that the message is moved from one sensor node to another by passing through a sequence of face edges. The objective is to discover the nearest node to the destination. As soon as it is terminated, the protocol sends the optimum sensor node on the edge. The regular greedy scheme acts well in condensed WSN environments, however, it collapses for uncomplicated structures [3, 29]. The performance simulation analysis results have indicated that there are some approaches to enhance the overall performance evaluation results of the GOAFR scheme. Moreover, it was found that GOAFR beats other well-known schemes like AFR, SPAN, and GPSR [3, 28, 29].

SPAN is a power-saving algorithm for multi-hop ad-hoc wireless networks that minimizes power expenditure without considerably reducing the capabilities of the WSN. The scheme is based on the investigation that when a zone of a common shared channel wireless network has enough nodes, only a small number of them need to be turned on at any time in order to route traffic for operational links [1–15].

This technique is of a distributed nature where nodes decide locally on the mode of operation. Such a decision is based on an approximation of the number of neighbors of each node that can gain from having it turned on and also on the extent of power remaining in it.

In SPAN, the system lifespan is enhanced because of the high idle-to-sleep energy-consumption ratio, which increases with the growth of the number of nodes of the network [51]. This scheme fits in well with 802.11 systems when operating in combination with the 802.11 energy-saving manner; it enhances average delay, capability, and system life [3, 51].

5.3   WSN routing protocols based on the nature of operation

In addition to the aforementioned taxonomy of routing protocols for WSNs, we can also classify these protocols based on the nature of operation of the WSN algorithms. Below is a brief description of the major protocols under this taxonomy.

5.3.1   Query-based routing approach

Here, the target nodes transmit a request for data from a node via the WSN and any sensor node that has the requested data transmits them. Every time a mediator/agent traverses a route with a path heading to an occurrence that it has not yet met, it produces a route state that heads to this occurrence or event.

If the agents find quicker paths or more inexpensive routes, then they enhance the paths in the related routing tables. Every sensor node keeps a record of its nearby nodes, an event table that is brought up to date every time recent events are found. Moreover, every sensor node may produce an agent in a stochastic manner. Here every agent holds an event table, which is coordinated with each sensor node that it goes to. This agent has an existence equal to a specific number of hops, following which it expires [32].

In general, a node will not produce a request unless it acquires a path to the necessary event. When there is no path available, the node sends a request in an arbitrary manner. Now, the node waits to find out whether the request went to the target node for a guaranteed amount of time, after which the node swamps the WSN if no reply is received from the target.

5.3.2   Multipath routing schemes

Here multiple paths are employed rather than only one route. This is done to improve the WSN performance. The strength of a routing scheme is assessed by the chance that a substitute route occurs between a sender and a target when the key route dies. Of course, this may be improved by providing several routes between the sender and the target on account of enhanced power use and network traffic production. Such substitute routes remain active via transmitting regular notices. Therefore, the WSN stability may be enhanced by raising the communication overhead needed to preserve the other routes. A scheme has been devised to move data over the route with nodes having the greatest remaining power [54]. The route can be modified at any time if a suitable route is located. The principal route will be employed till its power drops below the power of the standby route, when the standby route is employed. Hence the sensor nodes in the main route may not drain their power supplies via persistent usage of the same path, which can lead to an extended life. The infrequent usage of a group of sub-optimal routes has been devised to enhance the lifespan of the WSN [55]. Such routes are selected using the probability of how small the power usage of each route is. The route that has the maximum residual power when employed to send data in the network may perhaps be especially energy-costly as well. Hence, there is a compromise between reducing the overall energy spent and the remaining power of the network. A scheme to relax the residual energy of the route so as to choose a further power-efficient route has also been devised [47].

The authors in [56] proposed a multipath routing-based scheme to improve the reliability of the network, which is advantageous in order to deliver data in erratic settings. Having multiple routes between source and destination will definitely improve the reliability of the network. Nevertheless, this may raise the volume of traffic considerably. Therefore, we need to be careful and try to have a compromise between the volume of traffic and the trustworthiness of the WSN. An investigation of this compromise employing a redundancy function that relies on the multipath level and on the fading probabilities of the existing routes is found in [56]. The key concept behind this is to divide the earliest data packets into smaller packets and then transmit each of these small packets using one of the available multipaths. Based on the scheme, it was established for a specified largest node breakdown probability, employing a larger multipath level than a specific best estimate will raise the overall probability of breakdown. One example of multipath routing is directed diffusion scheme [31]. In this approach, a multipath routing protocol that discovers some moderately separate paths has been investigated. It was established that the utilization of multipath routing offers a substitute for power-efficient healing from stoppage in the network.

5.3.3   Coherent and non-coherent processing

There are two cases of data processing operations that were meant to be implemented in wireless sensor network systems: coherent and non-coherent data-processing-based routing [57]. In the latter, sensor nodes can locally treat the basic data before being transmitted to other sensor nodes in the network for more treatment. In the coherent routing technique, data are sent to combiners once some minor treatment is performed. This small processing may involve operations such as time marking and removal of replicas, among others. The coherent processing scheme is usually chosen in order to accomplish power-effective routing.

On the other hand, the non-coherent approach has small data traffic loading. Because the coherent-based approach produces stretched data flows, power efficiency can be obtained by employing the best path. There are three stages for the operation of the non-coherent approach [3, 57]: (a) recognition of target, collection of data, and preprocessing, (b) affirmation of membership, and (c) determination of central node.

Because the central node is chosen to make complex processing, it should have enough residual power, memory, and computing resources. Single and multiple winner schemes have been devised for non-coherent and coherent processing [57]. In the first scheme, only one collector sensor node is selected for compound processing cases. This selection is established on the residual power, memory, and computational capabilities of the selected sensor node.

At the completion of the single winner scheme operation, a minimum-hop spanning tree can deal with the WSN. The multiple winner scheme is basically a straightforward augmentation of the single winner technique. When all sensor nodes send data to the main collector sensor node, a huge amount of power may be used. To alleviate this situation and reduce the amount of energy spent, we can restrict the number of sources that can transmit data to the main collector sensor node. Normally, the multiple winner scheme has larger latency and overhead, and less scalability than the non-coherent processing networks.

In order to reduce the power consumption, we can restrict the number of nodes that may transmit to the key collector node. Each sensor node maintains a log of all the nodes that may send data rather than maintaining only the most potential candidates. Every node in the WSN will have a group of routes with the least power requirements to every other node. The single winner scheme can be employed to locate the sensor node, which requires the least power use. Hence, this latter node may function as the key node [57].

5.3.4   Quality-of-service (QoS)-based routing schemes

The QoS-based routing algorithms are designed to have a sense of balance between QoS and power spending, especially in offering good levels of QoS metrics such as low delay and jitter, low loss, and acceptable data rate at reasonable power level when sending data to the BS. In [1–30, 57], the sequential assignment routing (SAR) protocol was devised and this is believed to be the earliest routing scheme for wireless sensor networks to present the concept of QoS in routing. In this latter scheme, routing decisions are based on the following aspects: QoS on every route, energy supply, and precedence degree of the packet. In order to prevent sole path malfunction, a multi-route technique is employed and confined route re-establishment approaches are exploited. Hence, to establish multiple routes from the source sensor node, a tree rooted at the source to the base stations is created. Sensor nodes with low power levels or with no QoS guarantees are usually avoided while routing trees are constructed. When this process is completed, each sensor node will end up being a member of a multi-route tree. Hence, the SAR scheme is considered a table-driven multi-route technique that targets obtaining energy effectiveness and defect/fault tolerance. This scheme computes a weighted QoS metric as the result of the additive QoS metric and a weight factor related to the precedence degree of the packet. So SAR is targeted to reduce the mean weighted QoS measure during the entire lifespan of the WSN. Whenever the topology varies because of node malfunctioning, a route is re-calculated and can be available for use. In order to avoid any consequence that may result from the change in the network topology, a regular re-calculation of routes is initiated by the base station in order to face any modification in the network configuration. In order to come back from a failure, we use a handshake method that is based on a local route recovery technique between adjacent sensor nodes. This is performed via imposing routing-table reliability among all nodes whether upstream or downstream. Performance evaluation results using simulation analysis have shown that SAR provides lower energy spending than the minimum-energy metric scheme that concentrates on the power use of every packet regardless of its precedence [58]. This scheme keeps multiple routes from nodes to the base station. Of course, this guarantees fault tolerance and simple healing; however, the algorithm experiences the overhead of keeping tables and status at every node, particularly in large networks cases.

There is another scheme that comes under the QoS routing protocols; this is called the SPEED scheme. It offers smooth real-time end-to-end QoS assurance [58]. It requires every sensor node to keep data on its adjacent nodes and utilizes geographic transmitting in order to locate the routes. Moreover, it tries to guarantee a specific speed for every packet in the WSN in order to have every function approximate the end-to-end latency for the packets. This is found by dividing the range to the base station by the speed of the packet prior to establishing the admission resolution. It also can offer congestion prevention if the WSN is in congestion state. The latency approximation at every sensor node can be found by computing the time spent when an acknowledgement is obtained from an adjacent node as a reply to a sent packet [58].

It is worth mentioning here that the routing part in the SPEED scheme is termed stateless non-deterministic geographic forwarding (SNGF). The latter chooses the sensor node that can accommodate the speed constraint. Hence, if it does not meet the speed constraint, the relay fraction of the sensor node is examined so as to find the miss ratios of the neighbors of a node; nodes that could not offer the required speed. The obtained info is supplied to the SNGF part [3, 58].

5.3.5   Negotiation-based routing schemes

These protocols employ high-level data description so as to remove the unnecessary data communication via conciliation. The decisions on communication are based on the existing resources. Among the examples on the negotiation-based routing techniques are the SPIN family protocols [27] reviewed earlier, and the ones investigated in [13]. The key drive here is that the utilization of flooding to broadcast data may yield overlap between the transmitted data, therefore nodes may get replicas of the data. Such a phenomenon can use more power and extra processing overhead by transmitting the very same data by various sensor nodes. As explained earlier, the SPIN algorithm is meant to send the data of one sensor node to all others with the understanding that these sensor nodes are candidates to be BSs. Therefore, the key concept of the negotiation-based routing schemes is to overcome replica information and avoid unnecessary data from being transmitted to the following sensor or the BS by performing a sequence of negotiation messages before actual data are sent.

5.4   Summary

Despite the fact that numerous efforts have been exerted to solve the many issues in routing in wireless sensor networks, there are still many more challenges that obstruct effective solutions to these issues. One of these issues is the fact that the sensor nodes are usually deployed in unattended places. This is very different from the mobility applications, PDAs, and the Internet that directly and primarily interface with the human users. Secondly, a slight footprint characterizes sensors, as they are equipped with a power source that has a short lifespan; such nodes present stringent constraints on energy. This is also somewhat incomparable to the usual fix, reusable resources. Thirdly, transmissions between the sensor nodes are the main energy consumer in the WSN environment where a lot of energy is expended for sending a bit over 10 or 200 m, similar to that consumed by thousands to millions of operations.

It is assumed by most of the present routing schemes that the BS and the sensor nodes are stationary. Nevertheless, situations might happen wherein both the BS and the sensor nodes might need to be mobile. In such cases, an excessive drain of the energy from the nodes may occur in order to update the positions of the nodes and the command node regularly. Hence, there is a need for better routing protocols to effectively take care of the overhead of topology and mobility changes in such environments where the energy is limited.

The process of routing in WSNs is a rapidly growing area of research, although it is new and limited. Generally, the routing schemes have been categorized and several newly proposed protocols have been examined and compared with existing routing schemes. Each of these protocols has its own advantages and disadvantages based on the scenario of operation/application that has been taken into account and the factors that have been considered while evaluating them. Most of the discussed protocols are promising, but there are still many challenges left to be resolved in the area of routing in WSNs.

In this chapter we have reviewed the fundamentals of routing and challenges in WSNs. We identified the key design issues of routing in WSNs, which include: node deployment, energy consumption, data reporting, node heterogeneity, link heterogeneity, fault tolerance, scalability, network dynamics, transmission media, connectivity, coverage, data aggregation, and quality-of-service (QoS), among others. We also studied the major techniques for routing in WSNs. We classified these routing schemes into two main categories: network architecture-based routing schemes and the nature of operation-based routing schemes. We studied the main aspects of the popular techniques under each of these two broad routing protocol categories. Furthermore, we addressed the concepts for the design trade-off between key design considerations/factors such as power saving and communication overhead involved, along with the benefits and drawbacks of these routing protocols. Despite the fact that there are many good WSNs routing protocols today, we still need more efficient schemes, especially power-aware and secure schemes owing to the nature of these networks and their operating environments. It would be more effective to devise WSN routing schemes that are both energy aware and secure at the same time. This latter requirement is challenging as, if we want these schemes to be very secure, the security algorithm will require more computation, which leads to more power consumption from the battery [1–58].

References

[1] S. K. Dhurandher, S. Khairwal, M. S. Obaidat and S. Misra, "Efficient data acquisition in underwater wireless sensor ad hoc networks," *IEEE Wireless Communications*, Vol. **16**, No. 6, pp. 70–78, Dec. 2009.

[2] C. Ramachandran, M. S. Obaidat, S. Misra and F. Pena-Mora, "A secure, and energy-efficient scheme for group-based routing in heterogeneous ad-hoc sensor networks and its simulation analysis," *Simulation: Transactions of the Society for Modeling and Simulation International*, Vol. **84**, No. 2–3, pp. 131–146, Feb./March 2008.

[3] J. N. Karaki and A. E. Kamal, "Routing techniques in wireless sensor networks: a survey," *IEEE Wireless Communications*, Vol. **11**, No. 6, pp. 6–28, Dec. 2004.

[4] S. Singh, M. Woo, and C. Raghavendra, "Power-aware routing in mobile ad hoc networks," in *Proceedings of the Fourth ACM/IEEE International Conference on Mobile Computing*, 2005.

[5] D. Malan, T. Fulford-Jones, M. Welsh and S. Moulton, "Codeblue: an ad hoc sensor network infrastructure for emergency medical care," in *Proceedings of the International Workshop on Wearable and Implantable Body Sensor Networks*, 2004.

[6] S. Pattem, B. Krishnamachari and R. Govindan, "The impact of spatial correlation on routing with compression in wireless sensor networks," in *Proceedings of the Third International Symposium on Information Processing in Sensor Networks (IPSN)*, 2004.

[7] G. Lu, B. Krishnamachari and C. Raghavendra, "An adaptive energy-efficient and low-latency MAC for data gathering in sensor networks," in *Proceedings of the Fourth International Workshop on Algorithms for Wireless, Mobile, Ad Hoc and Sensor Networks (WMAN)*, 2004.

[8] M. Perillo and W. Heinzelman, "DAPR: a protocol for wireless sensor networks utilizing an application-based routing cost," in *Proceedings of the 2004 IEEE Wireless Communications and Networking Conference (WCNC)*, Vol. **3**, pp. 1540–1545, March 2004.

[9] J. Deng, Y. Han, W. Heinzelman and P. Varshney, "Balanced-energy sleep scheduling scheme for high density cluster-based sensor networks," in *Proceedings of the 4th Workshop on Applications and Services in Wireless Networks (ASWN)*, 2004.

[10] O. Younis and S. Fahmy, "Distributed clustering in ad-hoc sensor networks: a hybrid, energy-efficient approach," in *Proceedings of the Twenty-Third Annual Joint Conference of the IEEE Computer and Communications Societies (INFOCOM)*, 2004.

[11] S. Bhattacharjee, P. Roy, S. Ghosh, S. Misra and M. S. Obaidat, "Wireless sensor network-based fire detection, alarming, monitoring and prevention system for bord-and-pillar coal mines," *Journal of Systems and Software*, Vol. **83**, No. 3, pp. 571–581, March 2012.

[12] I. Akyildiz, W. Su, Y. Sankarasubramaniam and E. Cayirci, "A survey on sensor networks," *IEEE Communications Magazine*, Vol. **40**, No. 8, pp. 102–114, August 2002.

[13] A. Perrig, R. Szewzyk, J. D. Tygar, V. Wen and D. E. Culler, "SPINS: security protocols for sensor networks," *Wireless Networks*, Vol. **8**, No. 5, pp. 521–534, 2002.

[14] J. Kulik, W. R. Heinzelman and H. Balakrishnan, "Negotiation-based protocols for disseminating information in wireless sensor networks," *Wireless Networks*, Vol. **8**, No. 2–3, pp. 169–185, March/May 2002.

[15] C. Schurgers, V. Tsiatsis, S. Ganeriwal and M. Srivastava, "Optimizing sensor networks in the energy-latency-density design space," *IEEE Transactions on Mobile Computing*, Vol. **1**, No. 1, pp. 70–80, January/March 2002.

[16] J. Chang and L. Tassiulas, "Energy conserving routing in wireless ad hoc networks," in *Proceedings of the Nineteenth International Annual Joint Conference of the IEEE Computer and Communications Societies (INFOCOM)*, 2000.

[17] W. Heinzelman, A. Chandrakasan and H. Balakrishnan, "An application-specific protocol architecture for wireless microsensor networks," *IEEE Transactions on Wireless Communications*, Vol. **1**, No. 4, pp. 660–670, Oct. 2002.

[18] S. Ratnasamy and B. Karp, "GHT: a geographic hash table for data-centric storage," in *Proceedings of the First ACM International Workshop on Wireless Sensor Networks and Applications (WSNA)*, 2002.

[19] S. Intille, "Designing a home of the future," *IEEE Pervasive Computing*, Vol. **1**, No. 2, pp. 76–82, April 2002.

[20] D. Niculescu and B. Nath, "Trajectory-based forwarding and its applications," in *Proceedings of the Ninth Annual International Conference on Mobile Computing and Networking (MobiCom)*, 2003.

[21] W. Dargie and C. Poellabauer, *Fundamentals of Wireless Sensor Networks*, John Wiley & Sons, 2010.

[22] M. Zamalloa, K. Seada, B. Krishnamachari and A. Helmy, "Efficient geographic routing over lossy links in wireless sensor networks," *ACM Transactions on Sensor Networks*, Vol. **4**, No. 3, pp. 1–33, 2008.

[23] P. Levis, N. Patel, D. Culler and S. Shenker, "Trickle: a self regulating algorithm for code propagation and maintenance in wireless sensor networks," in *Proceedings of the 1st USENIX/ACM Symposium on Networked Systems Design and Implementation*, pp. 15–28, 2004.

[24] R. Hall, "An improved geocast for mobile ad hoc networks," *IEEE Transactions on Mobile Computing*, Vol. **10**, No. 2, pp. 254–266, Feb. 2011.

[25] L. Su, C. Liu, H. Song and G. Cao, "Routing in intermittently connected sensor networks," in *Proceedings of the 2008 IEEE International Conference on Network Protocols (ICNP)*, 2008.

[26] C. Intanagonwiwat, R. Govindan and D. Estrin, "Directed diffusion: a scalable and robust communication paradigm for sensor networks," in *Proceedings of ACM MobiCom 2000*, pp. 56–67, Boston, MA, 2000.

[27] W. Heinzelman, J. Kulik and H. Balakrishnan, "Adaptive protocols for information dissemination in wireless sensor networks," in *Proceedings of the ACM/IEEE Mobicom Conference (MobiCom '99)*, pp. 174–185, Seattle, WA, August, 1999.

[28] I. Stojmenovic and X. Lin, "GEDIR: loop-free location based routing in wireless networks," in *Proceedings of the 1999 International Conference on Parallel and Distributed Computing and Systems, Boston, MA, USA*, Nov. 3–6, 1999.

[29] F. Kuhn, R. Wattenhofer and A. Zollinger, "Worst-case optimal and average-case efficient geometric ad-hoc routing," in *Proceedings of the 4th ACM International Conference on Mobile Computing and Networking*, pp. 267–278, 2003.

[30] N. Sadagopan, B. Krishnamachari and A. Helmy, "The ACQUIRE mechanism for efficient querying in sensor networks," in *Proceedings of the First International Workshop on Sensor Network Protocol and Applications*, pp. 149–155, May 2003.

[31] D. Braginsky and D. Estrin, "Rumor routing algorithm for sensor networks," in *Proceedings of the First Workshop on Sensor Networks and Applications (WSNA)*, Atlanta, GA, October 2002.

[32] C. Schurgers and M. B. Srivastava, "Energy efficient routing in wireless sensor networks," in *Proceedings on Communications for Network-Centric Operations: Creating the Information Force-MILCOM*, McLean, VA, 2001.

[33] M. Chu, H. Haussecker and F. Zhao, "Scalable information-driven sensor querying and routing for ad hoc heterogeneous sensor networks," in *The International Journal of High Performance Computing Applications*, Vol. **16**, No. 3, pp. 293–313, August 2002.

[34] Y. Yao and J. Gehrke, "The Cougar approach to in-network query processing in sensor networks", *SIGMOD Record*, Vol. **31**, No. 3, Sep. 2002.

[35] W. Heinzelman, A. Chandrakasan and H. Balakrishnan, "Energy-efficient communication protocol for wireless microsensor networks," in *Proceedings of the 33rd Hawaii International Conference on System Sciences (HICSS '00)*, Vol. **2**, p. 10, January 2000.

[36] S. Lindsey and C. Raghavendra, "PEGASIS: power-efficient gathering in sensor information systems," *IEEE Aerospace Conference Proceedings*, Vol. **3**, pp. 1125–1130, 2002.

[37] A. Savvides, C.-C. Han and M. Srivastava, "Dynamic fine-grained localization in ad-hoc networks of sensors," in *Proceedings of the Seventh ACM Annual International Conference on Mobile Computing and Networking (MobiCom)*, pp. 166–179, July 2001.

[38] A. Manjeshwar and D. P. Agarwal, "TEEN: a routing protocol for enhanced efficiency in wireless sensor networks," in *Proceedings of the 15th International Workshop on Parallel and Distributed Computing: Issues in Wireless Networks and Mobile Computing*, pp. 2009–2015, April 2001.

[39] A. Manjeshwar and D. P. Agarwal, "APTEEN: a hybrid protocol for efficient routing and comprehensive information retrieval in wireless sensor networks," in *Proceedings of the 2002 International Parallel and Distributed Processing Symposium, IPDPS*, pp. 195–202, 2002.

[40] V. Rodoplu and T. H. Meng, "Minimum energy mobile wireless networks," *IEEE Journal Selected Areas in Communications*, Vol. **17**, No. 8, pp. 1333–1344, Aug. 1999.

[41] R. C. Shah and J. Rabaey, "Energy aware routing for low energy ad hoc sensor networks," in *IEEE Wireless Communications and Networking Conference (WCNC)*, Vol. **1**, pp. 350–355, Orlando, FL, March 2002.

[42] L. Li and J. Y. Halpern, "Minimum-energy mobile wireless networks revisited," in *IEEE International Conference on Communications (ICC) 2001*, Vol. **1**, pp. 278–283, 2001.

[43] L. Subramanian and R. H. Katz, "An architecture for building self configurable systems," in *Proceedings of IEEE/ACM Workshop on Mobile Ad Hoc Networking and Computing*, pp. 63–73, Boston, MA, August 2000.

[44] Q. Fang, F. Zhao and L. Guibas, "Lightweight sensing and communication protocols for target enumeration and aggregation," in *Proceedings of the 4th ACM International Symposium on Mobile Ad-hoc Networking and Computing (MOBIHOC)*, pp. 165–176, 2003.

[45] J. N. Al-Karaki, R. Ul-Mustafa and A. Kamal, "Data aggregation in wireless sensor networks – exact and approximate algorithms," in *Proceedings of IEEE Workshop on High Performance Switching and Routing (HPSR)*, pp. 241–245, Phoenix, Arizona, April 2004.

[46] Y. Xu, J. Heidemann and D. Estrin, "Geography-informed energy conservation for ad-hoc routing," in *Proceedings of the Seventh Annual ACM/IEEE International Conference on Mobile Computing and Networking*, pp. 70–84, 2001.

[47] Q. Li, J. Aslam and D. Rus, "Hierarchical power-aware routing in sensor networks," in *Proceedings of the DIMACS Workshop on Pervasive Networking*, May 2001.

[48] F. Ye, H. Luo, J. Cheng, S. Lu and L. Zhang, "A two-tier data dissemination model for large-scale wireless sensor networks," in *Proceedings of ACM/IEEE MOBICOM*, 2002.

[49] N. Bulusu, J. Heidemann and D. Estrin, "GPS-less low cost outdoor localization for very small devices," Computer Science Department, University of Southern California, Technical Report 00–729, Apr. 2000.

[50] S. Capkun, M. Hamdi and J. Hubaux, "GPS-free positioning in mobile ad-hoc networks," in *Proceedings of the 34th Annual Hawaii International Conference on System Sciences*, pp. 3481–3490, 2001.

[51] B. Chen, K. Jamieson, H. Balakrishnan and R. Morris, "SPAN: an energy-efficient coordination algorithm for topology maintenance in ad-hoc wireless networks," *Wireless Networks*, Vol. **8**, No. 5, pp. 481–494, Sep. 2002.

[52] Y. Yu, D. Estrin and R. Govindan, "Geographical and energy-aware routing: a recursive data dissemination protocol for wireless sensor networks," UCLA Computer Science Department Technical Report, UCLA-CSD TR-01–0023, May 2001.

[53] B. Karp and H. T. Kung, "GPSR: greedy perimeter stateless routing for wireless sensor networks," in *Proceedings of the 6th Annual ACM/IEEE International Conference on Mobile Computing and Networking (MobiCom '00)*, Boston, MA, Aug. 2000.

[54] J.-H. Chang and L. Tassiulas, "Maximum lifetime routing in wireless sensor networks," *IEEE/ACM Transactions on Networking*, Vol. **12**, No. 4, pp. 609–619, Aug. 2004.

[55] C. Rahul and J. Rabaey, "Energy aware routing for low energy ad hoc sensor networks," in *IEEE Wireless Communications and Networking Conference (WCNC)*, Vol. **1**, pp. 350–355, Orlando, FL, March 2002.

[56] S. Dulman, T. Nieberg, J. Wu and P. Havinga, "Trade-off between traffic overhead and reliability in multipath routing for wireless sensor networks," in *WCNC Workshop*, Vol. **3**, pp. 1918–1922, New Orleans, LA, March 2003.

[57] K. Sohrabi, and J. Pottie, "Protocols for self-organization of a wireless sensor networks," *IEEE Personal Communications*, Vol. **7**, No. 5, pp. 16–27, Oct. 2000.

[58] T. He, J. A. Stankovic, L. Chenyang and T. Abdelzaher, "SPEED: a stateless protocol for real-time communication in sensor networks," in *Proceedings of 23rd International Conference on Distributed Computing Systems*, pp. 46–55, Providence, RI, May 2003.

6 Transport protocols for wireless sensor networks

The traffic pattern in WSNs is convergent, and many-to-one. In other words, most of the traffic flow between the source sensor nodes and the sink node occurs through a multi-hop path consisting of intermediate sensor nodes. This characteristic traffic pattern results in a funnel-like structure between the source nodes and the sink node, as shown in Figure 6.1.

6.1 Transport protocol requirements for WSNs

Transport layer protocols provide two functions: congestion control and reliable data delivery or recovery from packet loss.

Reliable data transmission in WSN environment is a challenging task due to the following reasons [1].

(i) Sensors have limited computation and communication power, i.e. a sensor has limited processing power and short communication range.
(ii) Sensors are battery powered. So, energy conservation is an important issue.
(iii) Sensors are deployed close to the ground and this increases the unreliability of the communication channel due to signal attenuation, channel fading or shadowing.
(iv) Dense deployment of sensors increases channel contention and congestion.

6.1.1 Performance metrics

End-to-end reliable event transfer and event reliability
A sink successfully detects an event while a group of sensors senses that event and properly informs the sink about it. End-to-end reliable event transfer [2] is performed when a sink receives the first message that reports an event. If m_e is the first message that reports event e, then the probability of successful event transfer [2] can be expressed by equation (6.1):

$$Prob(success\ of\ m_e) = 1 - \prod_{s_i \in N'} Prob\{f(s_i, s_0) = 0\}, \qquad (6.1)$$

where $N' \subseteq N$ and N is the set of nodes which detects the event e, and $f(s_i, s_0)$ is the link state between node s_i and sink s_0. If E events occur within an update interval, then event reliability, which is the ratio of successfully delivered messages, can be expressed as [2]:

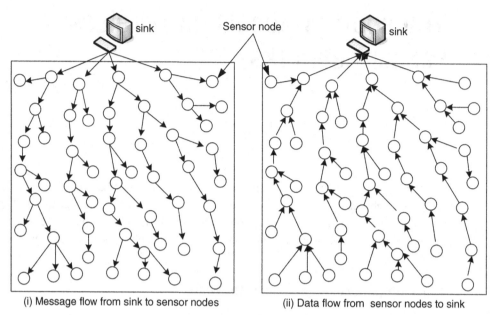

(i) Message flow from sink to sensor nodes   (ii) Data flow from sensor nodes to sink

Figure 6.1   Convergent, many-to-one traffic pattern in WSNs.

$$R(e) = \frac{\sum_{e-1}^{E} Prob(success\ of\ m_e)}{E}.$$

(6.2)

Node reliability

Node reliability [3] of a node i can be defined as

$$Ri(n) = \frac{number\ of\ packets\ whose\ source\ is\ i\ and\ received\ by\ sink}{total\ number\ of\ packets\ generated\ by\ i}$$

Congestion detection

Congestion degree [4] specifies the current congestion level at each sensor node. Congestion degree can be defined as:

$$d(i) = t_s^i / t_a^i,$$

(6.3)

where t_s^i denotes the average packet service time and t_a^i denotes the average packet inter-arrival time over a predefined time interval at each sensor node i [4].

Network efficiency

Network efficiency η is defined in [5] as the ratio of the total number of hops traveled by useful packets to the total number of packet transmissions in a network. A useful packet can be defined as a packet, which is ultimately delivered to the sink. All the retransmissions and the transmissions of dropped or corrupted packets are also taken into account in the total number of packet transmissions. The network efficiency is expressed by equation (6.4):

$$\eta = \frac{\sum_{u \in UP} hops(u)}{\sum_{p \in P} \sum_{h \in hops(p)} Txs(p,h)},$$

(6.4)

where UP denotes the set of useful packets and P is the set of all transmitted packets, $hops(p)$ denotes every hop taken by p, and $Txs(p,h)$ expresses the number of transmissions taken by p at each hop.

Node efficiency or imbalance
Node efficiency or imbalance ζ [5] is used to measure the performance of each node. Imbalance of node i can be calculated as

$$\zeta = \frac{\text{packets received at node } i \text{ from its children}}{\text{packets received at } i\text{'s parent from } i}.$$

(6.5)

Network fairness
Network fairness [5] implies how, fairly or equally, each node of the sensor network gets the chance to use the network resources and to transmit its data. Network fairness $\phi(i)$ of a node i can be calculated as:

$$\phi(i) = \frac{\left(\sum_{i=1}^{N} r_i\right)^2}{N \times \left(\sum_{i=1}^{N} r_i^2\right)},$$

(6.6)

where N is the total number of nodes and r_i is the average packet delivery rate of node i.

6.2   Internet transport protocols and their suitability for use in WSNs

In the traditional Internet, the intermediate nodes act as layer-3 devices (routers). In sensor networks, however, all nodes also have a transport layer, as shown in Figure 6.2 [6].

One of the most commonly used reliable transport protocols designed for the Internet is TCP. Traditionally, TCP was first designed to operate independently of the lower layers in conventional wired networks, which are more reliable than the wireless networks, in general, and WSNs, in particular. Additionally, some of the functional properties of WSNs make TCP unsuitable for use in these networks [7]. Firstly, the establishment of end-to-end connectivity between the source and the sink nodes in WSNs during a communication session is not usual. Secondly, sensor networks are characterized by high degrees of error-proneness associated with poor channel quality, low bandwidth of the medium, relatively frequent failure of sensor nodes, and congestion. Thirdly, each of the intermediate sensor nodes can store the packets for a relatively longer period of time

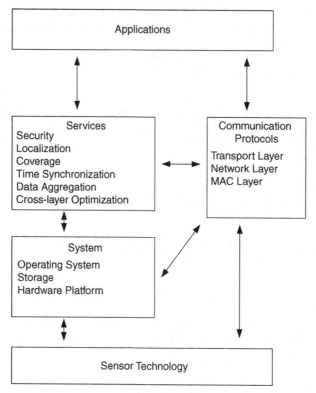

Figure 6.2   Layers and services of a node in WSN (adopted from [6], with minor modifications).

compared to the nodes in the Internet, and then send them to their next hop neighbor, when the channel becomes available. Finally, the resource limitations of the nodes in WSNs make TCP unsuitable for these networks.

6.3   Existing transport protocols for WSNs

6.3.1   Classification

Transport layer protocols for WSNs can be classified into two broad classes: (a) congestion control protocols, and (b) reliability protocols. Only a few of them provide both congestion control and reliability.

6.3.2   Congestion and flow control-centric protocols

The convergent, many-to-one traffic pattern commonly experienced in WSNs makes them highly prone to congestion.

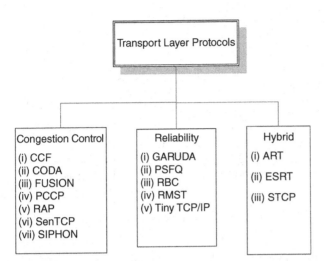

Figure 6.3   Classification of transport layer protocols.

6.3.2.1   Congestion detection and avoidance (CODA)

In CODA [8], congestion is detected based on the queue length of packets at the intermediate nodes. CODA comprises three mechanisms – congestion detection, open-loop hop-by-hop backpressure, and closed-loop multi-source regulation, as shown in Figure 6.4. In open-loop hop-by-hop backpressure, the source gets the backpressure signals depending on the local congestion state. In closed-loop multi-source regulation, the source gets an ACK from the sink and, when the congestion occurs, the sink stops sending ACKs to the source. As CODA is one of the most popular congestion control protocols, it is important to understand these three mechanisms in detail.

Congestion detection
CODA is designed to use the knowledge of present and past channel loading conditions, and the current buffer occupancy, in order to precisely determine the occurrence of congestion at each receiver. However, measuring the local loading conditions by listening to whether the shared transmission medium is congested due to traffic between other stations is not a wise approach in resource-constrained sensor networks. Therefore, CODA uses a sampling mechanism in order to periodically monitor the local channel and minimize costs. Following the detection of congestion, the nodes that detect it signal their upstream neighbors using a backpressure mechanism.

Open-loop hop-by-hop backpressure
A sensor node in a network running CODA uses this mechanism to broadcast back-pressure messages upstream when congestion is detected. This will trigger the nodes receiving the backpressure messages to decrease their data rates or drop packets based on the local congestion handling policy. Additionally, the upstream node receiving the

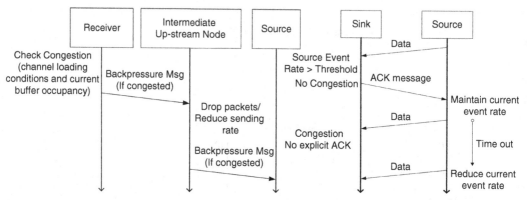

Figure 6.4   Open loop: hop-by-hop backpressure (left); closed loop: (multi-)source regulation (right).

backpressure message also makes a decision on whether or not it should propagate this message further upstream towards the source.

Closed-loop multi-source regulation
If there is persistent congestion over a longer duration of time, closed-loop multi-source regulation is used. This mechanism is used by a sink to determine and handle congestion over multiple sources. The source regulates and continues to transmit event data lower than the specified fraction of the maximum theoretical throughput of the channel being used. Closed-loop regulation is triggered only in the case when this throughput value is exceeded, as it is when the source is likely to contribute to congestion. In such a case, sink regulation is triggered. When such a case arises, the sink regulates the relevant sources by sending out an ACK-like constant, and slow time-scale feedback to them. These feedback messages allow these sources to maintain their current event rates. If, on the other hand, such a feedback message is not received at the source, the source reduces its own rate.

CODA controls the rate of flow of packets based on the additive increase and multiplicative decrease (AIMD) algorithm. This technique is energy efficient, but the reliability of the successful delivery of packets to the destination is not guaranteed because, on receiving backpressure signals, meaning that when the signals are received by the source node from the intermediate nodes, the nodes drop their packets based on the congestion parameters.

6.3.2.2   Congestion control and fairness (CCF)

CCF [9] controls congestion based on the packet service time by adjusting the transmission rate. CCF is designed to control congestion while ensuring fairness in the delivery of packets to the base station. The issue of fairness deals with ensuring that an equal number of packets is received from each sensor node in the network over a period of time. CCF has two major design considerations: (a) congestion control design, and (b) fairness design.

Congestion control design
CCF considers two types of congestion. In the first type, the rate of generation of packets is quite low and the simultaneous transmission of packets is independent of the rate of

their generation. As an example, all the sensor nodes could respond at the same time to a query sent by the base station. CCF suggests using phase shifting by introducing small amount of jitter in the data link layer, or even by performing some adjustments in the application layer.

The second type of congestion can be detected by monitoring the buffer/queue at the nodes. If a threshold is crossed, the downstream nodes are intimated to slow down their rates of transmission.

At the core of the congestion control solution in CCF is the computation of the number of downstream nodes, the measurement of the average rate at which packets can be sent by it, the per node rate disseminated by the parent, and the downstream propagation rate.

Fairness design

CCF uses the following two mechanisms for achieving fairness:

(a)   per-child packet queues, and
(b)   maintenance of per-child tree size.

In (a), one queue is maintained at the transport layer of every node for storing packets from each child node and another queue is maintained for storing packets generated by the node itself. One field in the header of each packet is designated to store the identifier of the last hop node. This helps to insert the received packet into the corresponding queue.

In (b), the subtree size of each child node is obtained and stored as a variable in the respective child node.

Fairness is implemented using two mechanisms: (i) probabilistic selection, and (ii) epoch-based proportional selection. In (i), the queue from which the next packet should be transmitted is probabilistically selected, and in (ii) the selection is made based on epochs, which are integer multiples of the total number of nodes in a subtree.

6.3.2.3   Priority-based congestion control protocol (PCCP)

Wang *et al.* [4] proposed PCCP as a faster and more energy-efficient congestion control algorithm than CCF. As suggested by its name, PCCP maintains a priority index, which is used as an indicator of the importance of each node. The packet inter-arrival time and the packet service time are used together to determine the degree of congestion. The node priority index and the degree of congestion values further help in imposing hop-by-hop congestion control.

Each node in PCCP is modeled to have a scheduler between the network and the MAC layers, as shown in Figure 6.5. This scheduler is tasked to maintain two queues – one for source traffic and the other for transit traffic. The weighted fair queuing (WFQ) or weighted round-robin (WRR) algorithms are used in order to impose fairness between the two traffic types. The same can also be applied for imposing fairness between all sensor nodes.

PCCP consists of the following three mechanisms: intelligent congestion detection (ICD), implicit congestion notification (ICN), and priority-based rate adjustment (PRA).

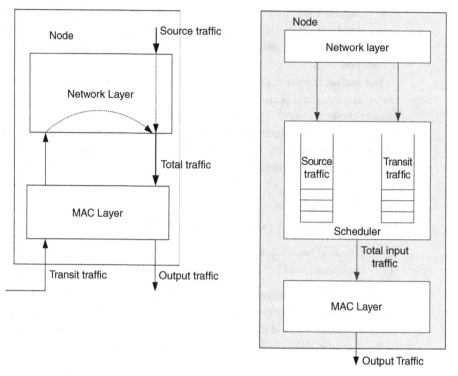

Figure 6.5   General node model (left); node model in PCCP (right) (adopted from [4], with minor modifications).

Intelligent congestion detection (ICD)

Unlike in TCP, in which congestion is detected at the endpoints, in PCCP, congestion is locally detected in the intermediate nodes, based on the mean packet inter-arrival time (t_a^i), which is the mean of the time between two successively arriving packets, and the packet service time (t_s^i), which is the time between the arrival of a packet at the MAC layer and the successful transmission of the last bit. ICD introduces the concept of congestion degree, which is defined as follows:

$$d(i) = \frac{t_s^i}{t_a^i}.\tag{6.7}$$

The congestion degree helps in estimating the current congestion level at each intermediate node. Equation (6.7) indicates that, for a node to experience congestion, the packet service time should be greater than the packet inter-arrival time. Thus, to avoid congestion, the packet service time should be smaller than the inter-arrival time. In other words, the rate at which the packets arrive at the node should be smaller than the rate at which the packets are serviced out of the nodes.

Implicit congestion notification (ICN)

PCCP uses the technique of ICN to piggyback congestion information in the header of the data packets. As it is a wireless channel, the congestion information is transmitted from the parent node to the child node, thereby avoiding the transmission of additional control messages, as is usually done in explicit congestion notification (ECN) techniques used in many traditional transport protocols for wired networks such as TCP.

The congestion information is stored in the header of the packets that are forwarded when triggered by either of the following two events:

(i)   when a threshold is exceeded by the number of packets forwarded by a node;
(ii) when a congestion notification is heard by a node from its parent node.

At every node i, PCCP proposes piggybacking t_a^i, t_s^i, the global priority, $GP(i)$, and the number of offspring nodes, $O(i)$. The global priority index is computed by summing the source traffic priority index and the global priority indices of the children nodes.

Priority-based rate adjustment (PRA)

Wang *et al.* [4] proposed that congestion could be avoided by adjusting the scheduling rate r_{svc}^i. In order to avoid congestion, r_{svc}^i should be lower than the MAC forwarding rate r_f^i. The additive increase and multiplicative decrease (AIMD) used in traditional transport control protocols such as TCP is not of much help in adjusting the transmission rate, as the congestion notification (CN) bit carries limited information. Therefore, it is important for the nodes to be notified as to exactly how much to increase or decrease the rate. The congestion degree, the priority index, and the global priority values, discussed earlier, help in providing more information for exact rate adjustment.

6.3.2.4   Trickle

Trickle [10] is a scheme designed to propagate code updates from the downstream nodes through the intermediate nodes in the multi-hop path towards the sink node. It has a periodical event in each node that suppresses broadcasting if the meta-data that it receives from its neighboring node exceed the threshold.

Trickle bases itself on the concept of "polite gossip." According to this concept, at periodic intervals of time, the sensor nodes broadcast meta-data to neighbors, only if they have not heard any similar meta-data from the other nodes in the recent past. These meta-data are only the summary of the code, and not the code itself. If a node receives meta-data from another node's gossip that are older than its own, it broadcasts an update in order to update the gossiper. One of the key features of Trickle is its ability to limit the number of packets by suitably adjusting the transmission rate. Consequently, only a small "trickle" of packets is necessary for keeping the nodes up-to-date.

Trickle was evaluated through simulations and an experimental test-bed setup. In one of the experimental studies, it was shown that Trickle is capable of reprogramming an entire network in 30 s, while imposing an overhead of only three packets per hour.

6.3.2.5   Fusion

Fusion [5] controls congestion by using three components that operate in a concerted manner in the different layers of the network protocol stack. These components are: hop-by-hop flow control, source rate limiting scheme, and prioritized MAC layer.

Hop-by-hop flow control

According to this mechanism, whenever some nodes experience congestion, the congestion information is rippled over to the downstream nodes by using the backpressure technique. A congestion bit is set in the packets that are broadcast with the congestion information. This helps in reducing the loss of packets due to congestion.

There are two components of congestion management through hop-by-hop flow control: congestion detection and congestion mitigation. The congestion detection mechanism operates by monitoring the size of the queue maintained at each sensor node. The congestion bit in the outgoing packets is set only when the available space in the queue falls below a certain limit. Congestion mitigation is performed by regulating the transmission of packets to the next-hop nodes, so that they do not overflow. A node that receives a packet with the congestion bit set temporarily stops the transmission of packets.

Source rate limiting scheme

This scheme was designed for reducing the unfairness towards source nodes from which packets have to traverse a large number of hops upstream towards the sink node. At the heart of this scheme is the token bucket algorithm, which is used to regulate the rate of transmission of packets at each sensor node. According to this scheme, whenever the parent of a node sends a certain number of packets to it, it accumulates one token. A sensor node sends only when the token count is more than zero.

Prioritized MAC layer

In general, medium access in wireless networks is controlled by CSMA-like protocols, which give an equal chance of success to all the participating stations. However, it was observed that, during congestion, the congested nodes will not be able to promptly send the congestion information to their neighbors, thereby leading to degraded performance of the network. Therefore, the prioritized MAC layer was designed in order to provide priority in medium access to backlogged nodes over the non-backlogged ones, thereby reducing the amount of buffer drops.

6.3.2.6   Siphon

Siphon [11] is one of the transport protocols which directly addresses the issue of overload traffic management using the concept of multi-radio virtual sinks (VSs). The VSs are used to remove data events from the sensor network when any symptom of traffic load occurs. They act as siphons to tunnel out traffic from regions experiencing traffic overload, as shown in Figure 6.6. This is analogous to opening safety valves in a pressurized fluid tank, when the threshold exceeds. The VSs are equipped with a higher

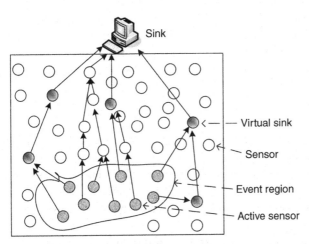

Figure 6.6   Virtual sinks.

capacity long-range radio interface such as the ones available in Wi-Fi or WiMAX, in addition to the existing low-powered radio interface required for them to connect to the sensor network. Siphon consists of a set of algorithms for performing virtual sink discovery and selection, congestion detection, and traffic redirection, as explained below.

Virtual sink discovery and selection

Siphon uses an in-band signaling mechanism for virtual sink discovery and selection. In this approach, a signature byte, which contains a virtual sink TTL (VS-TTL) field, is inserted into the control packets that periodically originate from a physical sink. This requirement can be suitably embedded into the existing routing protocol functionalities, which will not result in any significant additional overhead. The VS-TTL field has the hop count over which VS is advertised. Each sensor node contains the list of its neighbors that can lead it to discover its parent VS. Similarly, each VS maintains a list of its neighboring VSs. This connectivity is essential for the transmission of congestion information in the network.

Congestion detection

Siphon uses two types of congestion detection mechanisms: (a) node-initiated congestion detection, and (b) physical sink-initiated "post-facto" congestion detection.

In (a), the mechanism is similar to that used in CODA [8]. The knowledge of the present and past channel loading conditions, along with the knowledge of the current buffer occupancy, are used by the wireless receivers in order to determine local congestion levels being experienced by different sensor nodes. A theoretical upper bound of the channel throughput is pre-computed, and when this value is exceeded, the redirect algorithm is activated by a sensor node within the transmission range of the affected node, in order to funnel out, by using VSs, certain types of traffic from the neighborhood.

Unlike in (a), in which congestion detection is initiated by the node, in (b) the VSs are activated in a post-facto manner by the physical sinks, when congestion is inferred by them. As the physical sinks have a collection of data obtained from the different downstream source nodes, they are in an advantageous position to monitor the event data quality and the application fidelity. When the measured application fidelity reduces below a certain threshold, or when congestion is detected by the physical sinks, they activate the VSs.

Traffic redirection

Siphon proposes the use of a redirection bit in the network layer header. The redirection bit can be set either in an on-demand manner when congestion is detected, or it can be always set to be on. The latter approach can help in alleviating congestion more than the former, but it is not an energy-efficient approach, as it always keeps the secondary radio on without taking the neighborhood congestion conditions into consideration.

6.3.2.7   Learning automata-based congestion avoidance in sensor networks (LACAS)

LACAS [12], which is based on the concept of learning automata (LA), is a congestion avoidance scheme for sensor networks. It is assumed that an automaton, which is a simple autonomous machine (code) capable of making decisions, is equipped at every node in the network, as shown in Figure 6.7.

In LACAS, only the intermediate nodes during a transmission have their automata working for controlling congestion locally. In other words, at any time instant, if we observe the network topology, the automata stationed in the intermediate nodes, and not the ones in the source nodes, act as congestion controllers of data arriving from source nodes. Also, it can be noted that each of the nodes is independent in the network in its approach for controlling congestion.

For the input to the automaton at time $t = 0$, the authors considered that the number of actions associated with an automaton is limited to five, based on the rate at which an intermediate sensor node receives the packets from a source node. These actions are denoted as $\psi = \{\psi_1, \psi_2, \psi_3, \psi_4, \psi_5\}$, as shown in Figure 6.7. The rates, ψ, that are considered as inputs to an automaton stationed in a particular node, are based on the number of packets dropped till then in the concerned node. At any time instant, the most optimal action among the set of actions in an intermediate node is estimated by the number of packets dropped. To be specific, the rate of data flow into a node, for which there is the least number of packets dropped, is considered to be the most optimal action. At any time instant, the choice of an action by the automaton, i.e. the rate at which data should flow into the corresponding node, is rewarded or penalized by the environment. In the beginning, at $t = 0$, all of the actions have the equal probability (let, $P_{\psi_i}(n)$) of being selected by the automaton. It can be assumed that the automaton selects ψ_1 initially, based on the probability values of all the actions at time $t = 0$. The chosen action, which maps to a predefined rate of flow of data, then interacts with the environment. The environment examines the action ψ_1 and rewards/penalizes it based on the number of packets dropped at that node. If the action ψ_1 is rewarded, the probability value of ψ_1 is increased and the probability values of the other actions, i.e., $\psi_2, \psi_3, \psi_4, \psi_5$, are decreased as shown in equations (6.8a) and (6.8b).

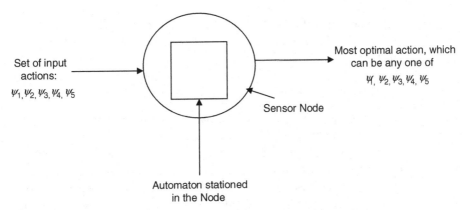

Figure 6.7   Diagram showing a node with an automaton stationed in it (adopted from [12], with minor modifications).

$$P_{\psi_1}(n + 1) = P_{\psi_1}(n) + \frac{1}{\lambda}\left(1 - P_{\psi_1}(n)\right), \tag{6.8a}$$

$$P_{\psi_{2,3,4,5}}(n + 1) = \left(1 - \frac{1}{\lambda}\right)P_{\psi_{2,3,4,5}}(n). \tag{6.8b}$$

On the other hand, if ψ_1 is penalized, the probability value corresponding to ψ_1 as well as for the rest of the actions, $\psi_{2,3,4,5}$, will remain unchanged. This is shown by equations (6.9a) and (6.9b) as follows.

$$P_{\psi_1}(n + 1) = P_{\psi_1}(n), \tag{6.9a}$$

$$P_{\psi_{2,3,4,5}}(n + 1) = P_{\psi_{2,3,4,5}}(n). \tag{6.9b}$$

The probability values associated with all the actions should be such that, at every time instant, the summation of their probabilities should equal unity, i.e. $\sum_{i=1}^{r} P_{\psi_i}(n) = 1$. At the next time instant, the automaton, based on the updated probability values, selects an action that interacts with the environment again, and is rewarded or penalized accordingly. The probabilities of all the actions are updated continuously, and this updating of probabilities continues until the most optimal action is selected, i.e. $\lim_{t \to \infty} P_{\psi_i}(n) = 1$, which means that the probability corresponding to the most desirable action tends to unity, as time tends to infinity. Let us assume that, as $t \to \infty$, the automaton selects action ψ_2 to be the most optimal action for the system. The sensor node emits the packets based on the rate corresponding to action ψ_2. Since this is the most optimal action selected by the automaton, the chance of the node getting congested reduces, and for this action ψ_2, the number of dropped packets reduces to the minimum.

6.3.2.8   Ant-based routing with congestion control (ARCC)

ARCC [13] utilizes the concepts of ant colony optimization (ACO) [14] to deal with congestion in a WSN, and finds an optimum path between a source and sink, considering the network performance issues such as throughput, fairness, and loss of packets.

The motivations behind ARCC are as follows.

(1)   The nodes of a WSN are often assigned different levels of priorities based on their role and location. As a consequence, a congestion control mechanism needs to assign weighted fairness to the nodes, depending on their priorities.

(2)   Owing to network dynamism, the number of nodes participating in data communication varies with time. So, a predetermined path computed by ACO algorithms may not always be a good fit. The enhancement carried by a congestion control mechanism may lead to a single path being declared as the most efficient one in minimum traffic times, as well as, in peak hours.

(3)   ACO routing protocols require to consider the quality-of-service (QoS) metrics to enhance the overall network performance.

(4)   Maintaining routing tables is an overhead for resource-constrained nodes of a WSN. Considering that fact, ARCC mitigates the overhead by running the ARCC algorithm every time, and thereby eliminates the book-keeping of past records.

In ad-hoc networks, due to the movement of nodes, a predetermined routing path may be detected invalid in later cases. To deal with node mobility, ARCC utilizes ACO along with congestion control every time a node requests communication with the sink or any of its parent nodes. Moreover, ARCC assumes that all the nodes of the network may not take active participation in the communication process. Nodes with limited battery power or with other constraints may behave selfishly at any time.

The proposed protocol employs a pheromone-based ACO approach to detect the optimum paths. The primary condition for optimality is used distance. Less energy is used for transmission through these paths than the other available paths. However, increasing the pheromone content along a path over a period of time leads to more nodes using the same path for communication, and this excessive communication through a path leads to congestion.

ARCC detects congestion at a node based on packet (ant) inter-arrival time and inter-service time. The inter-service time is defined as the time required to move across a node from the previous link to the next link. Congestion occurs at a node when the inter-service time exceeds a threshold value. After detecting congestion, a node notifies its upstream nodes, which means that the congestion information is communicated in the reverse direction from the detector node to the source. ARCC uses this notification for relieving congestion. After receiving congestion notification, the nodes avoid any additional transmission along the path. Moreover, other previously non-optimal paths are chosen as the candidates for the minimum path. Taking advantage of the broadcast nature of the wireless network, the congestion notifications can be piggybacked into the header of the data packets. This piggybacking process is referred to as implicit congestion notification (ICN) [15].

When an ant waits to be serviced in the queue of congested nodes, its pheromone deposit evaporates, and the rate of evaporation is directly proportional to the waiting time, signifying an overall delay due to the presence of congestion. In the meantime, the relieving algorithm utilizes a priority index attached to each node to cope with the congestion. Based on the attached priority index, an intermediate scheduler, attached between the MAC and network layer, estimates the appropriate traffic rate to be forwarded or blocked. Periodically, the pheromone contents of different paths are compared to detect a set of optimum paths at that time instant for a given network configuration. For each request, a transient routing table is prepared and actual transmissions take place through those estimated optimum paths.

6.3.3   Reliability-centric protocols

The notion of reliability in WSN is widely spread. In WSN, reliability [2, 16–20] may mean packet reliability, event reliability, end-to-end reliability, hop-by-hop reliability, upstream (sensors-to-sink) reliability, or downstream (sink-to-sensors) reliability. Packet reliability implies successful delivery of a packet from source to sink. Event reliability is achieved when a detected event is reported to the sink with a certain degree of accuracy. End-to-end reliability refers to the successful data delivery from source to destination, while hop-by-hop reliability means reliable data delivery from a source to its next hop. Upstream reliability means reliable delivery of data from sensor nodes to the sink, and downstream reliable delivery ensures delivery of data or query from the sink to all (or a subset of) sensor nodes.

6.3.3.1   Event-to-sink reliable transport (ESRT)

In WSN, for reliable detection of an event, a sink relies on aggregated data provided by multiple source nodes, not a single node. Hence, conventional end-to-end reliability is not required for WSN and may be achieved by unnecessary wastage of sensor resources. As the name suggests, ESRT [16] is a transport protocol for reliable event detection at the cost of minimum energy. Congestion control mechanism is also added to ESRT for achieving reliability and saving energy.

For monitoring an event reliably, or for reliable temporal tracking of an event, as mentioned by authors of ESRT [16], the sink evaluates the event features every t time units. This t represents the decision interval and depends on the application. The number of received data packets is used as the metric to evaluate the reliability of event features transportation from source nodes to sink. Two different kinds of event reliability are defined in ESRT. The number of received packets in the ith decision interval d_i at the sink is referred to as the observed event reliability, r_i. The number of received packets required for reliable event detection is referred to as the desired event reliability, R. In ESRT a normalized measure of reliability is used, $\eta = r/R$, and η_i refers to the normalized reliability at the end of decision interval d_i. The aim of ESRT is to maintain reliability within the range of $1 - \alpha \leq \eta_i \leq 1 + \alpha$, where α is the tolerance parameter, which depends on the application. The main task of ESRT is to maintain a reporting rate, f, the number of

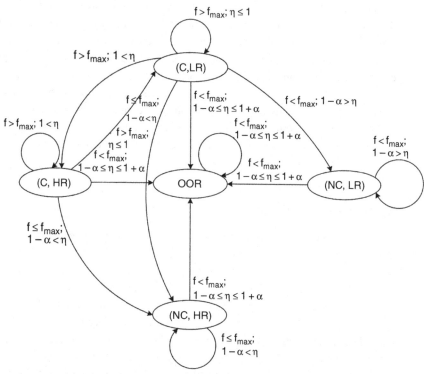

Figure 6.8   Network state and transition diagram of ESRT protocol (adopted from [16], with minor modifications).

packets transmitted per unit time by a source node, of all the source nodes, so that the required event reliability is achieved at the cost of minimum resource utilization.

In every decision interval d_i, ESRT identifies the state of the network by the reliability indicator η_i and the congestion detection mechanism. The current state of the network S_i is any one of the five states of the network, i.e. $S_i \in \{(NC, LR), (NC, HR), (C, HR), (C, LR), OOR\}$. The states and their transitions are described in Figure 6.8. ESRT estimates the reporting frequency f_{i+1} for next decision interval d_{i+1}, depending on current network state S_i, f_i, and η_i. ESRT tries to achieve and maintain the network state OOR.

(i)   (NC, LR)   *No congestion and low reliability:* $f < f_{max}$ and $\eta < 1 - \alpha$. This state may be the result of any one, or more, of the following reasons:
   (a)   hardware failure of any of the intermediate relay nodes,
   (b)   loss of packets due to link failures, and
   (c)   insufficient information reported by source nodes.
   ESRT focuses mainly on the third reason and ignores the first two. So, to improve reliability, the amount of source information is required to increase. ESRT increases the reporting frequency, f, aggressively to reach the desired level of reliability as soon as possible and the updated reporting frequency f_{i+1} becomes

$$f_{i+1} = \frac{f_i}{\eta_i}.$$

(ii) (NC, HR) *No congestion and high reliability:* $f \leq f_{max}$ and $\eta > 1 + \alpha$. In this state, the reliability level is more than sufficient. The source nodes consume more energy as they send more than adequate information. ESRT reduces the reporting frequency carefully, so that the energy of the source nodes can be saved and the required level of reliability is also maintained. The updated reporting frequency is:

$$f_{i+1} = \frac{f_i}{2}\left(1 + \frac{1}{\eta_i}\right).$$

(iii) (C, HR) *Congestion and high reliability:* $f > f_{max}$ and $\eta > 1$. In this state, ESRT reduces the reporting frequency to avoid congestion and save energy of source and intermediate nodes. The reduction of reporting frequency is done cautiously to maintain event-to-sink reliability. Multiplicative decrease is used to update the reporting frequency and it is:

$$f_{i+1} = \frac{f_i}{\eta_i}.$$

(iv) (C, LR) *Congestion and low reliability:* $f > f_{max}$ and $\eta \leq 1$. This is the worst case among the all-fine states. The achieved reliability is not adequate and the nodes experience congestion. ESRT reduces the reporting frequency aggressively to make the network state OOR. The reporting frequency is decreased exponentially and the updated reporting frequency is expressed as:

$$f_{i+1} = f_i^{(\eta_i/k)},$$

where k is the number of successive decision intervals.

(v) OOR *Optimal operating region:* In this state, the required reliability is achieved by expending minimum energy of nodes. Hence, the reporting frequency of next decision interval is the same as the current one.

$$f_{i+1} = f_i.$$

Congestion detection

ESRT employs a congestion-detection mechanism based on local buffer monitoring mechanism in sensor nodes. Let B be the buffer size of a sensor node, b_k and b_{k-1} the buffer occupation levels at the kth and $(k-1)$th reporting intervals, respectively, and Δb the buffer level increment at the end of the last reporting period, i.e.

$$\Delta b = b_k - b_{k-1}.$$

If a node observes that, for any kth interval, $b_k + \Delta b > B$, it assumes that it will experience congestion in the next interval. The node then notifies the sink by setting the congestion notification bit in the header of all the forwarded packets.

6.3.3.2 Reliable multi-segment transport (RMST)

RMST [17] is a selective NACK-based transport layer protocol designed to support directed diffusion [21]. It ensures guaranteed delivery and, if required, fragmentation and reassembly. Reliability mechanisms can be implemented in the MAC layer, transport layer, application layer, and/or any combination of these layers. MAC layer reliability provides hop-by-hop frame recovery, whereas the transport layer provides end-to-end guaranteed reliability, fragmentation/reassembly and route maintenance, and route repair mechanism. The application layer is also able to manage fragmentation/reassembly and end-to-end reliability.

RMST is designed as a filter, and operates above the diffusion stack of a node. The main responsibility of RMST is delivery of any or all fragments of a unique RMST-entity to all concerned sinks. A unique RMST-entity is a data set that may or may not be fragmented into multiple pieces, originating from the same source. Two distinctive transport services are provided: guaranteed delivery, and effective fragmentation and reassembling of messages. RMST operates in two modes: caching and non-caching, which are configurable at the run time.

In RMST, it is the receivers' responsibility to detect the loss of any fragment. The notion of receivers depends on the operating mode of RMST. In the non-caching mode, it is the sinks' responsibility to check the integrity of an entity; whereas, in the caching mode, any RMST caching node may initiate the recovery process of the missing fragments by requesting to its next upstream node along the path toward the source. To detect the loss, each RMST-entity is distinguished by an application-specific attribute (*RmstNo*) or a set of attributes. Each fragment of a data-entity is also marked by sequential fragment (*FragNo*) and the total number of fragments is also known. The loss of fragments can be classified into two categories: a *hole*, a missing fragment in a sequence of fragments, and an abruptly halted fragment flow. A timer-driven mechanism is used for loss detection. The sinks, in non-caching mode, and all the caching nodes, in caching mode, maintain watchdog timers. A watchdog timer is associated with each new flow or new data entity. A receiver or caching node detects the loss of an expected fragment if it does not arrive before timeout. In the case of hole detection, the caching node requests specifically for that missing fragment.

RMST-caching nodes send NACKs (negative acknowledgements) to their upstream nodes when they detect any loss of fragments. An RMST-receiver unicasts a NACK in the reverse direction of the reinforced path, from source to the sink. When a caching node detects a cache hit for a missing fragment, it forwards the fragment to its immediate downstream node in the reinforced path from source to sink. If an appropriate match is not found, the node forwards the NACK to its immediate upstream node in the reinforced path. The reinforced path is constructed from source to sink by directed diffusion. To deliver NACKs, an RMST builds a *back-channel* from sink to source. An RMST node checks the reinforcement messages and makes the sender of a reinforcement message as its immediate downstream node of the back-channel from sink to the source of the reinforcement message.

6.3.3.3   Reliable bursty convergecast (RBC)

Wireless sensor networks (WSNs) are deployed for either monitoring or surveillance purposes. The key purposes are collecting information and/or detection of interested events. In event-driven communications, there is a sudden increment of packets, generated and flown through the network, within a short span of time. This sudden increment of packets creates some challenges for reliable and real-time event detection. These high-volume data amplify the channel contention and, as a consequence, the packet collision probability is also enhanced. In a multi-hop network, the probability of packet collision is increased further.

The RBC protocol [19] applies window-less block acknowledgement and differentiated contention control mechanisms to overcome the challenges of reliable and real-time bursty convergecast. The window-less block acknowledgement scheme helps RBC to forward packets continuously in the presence of packet and acknowledgement loss, whereas the differentiated contention control mechanism reduces the channel contention by ranking the nodes on the basis of their queues and en-queued packets.

Window-less block acknowledgement

The packet queue, Q, of a sender S is structured by $(H+2)$ linked lists, where H denotes the maximum number of retransmissions at each hop. The arrangement of Q is shown in Figure 6.9. The linked lists form a virtual queue. The linked lists are marked as Q_0, $Q_1, \ldots, Q_{H+1}$. The virtual queues are ranked according to their indices, i.e. rank of Q_i is higher than Q_j if $i < j$.

The sender S puts a new packet to the tail of Q_0. Packets of Q_{j-1} are transmitted before the packets of Q_j. After transmission of a packet of Q_j, the packet is attached to the tail of Q_{j+1}. After the arrival of acknowledgement of a transmitted packet, the buffer holding the packet is freed and is added to the Q_{H+1}.

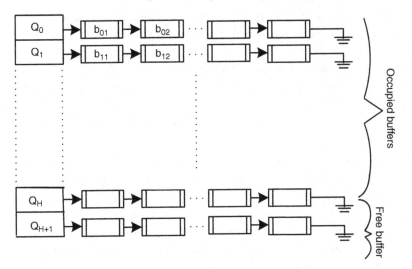

Figure 6.9   Arrangement of virtual queues within a node.

Queue buffers at senders are identified by unique IDs. When a sender S transmits a packet, p, it attaches the buffer ID, $b_{current}$, of that packet and also the buffer ID, b_{next}, of the next packet. After receiving p, a receiver R knows the buffer ID of the next expected packet. When R receives the next packet p' from S, it checks the buffer ID of p' with the b_{next} of the previous packet. If two buffer IDs are matched then no packet is lost, otherwise packet loss has occurred.

The receiver R uses block acknowledgement to acknowledge S. When R receives p_j, $p_{j+1}, \ldots, p_{j+m}$, m consecutive packets without any loss, it adds a 2-tuple $<b_j, b_{j+m}>$ to packet p_{j+m} before forwarding the packet, where b_i is the ID of the ith packet. The sender S checks or "snoops" the packet p_{j+m} while it is forwarded by R.

RBC uses counters for duplicate packet deletion. Duplicate packet detection and the deletion mechanism are necessary as in a lossy communication channel there is high probability of acknowledgement loss. A sender S maintains a counter for each virtual queue. The counter value is incremented whenever a new packet is stored in that queue. When the sender transmits a packet, it also appends the counter value with that packet. The receiver R preserves the counter value, piggybacked by the last packet. When R receives a new packet from the same buffer, it checks the counter value extracted from the newly received packet with the stored value. If the two values are equal, R considers the new packet as a duplicate packet and drops it; otherwise R accepts the packet as a new one.

Differentiated contention control

RBC uses two kinds of scheduling, intra-node and inter-node, to reduce channel contention and interference among packets of the same rank, as well as to balance the queue of the nodes. Window-less block acknowledgement is used to schedule the packets of a node. Within a node, packets of low-indexed queue have higher priority than the packets of high-indexed queue. For inter-node packet scheduling, nodes which have more high-ranked packets get the priority to transmit their packets first.

Nodes are prioritized according to their ranks for inter-node scheduling. The rank of a node i can be expressed as $rank_i$ and can be defined by a 3-tuple $<H - r, |Q_r|, ID_i>$, where r is the index of highest ranked non-empty queue of i, $|Q_r|$ is the number of packets in Q_r, and ID_i is the ID of node i. The first field is used so that the packets with high priority can be transmitted earlier. The second field enables a node whose queue length is more than its neighbors' to transmit earlier and the third field is used to break a tie between two nodes. The process of inter-node scheduling can be described as follows.

(i)   A node informs its neighbors by piggybacking its rank to the transmitted packets.
(ii) A node R compares its rank with the sender's, S, rank, known either by snooping or receiving a packet. If $rank_S$ is higher than $rank_R$, R will not transmit in the next $D_{RS} \times T_{pkt}$ time. T_{pkt} is the time required to transmit a packet at the MAC layer. T_{pkt} is estimated by the exponentially weighted moving average mechanism. $D_{RS} = 4 - i$, where the priority between R and S is decided by the ith element of their 3-tuple rank.

Timer management

Retransmission timer management has effects on the network performance. A small timeout value of the timer may cause unnecessary packet transmissions, whereas delay in packet delivery may increase due to a large timeout value of timers. In RBC, a sender adjusts its retransmission timer value according to the queue length of a receiver. After receiving a packet, p, a receiver, R, buffers the packet at the end of its Q_0. Before forwarding p, the receiver first transmits all other packets. So, the delay of forwarding the packet p by R depends on the length of its Q_0. As the length of Q_0 keeps changing, the delay of forwarding a packet also varies.

To adapt to the queuing condition of a receiver, R, a sender, S, needs to know the length, l_{Q_0}, of Q_0, the average delay, d_R, of forwarding a packet by R, and the deviation, δ, of d_R, and S gathers this information by snooping packets forwarded by R. Whenever S sends a packet to R, retransmission timer of S is initialized by

$$(l_{Q_0} + C) \times (d_R + 4\delta),$$

where C is a constant.

The receiver R detects loss of packet(s) when it receives a packet p from buffer b of sender S, while it is expecting a packet from buffer b' of S. R realizes that the packets from buffer b' to buffer b, excluding the packet from buffer b, are lost. In such cases, RBC uses block-NACK to acknowledge the sender. To inform the sender, the receiver R piggybacks a block-NACK $[b', b)$ to the next forwarded packet. S learns of the loss of the packet by snooping the packet piggybacking the block-NACK $[b', b)$. After snooping the block-NACK $[b', b)$ from R, S resets the retransmission timer to 0 for the respective packets. Then buffers of these packets are added to the tail of Q_{k-1} if currently the buffers are at Q_k. This way the priority of the lost packet(s) is enhanced and they can be retransmitted quickly.

6.3.3.4   Pump slowly, fetch quickly (PSFQ)

The PSFQ protocol [18] was designed as a simple, reliable, and scalable transport protocol that can be customized according to the WSN application needs. The protocol was developed with the broader goal of building a generic environment capable of reprogramming or re-tasking sensors according to the application needs. This goal is challenging to achieve, as the sensor networks are severely resource constrained. Further, as the number of sensor nodes increases, the re-tasking operations can become challenging.

In PSFQ, packets from the source nodes are pumped at a relatively slow rate. The nodes that experience packet loss are allowed to fetch the missing packets relatively quickly from the immediate neighbors that have copies of it. The losses are detected when a node receives a message with a higher sequence number than is expected.

PSFQ has three components [18]: message relaying (pump operation), relay-initiated error recovery (fetch operation), and selective status reporting (report operation), each of which is explained below.

Message relaying (pump operation)

An injected message is associated with the pump operation. The header of this message has four fields [18]: (i) the file identifier, (ii) the length of the file, (iii) the sequence number, and (iv) time to live (TTL). The pump operation helps in the timely dissemination of code fragments to the intended recipients in the re-tasking operations of the sensor network. It also helps in the flow control process to ensure that the re-tasking operations do not send data exceeding the capacity of the network for performing its regular sensing operations. The pump operation also helps to cut down on the transmission of redundant messages (particularly in the case of dense networks) and the propagation of loss to the downstream nodes.

Relay-initiated error recovery (fetch operation)

Once an error is detected at a node, due to the mismatch between the actual packet sequence number received by it and the one that is expected, a fetch operation is initiated.

In the fetch operation, the receiver node that detects the error requests the neighboring nodes to retransmit the packets. This is done by using a single fetch operation containing a batch of a number of message losses – a process known as "loss aggregation." It is based on the premise that losses in sensor networks often occur as a result of fading conditions, channel impairments, or other adverse physical phenomena, which result in loss of a group of packets and not a single packet. Therefore, PSFQ proposes the use of a single fetch operation dealing with a group of lost packets, and not a single lost packet. Further, it is often the case that not one neighbor, but multiple neighbors, would have to help the receiver node with parts of the messages that are lost. This is because different parts of the message could be residing with different neighbors.

A fetch operation has an associated NACK (negative acknowledgement) control message, which can help to identify a retransmission from the neighboring nodes. Among different fields in the NACK message, it also has a "loss window" field that helps in identifying its left and right boundaries.

A fetch timer is used along with the NACK messages sent out by the receiver node to its neighbors. The fetch timer helps in cases when no reply is received at the receiver from its neighbors in response to the NACK messages sent by it, or a partial set of missing parts of the lost messages are only received. In such cases, the NACK message is sent out repeatedly at set fetch timer intervals until all the required parts of the message are received.

Selective status reporting (report operation)

The report operation helps in sending out data delivery status feedback in the form of report messages. However, as most sensor network applications consist of a large number of nodes, it is not efficient that every node sends feedback. In order to minimize the feedback messages, in PSFQ, a feedback report message originates from the furthest target node back to the user through a multi-hop path consisting of intermediate sensor nodes that piggyback their report messages on the original report message. The header of the above-mentioned report message contains the identifier of the destination node that has to relay this message. Consequently, when this message arrives at the final destination it has a chain of node identifiers and sequence number pairs.

6.3.3.5   GARUDA

GARUDA [20] addresses the problem of reliable downstream, point-to-multipoint data delivery, i.e. delivery of data from the sink node to the source nodes. The protocol is designed for downstream delivery of packets and may not be suitable for upstream delivery. The name GARUDA was given to the protocol after the Hindu mythical bird that used to transport gods. GARUDA is reliable and is scalable with the increase in the size of the network, message characteristics, loss rate, and reliability semantics [20]. Park *et al.* [20] defined the reliability semantics according to the following classifications.

- Reliable delivery to all nodes in the field.
- Reliable delivery to a part of the field.
- Reliable delivery to minimal number of sensors that can cover the field.
- Reliable delivery to a probabilistic subset of the sensors in the field.

The main components of GARUDA are: (a) construction of loss-recovery servers (core), and (b) loss-recovery process. The core nodes are used to cache the packets and the non-core nodes are used to recover the lost packets.

The core is constructed by using a variant of the minimum dominating set (MDS) algorithm, which has often been used in scenarios that require reliable message delivery [22].

At the heart of the loss-recovery process is the construction of an A-map (availability map), and its exchange between the core nodes. The A-map helps to convey meta-level information regarding the availability of packets with the relevant bit set. An A-map with the corresponding bit set is received by a downstream core node. The loss-recovery process involves two stages (a) loss recovery from core nodes, in which, first, the core nodes recover from all lost packets, and (b) loss recovery from non-core nodes, which is executed when an A-map is overheard by a non-core node [20].

6.3.3.6 Asymmetric and reliable transport (ART)

ART [2] aims at providing event reliability, instead of per-message reliability, like some of the previously proposed protocols. The protocol is designed on the careful observation that there exist a lot of redundant message transmissions in WSNs. For instance, when an object is detected in a sensor field, all the sensors that detect the object transmit the message towards the sink node. Depending on the type of data aggregation mechanism used in the intermediate nodes in the multi-hop path, the copies of messages of the same event that travel towards the sink are reduced. But the sink node still gets multiple messages indicating the same event. Such multiple copies of a message corresponding to the same event do not necessarily offer more reliability. In fact, redundant transmissions of messages deplete the energy-constrained nodes of their energy. ART has been designed by keeping these observations in mind.

As we saw earlier, in WSNs there are two directions of message flows: sensors-to-sink (upstream) and sink-to-sensors (downstream)[1]. Accordingly, ART considers two types of reliable data transmissions: event reliability and query reliability. Tezcan and Wang [2] define event reliability to be achieved when at least a single message reporting every critical event is received at the sink node. Similarly, query reliability is defined to be achieved when every query that is sent by the sink is received by the sensors sensing the

---

[1] The term "asymmetric" in the protocol name is reflective of this fact.

portion of the terrain that needs to be sensed. The idea is to deliver the least possible number of messages while minimizing the energy consumption and latency in the delivery of the packets. End-to-end reliable data transfer in either the upstream or downstream direction normally happens when the first message corresponding to the event successfully arrives at the sink node.

ART further classifies the nodes in a sensor field into essential (E) nodes and non-essential (N) nodes. The E nodes are selected by running a weighted greedy algorithm at the sink node, based on the residual energy of the sensors. End-to-end reliable data transmission, in both the upstream and the downstream directions, is capacitated with the help of asymmetric acknowledgement (ACK) and negative acknowledgement (NACK) between the essential nodes and the sink node.

ART has the capabilities of congestion control. The classification of the nodes into E and N is utilized to mitigate congestion when it occurs. Congestion is monitored by the E nodes by monitoring the duration of ACK arrivals corresponding to event messages. If within a pre-configured timeout interval, no ACK is received by an E node, the data arrival at the N nodes is slowed down (or even temporarily stopped) by sending out congestion alarm messages. This makes the neighboring N nodes of an E node relatively passive. When an ACK is received again, a safe message is transmitted back, indicating the normal resumption of messages.

6.3.3.7   Collaborative transport control protocol (CTCP)

The CTCP protocol [7] is yet another transport protocol aimed at providing end-to-end reliability. It is a collaborative protocol, in which the concerned nodes in the network collaborate to detect and then mitigate congestion. The protocol is adaptive to different application needs by using a two-level mechanism of reliability variation.

CTCP is capable of delivery of packets to the application layer in the base station, even in the presence of failure and disruptions in the network. It explicitly takes reliability and energy-efficiency issues into account. Furthermore, as a congestion control protocol, it is capable of limiting the rate of forwarding of packets at nodes. There are two main functionalities of CTCP. They are explained below.

Hop-by-hop connection open and close

Prior to the transmission of the sensed messages, CTCP sends a packet (ABR), which contains the data flow identifier and the first sequence number, from the source node to the destination node, and the packet passes through every node in the path. On receipt of the packet at the base station, the base station reserves the buffers, performs the desired configuration tasks, and responds back to the source node with a response (RSP) message. The reliability requirements of the application and the connection identifier are also transmitted in the RSP message. Once the source node receives the RSP message, it starts transmitting data. After the source node completes sending the data, a closure (CLO) message is sent to the base station.

Controllably reliable delivery

Different applications have different reliability requirements. Additionally, CTCP mandates to offer generic transport functionalities to a wide range of applications. Therefore, the protocol must also be able to offer different levels of reliability as per the requirements of the applications. The changes in the reliability level are informed to the base station with the help of RSP packets.

As mentioned earlier, CTCP supports two levels of reliability. Reliability level 1 is used when the application data have some kind of redundancy inherent in them and can tolerate losses. This reliability level is helpful in saving energy. The usage of reliability level 2 increases the probability of data reaching their destination. This is because, when this reliability level is used, the failure of a node in the path does not interrupt the delivery of data.

6.3.4   Other protocols

6.3.4.1   Sensor TCP (STCP)

STCP [23] is a generic transport layer protocol applicable to WSN. Sensor networks support a wide range of applications such as target tracking and environmental monitoring. Applications differ by their characteristics and requirements. So, a generic transport layer protocol, independent of applications, MAC layer and network layer protocol, is capable of supporting different kinds of applications and their specific requirements. STCP is able to support several applications simultaneously in the same network and provides application-specific reliability and congestion detection and avoidance.

Transmission of sensed data is initiated by a sensor node by sending a session initiation packet to the base station. The purpose of sending the session initiation packet is to create an association or connection between source node and base station. The session initiation packet contains information such as the number of data flows generated from the source, type of each data flow, transmission rate, and required reliability of each data flow.

STCP supports both continuous data flow and event-driven data flow. The format of the session initiation packet is given in Figure 6.10. On receipt of the session initiation packet, the base station stores the information carried by the packet, sets the timer and parameters of each flow, and sends an acknowledgement. The ACK ensures the source node knows about the establishment of the association with the base station.

Sequence Number (16)		Flows (8)	Options (8)
Clock (32)			
Flow ID#1 (8)	Flow Bit (8)	Transfer Rate (8)	Reliability (8)
Flow ID#2 (8)	Flow Bit (8)	Transfer Rate (8)	Reliability (8)
Flow ID#N (8)	Flow Bit (8)	Transfer Rate (8)	Reliability (8)

Figure 6.10   Format of session initiation packet (adopted from [23], with minor modifications).

Data flow
As already mentioned earlier, STCP provides the means for continuous data flow and event-driven data flow.

Continuous data flow
In continuous data flow it is assumed that the clocks of sensor nodes and the base station are synchronized. The base station gathers information about transmission rate of each data flow originating from each source node from session initiation packets. From the collected transmission rate information, the base station is able to estimate the expected arrival time of a packet. It sends a NACK if it does not receive a packet within the expected time. The base station calculates the estimated trip time (*ETT*) on receipt of a packet from a source:

$$ETT = \text{current clock value at base station}$$
$$- \text{ clock value contained in the received packet.}$$

The base station may calculate the timeout value in two ways.

(i)   Timeout $= T + \varphi \times ETT$, where T is the time difference between two successive transmissions and the value of φ varies with *ETT*. If a packet arrives before the timeout value, the base station reduces the value of φ by 0.5. In the case of a lost packet or receipt of a packet after sending a NACK, base station increases the value of φ by 0.5.
(ii) In the second approach similar to proposed mechanism by [24], the value of timeout is calculated as:
   sampled_ETT = base station clock value $-$ packet's clock value,
   ETT_difference = sampled_ETT $-$ estimated_ETT,
   estimated_ETT = estimated_ETT + ($\alpha \times ETT$_difference),
   deviation = deviation + $\alpha \times$ (|ETT_difference| $-$ deviation),
   timeout = $\beta \times ETT$ + $\eta \times$ deviation.

Sensor nodes retransmit packets if a NACK is received.

Event-driven data flow
In event-driven data flow, clock synchronization is not required, as it is not possible for a base station to estimate the arrival of next packets. The base station informs successful reception of packets by sending ACKs. The source node buffers all the transmitted packets and retransmits those buffered packets whose ACKs will not arrive within a predefined time interval.

Reliability
Different approaches are taken to estimate the reliability of continuous data flow and event-driven data flow. For continuous flow, reliability is measured as the ratio of successfully received packets to the total transmitted packets. While the current reliability is more than the required reliability, the base station does not send a NACK even if a packet does not arrive within estimated time. The base station sends a NACK only if the

current reliability is lower than the required reliability. In the case of event-driven flow, the reliability is calculated as the ratio of total received packets to the highest sequence number of received packets. If the current reliability is more than required reliability a source node does not buffer a transmitted packet.

Congestion detection and avoidance
STCP uses an explicit congestion notification mechanism. Each STCP data packet contains a congestion notification bit in its header. Two threshold values, t_{lower} and t_{higher}, are maintained. An intermediate sensor node sets the congestion notification bit of a forwarded packet with a certain probability p if the buffer reaches t_{lower} value. But a node sets the congestion notification bit of each forwarded packet if the buffer exceeds the t_{higher} value. After detecting congestion, the sink informs the source of the congested path by setting the congestion bit of the ACK and transmitting it to the source node.

6.4   Summary

This chapter has illustrated the convergence of multi-hop traffic flow within a WSN. The details and suitability of numerous transport layer protocols were analyzed for a WSN. Some performance metrics that focus on enhancing the aspect of reliability, congestion control, and overall efficiency in wireless communication were discussed. We, then, classified the existing transport layer protocols for WSNs based on their characteristics. The congestion control TCP protocols, such as CODA, CCF, PCCP, Trickle, Fusion, Siphon, LACAS, and ARCC were illustrated in this context. TCP protocols that enhance the transmission reliability, viz. ESRT, RMST, RBC, PSFQ, GARUDA, ART, and CTCP were discussed next. We concluded this chapter by reviewing the hybrid protocol, STCP.

References

[1] J. Jones and M. Atiquzzaman, "Transport protocols for wireless sensor networks: state-of-the-art and future directions," *International Journal of Distributed Sensor Networks*, Vol. **3**, No. 1, pp. 119–133, 2007.

[2] N. Tezcan and W. Wang, "ART: an asymmetric and reliable transport mechanism for wireless sensor networks," *International Journal of Sensor Networks*, Vol. **2**, Nos. 3–4, pp. 188–200, 2007.

[3] M. A. Rahman, A. El Saddik and W. Gueaieb, "Wireless sensor network transport layer: state of the art," in *Sensors*, LNEE 21, S. C. Mukhopadhyay and R. Y. M. Huang (Eds.), Springer, 2008, pp. 221–245.

[4] C. Wang, K. Sohraby, V. Lawrence, L. Bo and Y. Hu, "Priority–based congestion control in wireless sensor networks," *IEEE International Conference on Sensor Networks, Ubiquitous, and Trustworthy Computing (SUTC'06)*, 2006.

[5] B. Hull, K. Jamieson and H. Balakrishnan, "Mitigating congestion in wireless sensor networks," in *Proceedings of ACM Sensys'04*, Baltimore, MD, Nov. 3–5, 2004.

[6] J. Yick, B. Mukherjee and D. Ghosal, "Wireless sensor network survey," *Computer Networks (Elsevier)*, Vol. **52**, No. 12, pp. 2292–2330, 2008.

[7] E. Giancoli, F. Jabour and A. Pedroza, "CTCP: reliable transport control protocol for sensor networks," in *Proceedings of the International Conference on Intelligent Sensors, Sensor Networks and Information Processing (ISSNIP)*, Sydney, Australia, December 2008, pp. 493–498.

[8] C.-Y. Wan, S. B. Eisenman and A. T. Campbell, "CODA: congestion detection and avoidance in sensor networks," in *Proceedings of ACM Sensys '03*, Los Angeles, CA, Nov. 5–7, 2003.

[9] C. T. Ee and R. Bajcsy, "Congestion control and fairness for many-to-one routing in sensor networks," in *Proceedings of ACM Sensys '04*, Baltimore, MD, Nov. 3–5, 2004, pp. 148–161.

[10] P. Levis, N. Patel, D. Culler and S. Shenker, "Trickle: a self regulating algorithm for code propagation and maintenance in wireless sensor networks," in *Proceedings of 1st Symposium on Networked Systems Design and Implementation*, San Francisco, CA, Mar. 29–31, 2004.

[11] C.-Y. Wan, S. B. Eisenman, A. T. Campbell and J. Crowcroft, "Siphon: overload traffic management using multi radio virtual sinks in sensor networks," in *Proceedings of ACM Sensys '05*, San Diego, CA, Nov. 2–4, 2005.

[12] S. Misra, V. Tiwari and M. S. Obaidat, "LACAS: learning automata-based congestion avoidance scheme for healthcare wireless sensor networks," *IEEE Journal on Selected Areas in Communications*, Vol. **27**, No. 4, pp. 466–479, May 2009.

[13] S. K. Dhurandher, S. Misra, H. Mittal, A. Agarwal and I. Woungang, "Using ant-based agents for congestion control in ad-hoc wireless sensor networks," *Cluster Computing (Springer)*, Vol. **14**, pp. 41–53, 2011.

[14] M. Dorigo, A. Colorni and V. Maniezzo, "The ant system: optimization by a colony of cooperating agents," *IEEE Transactions on System, Man, and Cybernetics–Part B*, Vol. **26**, No. 1, pp. 1–13, 1996.

[15] C. Wang, B. Li, K. Sohraby, M. Daneshmand and Y. Hu, "Upstream congestion control in wireless sensor networks through cross-layer optimization," *IEEE Journal on Selected Areas in Communications*, Vol. **25**, No. 4, pp. 786–795, 2007.

[16] Y. Sankarasubramaniam, O. B. Akan and I. F. Akyildiz, "ESRT: event-to-sink reliable transport in wireless sensor networks," in *Proceedings of the 4th ACM International Symposium on Mobile Ad Hoc Networking & Computing (MobiHoc '03)*, ACM, New York, NY, USA, 2003, pp. 177–188.

[17] F. Stann, and J. Heidemann, "RMST: reliable data transport in sensor networks," in *Proceedings of the First IEEE International Workshop on Sensor Network Protocols and Applications*, May 2003, pp. 102–112.

[18] C.-Y. Wan, A. T. Campbell and L. Krishnamurthy, "PSFQ: a reliable transport protocol for wireless sensor networks," in *Proceedings of the First ACM International Workshop on Wireless Sensor Networks and Applications (WSNA '02)*, Atlanta, Georgia, USA, Sept. 2002.

[19] H. Zhang, A. Arora, Y. Choi and M. G. Gouda, "Reliable bursty convergecast in wireless sensor networks," *Computer Communications*, Vol. **30**, No. 13, pp. 2560–2576, 2007.

[20] S.-J. Park, R. Vedantham, R. Sivakumar and I. F. Akyildiz, "GARUDA: achieving effective reliability for downstream communication in wireless sensor networks," *IEEE Transactions on Mobile Computing*, Vol. **7**, No. 2, pp. 214–230, Feb. 2008.

[21] C. Intanagonwiwat, R. Govindan, D. Estrin, J. Heidemann and F. Silva, "Directed diffusion for wireless sensor networking," *IEEE/ACM Transactions on Networking*, Vol. **11**, No. 1, pp. 2–16, Feb. 2003.

[22] R. Sivakumar, P. Sinha and V. Bharghavan, "CEDAR: a core-extraction distributed ad hoc routing algorithm," *IEEE Journal of Selected Areas in Communications*, Vol. **17**, No. 8, pp. 1454–1465, 1999.

[23] Y. G. Iyer, S. Gandham and S. Venkatesan, "STCP: a generic transport layer protocol for wireless sensor networks," in *Proceedings of 14th International Conference on Computer Communications and Networks (ICCCN 05)*, San Diego, California USA, Oct. 17–19, 2005, pp. 449–454.

[24] V. Jacobson, "Congestion avoidance and control," in *Proceedings on Communications Architectures and Protocols (SIGCOMM '88)*, V. Cerf (Ed.), ACM, New York, NY, USA, pp. 314–329, 1988.

7 Localization and tracking

The term "localization" in wireless sensor networks (WSNs) refers to determining the location of a device in the absence of additional infrastructure, such as satellites. As such the location discovery service enables a device to know its location. The use of additional devices is not suitable for wireless networks, especially the resource-constrained ones, because of the power-hungry nature of such devices. For example, the global positioning system (GPS) uses the distance measurements from satellites to calculate the location of a device. Additionally, GPS cannot be used in indoor localization. Traditionally, the localization problem has its origin in the field of robotics, where it is necessary to locate a robot. For navigation of a robot through a terrain, it is important to have the location information.

The information provided by wireless sensor networks (WSNs) is highly correlated with respect to space and time. In applications such as target tracking [1], geographic routing [2, 3], environmental monitoring [4], structural health monitoring [5], and forest fire monitoring [6], the location of information generating sensors is very important. As an example, in forest fire monitoring applications, the location of fire is required to be tracked in order to take preventive measures. Further applications include the optimization of medium access and improvement of location-based geographic routing protocols. However, being a network of small low-powered devices, the localization problem for WSNs is very different from other networks. Also, the localization of the nodes is important for tagging the sensor information with nodes' location information.

In WSNs, the nodes determine their geographic positions by using externally aided localization or self-localization techniques. In such networks, most of the nodes are static, and generally nodes determine their positions in the network initialization phase itself. However, there may be application requirements to know the location information on-demand. Some popular methods used in WSNs are trilateration and triangulation, where relative measurements from three or four reference nodes are utilized to calculate a node's own location. Few others to name are bounding box, multidimensional scaling, and hop-count-based approaches. Some of these are discussed in this chapter.

Target tracking is another well-studied problem area in different wireless networks. Tracking of a target requires knowledge about its location at different times. In robotics and navigation, different target tracking approaches are proposed in the literature. Tracking has wide applications in the military field where the location determination of an enemy vehicle is necessary. General common characteristics of various tracking systems are target location reporting and collaborative processing to remove redundancy.

As mentioned earlier, it is worth emphasizing that localization – both node localization and target localization – are necessary for successful target tracking.

Different problems related to target tracking are studied, such as target number counting, and target size or area detection. In sensor networks, as the battery power of a node is limited, only a few nodes that can sense the target are kept in the active state. In this regard, different approaches are studied to optimally select the active nodes. The target tracking schemes related to WSNs are also required to maintain a balance between various network resources such as power consumption, communication bandwidth, and protocol overhead [7]. However, using WSNs for target tracking also offers a few advantages, such as increased observation quality and tracking accuracy, and system robustness.

7.1   Localization

In WSNs, it is possible for a node to determine its position within a few meters of accuracy, if a node is equipped with the global positioning system (GPS). However, in practice, the use of GPS in WSNs involving thousands of nodes is not feasible. The first reason is that a GPS receiver is expensive and the cost of deployment of WSN increases if GPS is built inside every node. The second reason is that the use of GPS with every node is not an energy-efficient solution in WSNs, as these networks are intrinsically energy constrained. GPS also does not work in indoor environments, and environmental factors such as large buildings affect GPS performance [8].

In order to understand the problem of localization, let us consider that a sensor network is deployed in a region, with N randomly deployed nodes. Let us assume that the position of a node i is denoted as P_i, where $i = 1, 2, \ldots, N$. The coordinate of P_i is presented as $P_i = (x_i, y_i, z_i)$. In a two-dimensional space, $z_i = 0$. Some of the nodes know their positions by using GPS or some other methods. These nodes are known as beacon nodes, anchor nodes, or landmarks. The nodes that do not know their positions at the beginning are known as unknown nodes. Let the communication range of all types of nodes be R. If a node can directly communicate with any other node, then the distance between these two nodes is less than or equal to r, and these two nodes are the neighboring nodes. The localization problem is abstracted as follows. A WSN is represented by a graph $G = (V,E)$, $|V| = N$, and there exists an edge between two nodes, if they are able to communicate directly with one another. B indicates the set of beacon nodes with (x_b, y_b) for all $b \in B$ is given where $B \subseteq V$. The positions (x_u, y_u) for all unknown nodes $u \in U$ are required to be determined [9]. See Figure 7.1.

General methods of localization

Two issues are related with localization – the first is how to define the coordinate system, and the second is how to calculate the distance between two nodes [10]. Unknown nodes calculate their location by referencing certain number of anchors by using various ranging and direction methodologies [11].

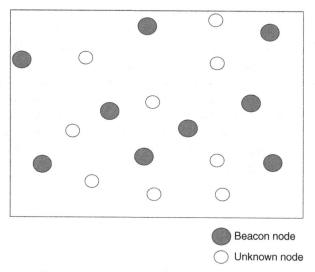

Beacon node

Unknown node

Figure 7.1   Example of wireless sensor network topology.

7.1.1   Distance estimation techniques

7.1.1.1   Received signal strength indication (RSSI)

Every sensor node has a transceiver. This transceiver can be used to find the range between two nodes. There are two types of range computations; one is based on hop count, and another is based on received signal strength indication (RSSI).

The energy of a radio signal decreases when the signal propagates through a medium. The decay of signal strength is proportional to the square of the distance traveled by it. So, a receiver can estimate its distance from the source by knowing the signal strength at the source end and the strength of the signal at the receiver end. A relationship between the received power and distance is obtained from the following equation [12].

$$P_r = kd^{-\alpha}. \tag{7.1}$$

In equation (7.1), P_r is the power of the received signal, k is a constant that depends on the frequency and transmitted power, d is the distance between the transmitter and the receiver, and α is the attenuation exponent. However, RSSI measurements contain noise up to several meters in a few areas such as indoors. Noise occurs due to environmental factors, such as the presence of a tree or wall, which can reflect radio signals.

7.1.1.2   Radio hop count

Though RSSI is error prone, radio can still be used to calculate the distance between two sensors. If two nodes can directly communicate with each other, then it can be observed that the distance between two nodes is less than or equal to R, where R is the maximum radio range of a node. A WSN can be represented as an unweighted graph, in which the sensors are represented by vertices and there is an edge between two nodes, if the Euclidean distance between these two nodes is less than or equal to R. The length of shortest path between

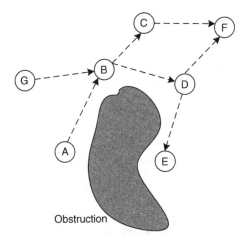

Obstruction

Figure 7.2   Example of hop count. Hop count of AE =3 and hop count of AF =3. Distance AE is less than distance AF (adopted from [10], with minor modifications).

nodes s_i and s_j is h_{ij}, where h_{ij} is the hop count between nodes s_i and s_j. If the distance between s_i, s_j is denoted by d_{ij}, then d_{ij} is less than or equal to $h_{ij} \times R$. In the ideal case, the Euclidean distance between nodes s_i and s_j is equal to $h_{ij} \times R$. But the distance between any two nodes depends on their spatial distribution. The expected distance, which is covered per communication hop, is denoted as d_{hop}. If n_{local} is the expected number of neighbors, then the value of d_{hop} can be calculated by equation (7.2) [10].

$$d_{\text{hop}} = R \left(1 + e^{-n_{\text{local}}} - \int_{-1}^{1} e^{-\frac{n_{\text{local}}}{\Pi} (\arccos t - t\sqrt{1 - t^2})} \, dt \right). \qquad (7.2)$$

Then, d_{ij} can be expressed as $d_{ih} \approx h_{ij} \times d_{\text{hop}}$. The distance between any two nodes is calculated as integral multiples of d_{hop}. This justifies an inaccuracy of almost $0.5R$ in every measurement. Further, environmental obstacles may influence the connectivity of the graph. An example is shown in Figure 7.2.

7.1.2   Time difference of arrival (TDOA)

In the TDOA approach, each node is equipped with a speaker and a microphone. In the transmitter initiated approach, as stated in [10], the transmitter first sends a radio message. After sending the message, it delays for some fixed interval of time, t_{delay}, and then generates some fixed patterns of sound by its speaker. When a receiver listens to the radio message, it records the receiving time, t_{radio}, and waits for the sound signal. It records the receiving time of sound signal as t_{sound}.

The receiver then calculates the distance d from the sender, by using equation (7.3) [10]:

$$d = (s_{\text{radio}} - s_{\text{sound}}) \times (t_{\text{sound}} - t_{\text{radio}} - t_{\text{delay}}), \qquad (7.3)$$

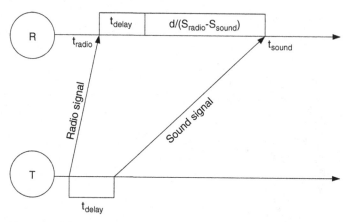

Figure 7.3   Example of time difference of arrival (TDOA) (adopted from [10], with minor modifications).

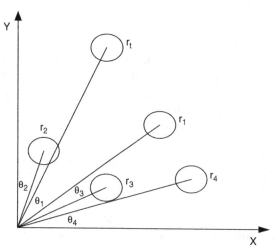

Figure 7.4   Example of time difference of arrival (TDOA) where a transmitter calculates its position by using location information from four receivers (adopted from [8], with minor modifications).

where s_{radio} and s_{sound} are the speed of radio signal and sound signal, respectively, as can be seen in Figure 7.3.

In [8], another approach is described, in which a transmitter sends signal to multiple receivers with known locations. The TDOA of a pair of receivers i, j is given by equation (7.4) [8], and the scenario is as explained in Figure 7.4:

$$\Delta t_{ij} \triangleq t_i - t_j = \frac{1}{c} \left(\|r_i - r_t\| - \|r_j - r_t\| \right), \ i \neq j, \tag{7.4}$$

where t_i is the receiving time of the signal at receiver i, c is the signal propagation speed, and $\| . \|$ is the Euclidean distance.

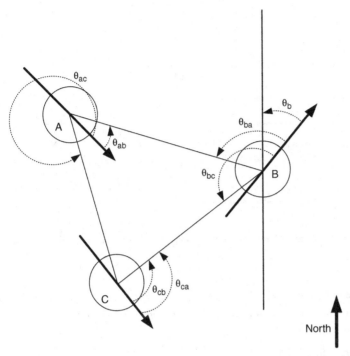

North

Figure 7.5   Example of angle of arrival (AOA) (adopted from [13], with minor modifications).

7.1.3   Angle of arrival (AOA), digital compasses

In the AOA approach [13], nodes are capable of sensing the angle of arrival of a received signal. AOA sensing requires an array of antennas or multiple ultrasound receivers to be equipped with the nodes. Each node in the network measures all angles based on its main axis. After deployment, the axis of a node directs arbitrarily. In Figure 7.5, the axis of a node is represented by an arrow passing through the center of the node. The meaning of "bearing" is an angle with respect to another object. Using AOA, each node can estimate the bearing to the neighbor nodes with respect to its own axis. A radial is a reverse bearing or the angle under which an object is seen by another object. The "heading" is the absolute bearing of each node and it directs to the North. In Figure 7.5, for node B, the bearing to node C is $\angle \theta_{ba}$, while the radial from C is $\angle \theta_{cb}$, and the heading is $\angle \theta_b$. A node can find the angle between its own axis and direction of incoming signal by interacting with its neighbors. Node A finds two incoming signals at $\angle \theta_{ab}$ and $\angle \theta_{ac}$, so that it can compute $\angle CAB = \angle \theta_{ac} - \angle \theta_{ab}$. Node A can also infer its heading or orientation from B's orientation. The heading of A is $2\pi - (\angle \theta_{ba} + \pi - \angle \theta_{ab}) + \angle \theta_b$. If an unknown node is capable of calculating AOA of a signal from at least two beacon nodes, then it can estimate its position by the triangulation process. Although methods of finding AOA have high accuracy, the problem of using this approach is that extra hardware is required for finding AOA. Each node requires multiple receivers. A digital compass gives the

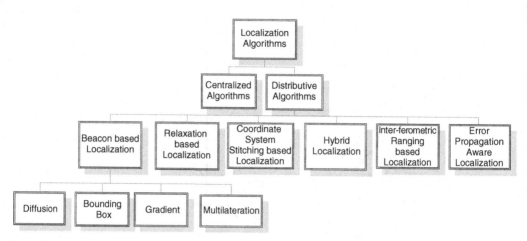

Figure 7.6   Classifications of localization algorithms (adopted from [9] with minor modifications).

global orientation of a node. As in Figure 7.5, the orientation of node B is $\angle\theta_b$ and node A can estimate its global orientation from the orientation of node B.

7.1.4   Localization algorithms

Localization algorithms can be categorized as centralized and distributive. In centralized algorithms, a single powerful node calculates the positions of the unknown nodes. An unknown node sends its measured information, such as beacon position and distance, to the central node, and the latter node sends back the estimated position to the former node. The problem with the centralized approach is that lots of packets are exchanged among the central and the unknown nodes and, as a result, scalability is a real issue. If the network size increases, then the power consumption due to communication also increases. In distributive localization schemes, an unknown node estimates its own location. The taxonomy of localization algorithms is shown in Figure 7.6.

7.1.4.1   Centralized algorithms
MDS-MAP

MDS-MAP [14] is a centralized localization algorithm. A WSN is represented as an unweighted graph, where nodes are represented by vertices and there is an edge between two nodes, if the distance between them is less than or equal to the communication range of radio. The connectivity graph is assumed to be a connected graph that is a path between any pair of nodes. The algorithm first produces a relative map of the network and then the relative map is transformed into absolute positions. The task of finding a relative map has two phases. In the first phase, the neighbors of each node are discovered and a connectivity graph is created. In the second phase, this graph is mapped onto a two- or three-dimensional plane. In an absolute map, the geographic coordinates of each node are determined. Let us assume that there are some nodes with known positions in the three-dimensional space. The straight-line distance between any pair of nodes is known. Multidimensional scaling (MDS) can be used to map the nodes on a two-dimensional plane. In this mapping, nodes are placed on the basis of three-dimensional distances among them. MDS starts with proximity matrices, which are derived from points of multidimensional spaces. Then, MDS is used to determine the placement of points on a low-dimensional space, generally two- or three-dimensional, where the distances among the points maintain original similarities.

We have the following steps in the MDS-MAP algorithm.

Step 1 Ranging data from the network are gathered, and a sparse matrix R is calculated. Here, r_{ij} is the range between nodes i and j, or zero if no range is collected.

Step 2 Shortest paths between all pairs of nodes in the region of consideration are determined. The shortest path distances are used to calculate the distance matrix, D, for MDS.

Step 3 Classical MDS is applied on the distance matrix D. The two-dimensional (or three-dimensional) relative map is constructed by retaining the first two (or three) largest eigenvalues and eigenvectors.

Step 4 If there exists a sufficient number of anchor nodes, then the relative map is transformed into an absolute map on the basis of absolute coordinates of anchor nodes.

Adaptive beacon placement
Beacon nodes or anchor nodes know their geographic location and act as a reference point in the localization algorithms [15]. In distance-based approaches, the density and placement of beacons affects the quality of localization (see Figure 7.7). Every node needs to hear a minimum number of beacon nodes and those beacons should be non-collinear. A uniform and dense distribution of beacons is not efficient, although it may appear that distribution improves the quality of localization. The hardware of beacons

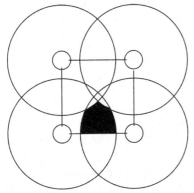
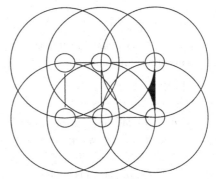

2*2 Grid of beacons
Fewer and larger localization areas

3*2 Grid of beacons
More and smaller localization areas

Figure 7.7 Beacon density vs. granularity of localization regions (adopted from [15], with minor modifications).

is costly; so, a larger number of beacons means increased overall cost of the WSN. A high density of beacons means that the probability of collisions due to the transmission also increases. So, a limited number of beacons are required for reducing the number of collisions, saving energy and, hence, prolonging the lifetime of the network. Environment (or terrain) is another crucial factor.

Beacons are placed at known position (X_B, Y_B), and they periodically transmit with a time period t. Clients listen for time period $t >> T$. If the number of messages received by a client from a particular beacon exceeds a threshold value, then that client is connected with the beacon. A client then estimates its position (X_E, Y_E), as it is the centroid of all connected beacons.

If the actual position of clients is (X_A, Y_A), then the localization error LE is equal to [15]:

$$LE = \sqrt{(X_E - X_A)^2 + (Y_E + Y_A)^2}.$$

As the density of beacon nodes increases, the size of localization area becomes finer, and hence, the localization error decreases. The centroid approach introduces some error. According to [15], if the nominal beacon transmission range is R, the distance between two adjacent beacons is d, and for a range overlap ratio (R/d) of 1, the maximum error is bound to $0.5d$. This error rate decreases with increasing value of the range overlap ratio.

In [15], the authors proposed a localization approach, in which localization improves incrementally by either adjusting the position of beacon nodes, or adding some new beacons. This improvement is based on existing localization at any time instant. GPS-equipped mobile robots are used for localization. These robots also carry a radio, similar to sensors, to estimate the localization error at any point in the terrain. Robots also carry additional beacons. Robots estimate the proper position to deploy the beacons on their calculation of localization errors.

The authors assumed that the area is a square, where the length of a side is S. Each robot's range is s and transmission range of beacons is R. Three beacon placement algorithms were proposed, namely – Random, Max, and Grid [15].

Random
Step 1   A point (X_r, Y_r) is chosen randomly.
Step 2   A beacon is placed on that point.
This algorithm is used to calculate localization error of other algorithms.

MAX
Step 1   The terrain is divided into $s \times s$ squares.
Step 2   Localization error at each point is calculated. The coordinate of each point is $(i*s, j*s)$, where $0 \le i, j \le S/s$.
$DP_T = \left(\frac{S}{s} + 1\right)^2$ number of data points in the terrain.
Step 3   New beacon is added to point (X, Y) which has the maximum localization errors among all points.

MAX is influenced by propagation effects or random noise. As a result, at a certain point, the localization error is high, while the localization error of a nearby point may remain low.

Grid

In Grid, a candidate point is estimated by calculating the cumulative localization error over each grid for several overlapping grids in the terrain.

Step 1   The terrain is divided into $s \times s$ squares.

Step 2   Localization error at each point is calculated. The coordinate of each point is $(i*s, j*s)$ where $0 \le i, j \le S/s$.

Step 3   The terrain is divided into N_G partially overlapping grids as follows.

   Step 3.1   Each grid has a side, $gridSide = 2R$.

   Step 3.2   The center of grid $G(i; j)$ is

$$G_C (i, j) = (X_C (i, j), Y_C (i, j)),$$

   where $1 \le i, j \le \sqrt{NG}$,

$$X_C(i,j) = \frac{gridSide}{2} + \frac{(i-1)(S - gridSide)}{\sqrt{NG} - 1}, \text{ and}$$

$$X_C(i,j) = \frac{gridSide}{2} + \frac{(j-1)(S - gridSide)}{\sqrt{NG} - 1}.$$

Step 4   For each grid $G(i; j)$, the cumulative localization error $LE(i; j)$ is calculated for all the points measured in Step 2 that lie in the grid $G(i; j)$. The number of data points per grid is

$$DP_G = DP_T \times \frac{side^2}{(2.R)^2},$$

where $DP_T = \left(\frac{S}{s} + 1\right)^2$ is the number of data points in the terrain.

Step 5   The new beacon is added at the center $G_C(i; j)$ of the grid $G(i; j)$ with the maximum cumulative localization error.

7.1.4.2   Distributive localization algorithms

Localization algorithms should mainly fulfill three conditions. They should be self-organizing, robust, and energy efficient [16]. Sensor nodes are distributed randomly at the time of network installation when the sensors are placed in an uncontrolled manner, for example, when sensors are deployed from an airplane. So, sensors should self-organize themselves in such scenarios. Algorithms should be robust against node failure, and energy efficient, as battery power is a critical resource.

Nodes without GPS may find their locations in three phases [16].

Phase 1. Finding the distance between anchor nodes and node itself.

Phase 2. Calculating the location from the calculated distances.

Phase 3. Refinement of the location using information of neighbors.

Beacon-based distributed algorithms

(a)   Diffusion

In the diffusion technique [10], an unknown node first finds the positions of its neighboring nodes. Then, the unknown node estimates its position as the centroid of its neighbors. In the approach proposed in [17], the nodes find all the neighboring beacon nodes and determine their positions as the average of the beacon positions. This technique is simple, but is error-prone. If beacon density is low, or if neighboring beacons are not uniformly distributed in that area, then the calculated positions are inaccurate.

In another type of approach [18], each node estimates its position as the centroid of its neighbors, beacons and the unknown nodes. Unknown nodes run this process until the result converges. The steps of this algorithm are as follows.

Step 1   All unknown nodes initialize their position as (0, 0).
Step 2   An unknown node finds the positions of all its neighbors.
Step 3   An unknown node estimates its position as the average of all its neighbors' position.
Step 4   Steps 2, 3 are repeated until the result converges.

The accuracy of this process also depends on the number of beacon nodes and their distribution. If the node distribution is not uniform, then the calculation is error prone. Additionally, the second method is energy consumable because nodes interchange their calculated position several times.

(b)   Bounding box
In the bounding box approach [19], each node listens to its neighboring beacon nodes. Let us assume that the position of a beacon is (x_b, y_b), and the communication range is r. If an unknown node hears the beacon, then it is located within a box, whose two corners are $((x_b - r), (y_b - r))$ and $((x_b + r), (y_b + r))$. It can be expressed as:

$$x \in [x_b - r : x_b + r]; \; y \in [y_b - r, y_b + r].$$

The position of the node is within the intersection of all the bounding boxes corresponding to all the neighboring beacons. The intersection is itself a box whose minimum and maximum values can be calculated from the bounding boxes of neighbors. The center of gravity is the estimated position of the unknown node (Figure 7.8).
The intersection area can be calculated as

$$[\max(x_i - r), \max(y_i - r)] \times [\max(x_i + r), (y_i + r)],$$

where $i = 1, 2, 3 \ldots n$, and n is the number of neighboring beacons.

(c)   APIT
APIT is a range-free area-based localization scheme [20]. As with other localization schemes, two types of nodes, beacons and unknown nodes, are used here. APIT segments the whole area into triangular regions among beacon nodes. A node may be present inside or outside these triangles. A node can estimate its location by narrowing down the area by finding which triangles it is inside and outside (see Figure 7.9).
The method used to estimate the possible narrow area in which the node can be found is known as the point-in-triangulation test (PIT). In this test, a node chooses any three beacon nodes and tests whether it is inside the triangle made by chosen beacons or not, by comparing the signal strength of non-beacon neighbors. APIT repeats the PIT test with different combinations of neighboring beacons until all the combinations are exhausted or the desired accuracy is achieved.

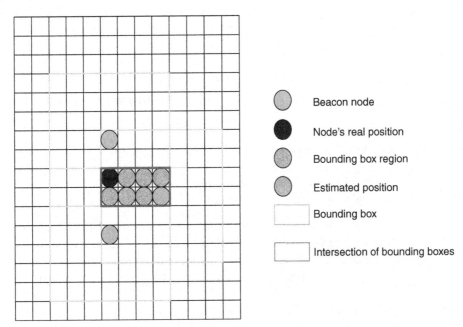

Figure 7.8 An example of intersection of bounding boxes (adopted from [19], with minor modifications).

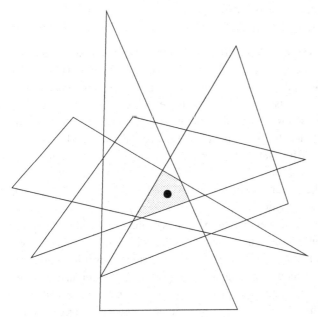

Figure 7.9 Area-based APIT algorithm overview (adopted from [20], with minor modifications).

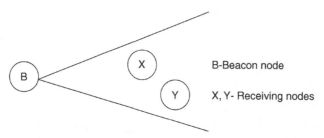

Figure 7.10   Signal strength variation with distance (adopted from [20], with minor modifications).

After PIT testing, APIT calculates the center-of-gravity (COG) of the intersection of all the triangles to estimate the node position. The algorithm has four steps.

Step 1   A node receives messages from all beacons where the distance between the node and any beacon is less than r, where r is the communication range of a node.

    Step 1.1   After receiving messages from beacons, each node maintains a table where entries are beacon ID, beacon position and received signal strength.

    Step 1.2   Nodes exchange the table with all neighboring nodes and build a super table using the information from all neighbors.

Step 2   PIT testing is done for a predefined number of beacon combinations.

Step 3   This step is called APIT aggregation, in which the intersection area of all the triangles, estimated at Step 2 is found.

Step 4   The COG is calculated.

During the PIT test, a node determines whether it is inside a triangle made by three beacons or not. Let us assume that the three beacons are A, B, and C. If the node is inside the triangle $\triangle ABC$ and it moves to a new position, then it moves nearer to beacons A, B, or C. If the node is outside $\triangle ABC$, then there exists a direction such that, if the node shifts, then it does so towards the triangle or away from it. If there exists a point p nearer to the node such that at that point p the node is closer/further to all of A, B, or C, then the node is outside $\triangle ABC$. But in practice a node may not be able to move generally. So, Approximate PIT (APIT) tests this differently. If the signal strength of beacon B at node X is more than the signal strength of B at Y, then Y assumes that X is nearer to B. This situation is shown in Figure 7.10.

In the APIT test, if any neighbor of node N is nearer to or distant from all three beacons A, B, and C, simultaneously, then N assumes that it is outside triangle $\triangle ABC$. Otherwise, it assumes that it is inside the triangle. This scenario is shown in Figure 7.11.

The APIT aggregation point is done by the grid-SCAN algorithm. In this algorithm, the whole area is divided into grids. The length of the grid side depends on the required accuracy. For each APIT in decision (when a node is within a region), the values of the grid regions that are inside the triangle under consideration are incremented. For each APIT outside decision (when a node resides outside of a triangle) the values of the corresponding grid regions are decremented. After calculating all the triangular regions, the information is used to find the maximum overlapping area to find center-of-gravity (COG) (Figure 7.12).

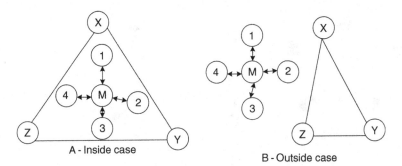

Figure 7.11   Approximate PIT test (adopted from [20], with minor modifications).

0	0	0	0	0	0	0	0	0	0
0	0	1	0	0	0	0	0	0	0
0	0	1	1	1	1	0	0	0	0
0	0	1	2	1	1	0	0	0	0
0	0	2	2	2	1	0	0	0	0
0	1	2	2	2	1	1	0	0	0
1	2	2	2	2	2	−1	0	0	0
1	1	1	1	1	1	−1	−1	−1	0
0	1	1	1	1	1	−1	−1	−1	−1
0	0	0	0	0	0	−1	−1	−1	−1

Figure 7.12   SCAN approach (adopted from [20], with minor modifications).

(d)   Multilateration

Multilateration is a distributive process of localization. The nodes which know their positions are known as beacon nodes, and the nodes which do not know them are known as the unknown nodes. Beacon nodes broadcast their location to their neighbors. Neighboring unknown nodes estimate their location from a received broadcast beacon message. Once an unknown node estimates its location, it transforms itself in a beacon node from an unknown node. The newly transformed beacon node now broadcasts its location to its neighbors. This process is repeated until all the unknown nodes satisfying the requirements for multilateration estimate their positions (Figure 7.13).

Atomic multilateration is a process in which an unknown node can estimate its position if it is a one-hop neighbor of at least three beacon nodes. These three beacon nodes should not be collinear points.

The error of measured distance between an unknown node and its ith beacon node is the difference between the measured distance and the estimated Euclidean distance [21] (equation (7.5)); (x_u, y_u) is the estimated coordinate point of unknown node u for $i = 1, 2, 3, \ldots, N$, where N is the total number of beacon nodes, s is the signal

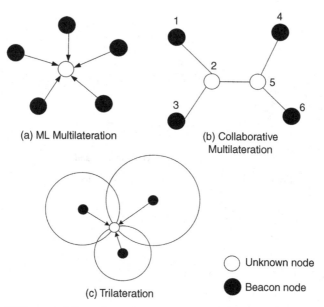

(a) ML Multilateration

(b) Collaborative Multilateration

(c) Trilateration

○ Unknown node

● Beacon node

Figure 7.13   Different multilateration techniques (partly adopted from [21], with minor modifications).

propagation speed, and t_{iu} is the signal propagation time from beacon i to unknown node u:

$$f_i(x_u, y_u, s) = st_{iu} - \sqrt{(x_i - x_u)^2 + (y_i - y_u)^2}. \qquad (7.5)$$

If a node has three neighboring beacon nodes, then from a set of three equations of the form (algorithm ML1), it can uniquely determine its location.

Iterative Multiplication is an algorithm that uses the atomic multilateration technique to estimate the locations of the unknown nodes iteratively. The iterative multilateration algorithm can be used in a centralized or distributive manner.

> *Algorithm ML1* Iterative Multilateration Algorithm [21]
> boolean iterative Multilateration (G)
> > (Max-Beacon-Node, Beacon-Count) ← unknown node with most beacons
> > while Beacon-Count ≥ 3
> > > setBeacon (Max-Beacon-Node)
> > > (Max-Beacon-Node, Beacon-Count) ← unknown node with most beacons

When a central node uses the centralized version of the iterative multilateration algorithm, the node must have global knowledge of the network. The input of the algorithm is graph G, which is known as the network connectivity graph. The weight of an edge of G denotes the separation of two adjacent nodes. The algorithm first estimates the location of unknown nodes with maximum number of beacons using atomic multilateration. After estimating its location, the unknown node becomes a beacon itself. The algorithm repeats until there exists an unknown node with three or more beacons.

Collaborative multilateration: In an ad-hoc network, nodes as well as beacons are deployed randomly. There is high possibility that, for some nodes, conditions for atomic multilateration are not fulfilled. That is, some node may not have three or more than three beacons. An example is shown in Figure 7.13b. When this happens, a node tries to estimate its position by using the information of more than one hop beacons. This process is known as the collaborative multilateration process. In Figure 7.13b, nodes 2 and 5 are the unknown nodes and nodes 1, 3, 4, and 6 are the beacon nodes. So, to find the location of nodes 2 and 5, collaborative multilateration processes should be used. In collaborative multilateration, an ad-hoc network is represented by graph $G = (V, E)$, where V represents the set of vertices and $|V| = n$, number of nodes in the network, and E is the set of edges. The beacon nodes are denoted by a set B where $B \subseteq V$ and unknown nodes are represented by a set U, where $U \subseteq V$. The nodes participating in collaborative multilateration form a sub-graph G' of G. Every edge of G' connects two participating nodes and can be represented by equation (7.6) [21]:

$$f(x_u, y_u) = D_{iu} - \sqrt{(x_i - x_u)^2 + (y_i - y_u)^2}, \tag{7.6}$$

where (x_u, y_u) is location of unknown node u, $\forall u \subseteq U$, $i \subseteq B$ or $i \subseteq U$, and $u \subseteq U$. A node is participating if it is a beacon node or it has at least three participating neighbors. In Figure 7.13b, if collaborative multilateration starts from node 2, then node 2 should have three participating neighbors. Neighbors of node 2 are nodes 1, 3, and 5. Nodes 1 and 3 are beacon nodes; so they are participating nodes. Node 5 has two beacon neighbors, nodes 4 and 6. Node 5 is also connected to node 2, thereby making both of them participating. A participating node pair is a beacon–unknown node pair or unknown–unknown node pair such that both unknown nodes are participating nodes.

We discuss a few schemes where trilateration was used. Some of these schemes were proposed for underwater sensor networks (UWSNs). However, by slight modification, they may be applied to terrestrial WSNs also.

• *Confidence-based iterative localization (CIL)* [22]   Multilateration-based approaches suffer a few basic challenges such as poor range estimation, noisy environment, and fluctuation in wireless communication. Following these challenges, the authors study the quality of trilateration (QoT) to put a figure on the relationship between the objects and ranging noise [22]. Based on this, confidence-based iterative localization (CIL) was proposed.

In CIL, the nodes assign a confidence value, which reflects its accuracy to localize. Initially, the highest confidence value is assigned to the beacon nodes and the unlocalized nodes are assigned with lowest values.

Confidence calculation   As the name suggests, this is a measure of the trust value of the localization result. Assume a trilateration t of node s such that $t = Tri(s, \{s_j, j = 1, 2, 3\})$. The real location of the node is $p(s)$ and the calculated location based on these beacon nodes (s_1, s_2, s_3) is $p_t(s)$. Assume that $d(s, s_i)$ denotes the distance between s and s_i, and it follows a distribution $f_{s,s_i}(x), x \in [0,+\infty)$. Then, confidence is defined as

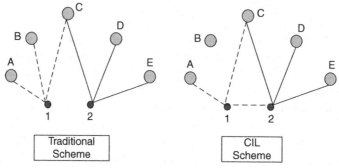

Figure 7.14   Comparison between traditional approaches and CIL (adopted from [22], with minor modifications).

$$C_t(s) = Q(t) \cdot \prod_{i=1}^{3} C(s_i),$$ where $Q(t)$ is the quality of trilateration t based on the definition of $Disk(p, R)$. $Disk(p, R)$ is defined as the disk area which is centered at p with a radius of R. $Q(t)$ is defined as [22],

$$Q(t) = \Pr\left(p \in Disk\left(p(s), R\right)\right)$$
$$= \int_p f_t(p) dp, \; p \in Disk\left(p(s), R\right).$$

From the definition of confidence of a node, it is concluded that confidence of a node s cannot be higher if its reference node s_i has a relatively lower confidence value. Otherwise, the confidence value of s depends on $Q(t)$. A node decides its confidence as follows:

$$Q(t) = \begin{cases} 1, \; \textit{if s is a beacon} \\ \max_{t \in CT(s)} (C_t(S)), \; \textit{otherwise} \end{cases}, \text{where } CT(s) \text{ is the candidate trilateration for node } s.$$

Location refinement   After calculating the confidence, each node continuously tries different possible trilaterations with the available reference nodes. Whenever a node finds a better localization option in terms of confidence value, estimation is done again to refine the location. For example, in Figure 7.14, two nodes 1 and 2 are taken. Node 1 refines its location based on the location update of node 2, as node 2 has a better confidence value. However, refining location increases accuracy at the cost of increased communication and computation overhead.

● *High-speed AUV-based silent localization (HASL)* [23]
This scheme utilizes the flexibility of a mobile beacon provider, an autonomous underwater vehicle (AUV). The main goal of this scheme is to reduce the effects of passive node mobility in increasing localization error of underwater sensor nodes. Also, maintaining energy-efficiency of the sensor nodes during localization process was another concern. Thus, the authors make use of the "silent" messaging technique. Three AUVs were assumed to cover the deployment region broadcasting their locations. The trajectory may be preplanned based on the deployment scenario. The AUVs move at

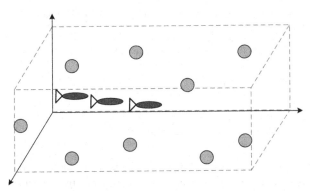

Figure 7.15   AUV-based localization scenario [23].

a fixed depth maintaining their velocity. Sensor nodes employ trilateration technique to estimate their coordinates (Figure 7.15).

The three AUVs are time-synchronized, i.e. they send beacons at the same time in the first phase of the algorithm. The beacon messages contain the following fields – AUV id, x_A, y_A, z_A, and timestamp; where x_A, y_A, and z_A are the location of the AUV at that time. This synchronized beacon sending ensures reduced delay between the reception of three beacons. Localization error due to node mobility is reduced in this fashion. A sensor node calculates its own coordinates after receiving beacons from three AUVs indicating same timestamp. This set of beacons is called effective beacon messages. However, the sensor nodes are able to acquire their z-coordinate value with the assistance of a pressure sensor attached to them. Let the three AUV position locations be (x_{A1}, y_{A1}, z_{A1}), (x_{A2}, y_{A2}, z_{A2}), and (x_{A3}, y_{A3}, z_{A3}), respectively, at time t_1. The unlocalized node's location then can be determined by [23]:

$$(x - x_{A1})^2 + (y - y_{A1})^2 + (z - z_{A1})^2 = d_1^2$$
$$(x - x_{A2})^2 + (y - y_{A2})^2 + (z - z_{A2})^2 = d_2^2$$
$$(x - x_{A3})^2 + (y - y_{A3})^2 + (z - z_{A3})^2 = d_3^2,$$

where, d_1, d_2, and d_3 may be calculated by RSSI measurement of the received beacons.

● *Mobility-assisted Localization (MobiL)* [24]

An inherent feature of underwater sensor networks (UWSNs) is that the nodes move according to the underwater currents. Because of the effect of passive node mobility, the location estimation error in sensor nodes increases. The greater amount of messaging between the unlocalized node and the reference node increases the protocol overhead, energy consumption, and delay, and thus increases the estimation error in the presence of node mobility.

This scheme presents a localization method for the sensor nodes in UWSNs. In this scheme, the localization process starts from the surface-based anchor nodes. The overall procedure is divided in two phases – mobility prediction, and ranging and localization. In the first phase, the authors used a velocity-estimation technique proposed by [25, 26]. Sensor nodes update their velocity based on the velocities of their neighbors. This is considered based on the fact that the movement of underwater nodes shows a group-like behavior:

$$v_j^x(d) = \sum_{i=1}^{n} \xi_{ij} v_i^x(d)$$

$$v_j^x(d) = \sum_{i=1}^{n} \xi_{ij} v_i^x(d)$$

The second phase utilized trilateration-based localization to localize the nodes. "Silent" messaging was considered to keep the energy consumption of the sensor nodes low. As the underlying localization procedure considered was trilateration, a node requires three beacons from three reference nodes. Let us assume that node S receives three beacons at time t_1, t_2, t_3. The displacement of node S is from (x, y, z) to (x', y', z') to (x'', y'', z'') in the meantime. The assumption is that the node can determine its position in the z-coordinate and have a negligible movement in the z-direction, i.e. $z = z' = z''$. Thus, using the velocity of node S, $y' = y_1 + (t_2 - t_1) \times v_S^y$. Similarly, x'' and y'' are calculated:

$$(x - x_1)^2 + (y - y_1)^2 + (z - z_1)^2 = d_1^2$$
$$(x' - x_2)^2 + (y' - y_2)^2 + (z - z_2)^2 = d_2^2$$
$$(x'' - x_3)^2 + (y'' - y_3)^2 + (z - z_3)^2 = d_3^2.$$

Replacing these values in the above equation, the location of node S can be calculated.

Relaxation-based localization
Anchor-free localization (AFL)
Anchor-free localization (AFL) is a concurrent and anchor-free scheme to solve the localization problem [27]. The term concurrent means that all nodes estimate and refine their location estimation simultaneously. In anchor-free localization, nodes try to estimate positions from local distance information where no node has any location information using GPS or any other method. The estimated coordinate system is not unique and it can be mapped on a global coordinate system in various ways by rotating, flipping, or translating. In this method, according to its authors, each node is assumed to be a "point mass" and these nodes are connected with "strings." Force-direction relaxation methods were used in this localization scheme which attains a minimum-energy configuration of the nodes.

The overall method of AFL is divided in two phases. In the first phase, the topology of the network is discovered and the output is a "fold-free graph." Each node n_i has a unique ID represented as ID_i. The hop count between two nodes is also calculated along the shortest possible path between these two nodes. It is denoted as h_{ij} for the hop count between nodes n_i and n_j. Initially, five nodes are to be selected from the graph. Let n_1 be the first node selected, based on the condition that it is located at the periphery of the graph. Then, n_2 is selected such that it is located at the maximum possible hop count away from node n_1. Similarly, n_3 and n_4 are selected. Here, n_3 is the maximum hop away from both n_1, n_2, and n_4 is the maximum hop away from n_3 and equidistant from both n_2, n_3.

Finally, node n_5 is selected such that it is equidistant from each of nodes n_1, n_2, n_3, n_4. In this fashion, for a node n_i, the polar coordinate (p_i, θ_i) is found, based on the information of h_{1i}, h_{2i}, h_{3i}, h_{4i} and the radio transmission range R. It can be calculated as [27],

$$p_i = h_{5i} \times R$$
$$\theta_i = \tan^{-1}\left(\frac{h_{1i} - h_{2i}}{h_{3i} - h_{4i}}\right).$$

In the second phase, a mass-spring-based optimization technique is used to correct and balance the localized error. For this, each node n_i estimates the distance d_{ij} to node n_j and also knows the measured distance r_{ij} to this node n_j. A unit vector in the direction p_i to p_j is denoted as v_{ij}. The force F_{ij} in this direction is then calculated as [27],

$$F_{ij} = v_{ij}(d_{ij} - r_{ij}).$$

The resultant force on node n_j is,

$$F_i = \sum_{i,j} F_{ij}.$$

In the next step, the total energy of node n_i is calculated,

$$E_i = \sum_j (d_{ij} - r_{ij}).$$

The energy E_i of each node n_i is reduced when it moves in the direction of the force F_i. The algorithm converges when F_i and E_i for each node are zero.

The algorithm is as follows.

Step 1 A node n_1 is chosen at the periphery of the graph.

Step 2 A node n_2 is selected such that n_2 is maximum hop count away from n_1.

Step 3 A third node n_3 is selected such that n_3 is maximum hop count away and equidistant from both nodes n_1 and n_2.

Step 4 Node n_4 is selected in such a way that it is maximum hop count away from node n_3 and equidistant from nodes n_1 and n_2.

Step 5 Node n_5 is selected as it is equidistant from each of nodes n_1, n_2, n_3, and n_4.

Step 6 For each node n_i, the hop counts h_{1i}, h_{2i}, h_{3i}, h_{4i}, h_{5i} are estimated from chosen reference points.

Step 7 For each node n_i, the approximate polar coordinate (p_i, θ_i) is estimated by using the hop counts and radio range R

$$p_i = h_{5i} \times R$$
$$\theta_i = \tan^{-1}\left(\frac{h_{1i} - h_{2i}}{h_{3i} - h_{4i}}\right).$$

Step 8 A local optimization technique is performed on current estimated coordinate by the nodes.

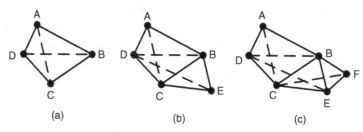

(a)   (b)   (c)

Figure 7.16   Example of robust quad chaining (adopted from [28], with minor modifications).

Coordinate system stitching-based localization

Robust distributed network localization with noisy range measurements

In this approach, the nodes first estimate the distances of all one-hop neighbors [28]. This distance information is then exchanged among the neighbors. Using the shared distance information, each node localizes itself and its neighbors and then the nodes are organized into clusters. A cluster refers to a node and its one-hop neighbors. During this process, the nodes form a local coordinate system as they do not have any knowledge about the global one. To transform the local coordinate system into global, the overlapped clusters are merged or stitched. All sets of four nodes that are fully connected are found. These quadrilaterals are taken as the smallest sub-graph and called "robust quad." The relative positions of the nodes of a robust quad are unambiguous even in the presence of measurement noise. Two robust quads are "chained" if they have three common nodes.

In Figure 7.16, ABCD is a robust quad because all nodes are completely connected and their relative positions are unambiguous. In the next stage, node E joins the quad by using known positions of ABD. F is joined by using the same procedure where ADFE is a robust quad.

The algorithm has three phases, which is described as follows.

Phase 1   Cluster localization: In the first phase, each node becomes the center of a cluster and finds all its neighbors and estimates the relative location of all the neighbors. All robust quads within a cluster are identified and the largest sub-graphs with overlapped quads are also identified. The position of a node within a cluster is then computed using chaining of robust quads and trilateration. This phase is shown in Figure 7.16.

Phase 2   Cluster optimization (optional): In this phase, the estimated positions are refined using numerical optimization techniques such as spring relaxation or Newton–Raphson. Any error that accumulates in the computation is reduced in this fashion. However, this is an optional phase, which may be omitted to gain maximum efficiency. One of the advantages of this phase is that no additional overhead is added to the protocol.

Phase 3   Cluster transformation: Transformations among the local coordinate systems of neighboring clusters are computed. This is done by selecting the set of nodes in common between two clusters.

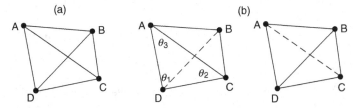

Figure 7.17 Robust quadrilateral and its decomposition in four triangles (adopted from [28], with minor modifications).

Quadrilaterals are taken as unit sub-graph for localization, as they can be used for unambiguous localization. As in Figure 7.17, four nodes are connected using six distance measurements. If no three nodes are collinear, then the relative positions of four nodes are unique with respect to graph operations such as rotation, translation and reflection, where b is the length of shortest side, θ is the smallest angle and d_{min} is the threshold value. If all four triangles of a quadrilateral are robust then the quadrilateral itself is also robust. So, a robust quadrilateral is a fully connected quadrilateral whose four sub-graphs are also robust.

The robust quadrilateral is formed based on few conditions; however, these conditions are not sufficient to guarantee the unique graph realization when distance measurements are noisy [28]. For this reason, further graph decompositions are done. This is shown in Figure 7.17a, where the quadrilateral is further decomposed into four triangles $\triangle ABC$, $\triangle BCD$, $\triangle ACD$, $\triangle BCD$. If each triangle satisfies equation (7.7), then it is called a robust triangle:

$$b\sin^2\theta = d_{min}. \tag{7.7}$$

In this equation, b is the length of the shortest side of the quadrilateral, θ is the smallest angle, and d_{min} is a threshold chosen based on the measurement noise.

Hybrid localization
Localization in sensor networks with limited number of anchors and clustered placement
This is a distributed localization scheme [29] composed of two different techniques – multidimensional scaling (MDS) [30] and proximity-distance map (PDM) [31]. The localization scheme works with a limited number of anchors and clusters placed. Initially, only a few nodes, called as primary anchors, are attached with GPS receivers. Secondary sensors are selected as a subset of the ordinary nodes. These secondary sensors are localized using MDS and the other remaining nodes are localized by using PDM.

The steps of the scheme are as follows [29].

Step 1   The secondary anchors are selected in this step: k_p is the number of primary anchors and k_s is the number of secondary anchors for each of the primary anchors. Primary anchors send an invitation packet containing its unique ID, counter, and the number k_s. Initially, the counter is set to zero. An ordinary node performs Bernoulli trial with success rate of p upon receiving this message. If the outcome is true, then it becomes a secondary anchor. Thus, the total number of anchors in the network is $p = k_p \times (1 + k_s)$.

Step 2   The primary anchors send packets containing the coordinate and proximity of the packet, i.e. the hop count of the packet. Secondary anchors do the same, except that the coordinate value in the packet is left blank.

Step 3   All of the nodes receive a packet containing the proximity value. If a node receives more packets, then it stores it only for lower proximity value.

Step 4   The proximity value is exchanged between the anchor nodes. After knowing the proximity value for any pair of anchor nodes, secondary anchors localize themselves using MDS.

Step 5   Proximity distance mapping T is calculated using the proximity matrix P and geographic distance matrix L: $T = LP^T (PP^T)^{-1}$.

Step 6   Ordinary sensor nodes calculate their positions based on the stored proximity vector p_s and the position information of the anchors.

The advantage of this scheme is its reduced complexity over MDS. MDS requires $O(n^3)$ computations for n number of nodes and PDM requires $O(m^3)$ calculations for m anchors. This scheme performs with $O(p^3)$ complexity.

Interferometric ranging based localization
Radio interferometric positioning system (RIPS)
Radio interferometry is based on the concept of interference between pairs of two senders and two receivers [32]. By further analyzing the interferometric signal, the location of two senders or two receivers can be calculated accurately. The radio interferometric ranging is shown in Figure 7.18. The idea given by the authors of this work [32] was that

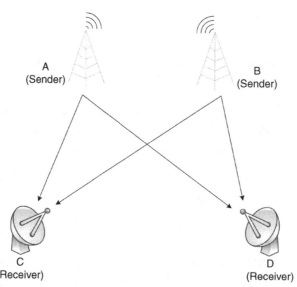

Figure 7.18   Radio interferometric ranging (adopted from [33], with minor modifications).

if both the senders use almost the same frequency, then the resultant signal will have a low envelop, which may be easily detected by the available hardware of the sensor node. However, due to lack of synchronization between the nodes, there is a phase offset between the received signals, which is a function of the relative positions of the sender–receiver pair and the communication frequency:

$$phase\ offset = 2\pi \frac{d_{AD} - d_{BD} + d_{BC} - d_{AC}}{\lambda_{carrier}},$$

where d_{ij} denotes the distance between i and j, and $\lambda_{carrier}$ is the wavelength.

However, interferometric ranging based localization is an NP-complete problem [33]. Also, to measure the inter-node distance, phase offset is calculated at different frequencies requiring multiple transmissions. The selection of a pair of sender–receiver nodes provides range estimation between the two nodes. Therefore, finding such pairs is a problem in itself. Genetic algorithms (GA) were used as the optimization tool. By means of optimization, the goal is to find the relative positions of the nodes, with the set of nodes as an unknown location and the set of ranges as given.

Error propagation aware localization
An error propagation aware algorithm for precise cooperative indoor localization [34] was proposed by Alsindi *et al*. The error propagation aware (EPA) algorithm considers path loss (PL) and distance measurement error (DME) in node localization. The major contribution of this scheme was to control error propagation in node localization. For this, the authors first modeled the ranging error specific to an indoor application, and proposed a method for the anchors to be capable of identifying the extent of the error.

Stage 1 Channel sensing: Anchor nodes broadcast packets with unique ID, coordinate (x, y) and position error variance (σ_p^2). Nodes sense the channel and collect TOA information about the anchors.

Stage 2 Position statistics: σ_p^2 for the GPS-powered anchors is assumed to be zero, as they are able to calculate their position accurately. Anchors also form a covariance matrix that characterizes the extent of an error.

Stage 3 Weighting matrix: Ranging variance (σ_r^2) is computed from the power of the detected direct paths. Using these metrics, the weighting matrix is formed as follows:

$$W = W_r + W_p$$
$$W_r = diag(\sigma_{r1}^2, \ldots \sigma_{rn}^2)\ \text{and}\ W_p = diag(\sigma_{p1}^2, \ldots \sigma_{pn}^2).$$

Range estimation error is assumed to be larger than the position estimation error $(\sigma_{ri}^2 > \sigma_{pi}^2)$.

Stage 4 Multilateration through WLS: A node calculates its position using the weighting matrix into the weighted least square (WLS) algorithm [35]. The covariance matrix is also calculated in this stage.

Stage 5   Confidence testing: A few conditions are checked in a node before it becomes an anchor, to reduce the error propagation.

Stage 6   State transformation: After successfully passing the confidence tests, a node becomes an anchor. The node then starts broadcasting packets to its neighbors. The process is repeated until all nodes get to know their position and become an anchor.

7.2   Target tracking

Tracking a target with the help of deployed sensor nodes creates a few possible applications in both the civilian and military domains. The applications of target tracking include surveillance, habitat monitoring, disaster relief, traffic monitoring, and intruder detection [36]. There are a few advantages of applying target tracking in WSNs. For example, improved qualitative measurement, accurate and timely signal processing, and increased robustness [37]. However, there exist several challenges associated with it. Some of these challenges are for using WSNs such as limited battery power, low bandwidth, short communication range of nodes, and limited processing capability. The target tracking approaches mainly focus on finding a balance between the energy consumption of the sensor nodes and tracking accuracy. Distributed control over a large deployment area is also required in target tracking applications. The problem with the centralized approaches is that they are vulnerable to a single point of failure, and, moreover, they are not scalable to large number of nodes. However, distributed approaches increase computation and communication cost in the networks.

Typically tracking a target is based on three steps – node localization, target localization, and target location update. Now, based on the number of targets to track, the different existing approaches may be divided in two categories – single target tracking and multi-target tracking. In Figure 7.19, a single target tracking scenario is shown, in which the sensor nodes alongside the target trajectory are activated and the remaining nodes remain in the sleep state.

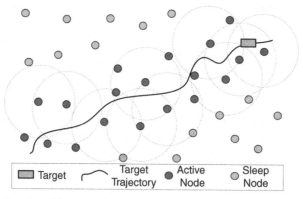

Figure 7.19   Target tracking scenario.

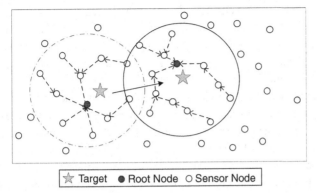

| ☆ Target   ● Root Node   ○ Sensor Node |

Figure 7.20   Convoy tree reconfiguration in DCTC with target movement (adopted from [38], with minor modification).

7.2.1   Single target tracking

The proposed solutions are classified into five different approaches – tree-based tracking, cluster-based tracking, prediction-based tracking, mobicast message-based tracking, and hybrid tracking methods. In the following one protocol of each class is discussed.

7.2.1.1   Tree-based tracking

DCTC (dynamic convoy tree-based collaboration for target tracking)
As the name suggests, this scheme introduces a concept called dynamic convoy tree-based collaboration (DCTC) for detection and tracking of mobile targets [38]. The convoy tree is formed with the sensor nodes around the mobile target, and it is dynamically maintained by adding or removing nodes as the target moves. The initial convey tree is formed when a target is first detected. One root is selected that collects more information from the sensor nodes to refine the information. Later, as the target moves, some nodes in the tree are no longer needed and so they are removed from the convoy tree. The root predicts the future movement direction of the target and the nodes in that area are activated (Figure 7.20).

The root of the convoy tree also needs to be changed as the movement of the target progresses. This helps to optimize the communication overhead between the nodes. Convoy tree reconfiguration is formulated as an optimization problem. The optimal solution is based on dynamic programming (o-DCTC) with maximum tree coverage and minimum cost. There are two methods proposed for expansion and pruning of the tree, namely the conservative scheme and the predictive scheme. Two tree reconfiguration schemes are sequential reconfiguration and localized reconfiguration.

Tree expansion and pruning
As the target progresses in the sensor field, in each time interval, some nodes are added to the tree and some nodes are excluded.

In the conservative scheme, the distance between the current target location and a node is the qualifying criterion for a node to be present in the tree. The distance should be less than $v_t + \beta + d_s$, where v_t is the current velocity of the target, β is a parameter, and d_s is the radius of the monitoring region. This target is assumed to move in any direction with velocity less than $v_t + \beta$. An increased value of β provides better coverage. However, if the value of β is too large, then the tree contains redundant nodes, which increases the overall power consumption as more nodes will be in the active mode.

A prediction-based scheme aims to reduce the redundant nodes in the convoy tree. Some assistance of a prediction scheme is required. The distance between the nodes and the predicted location of the target should be less than d_s for the nodes to be added in the tree. However, the coverage of the tree is reduced if the prediction is not accurate. After identifying the nodes to be removed from the tree, the root notifies the grid about this. The grid again sends prune messages to these nodes and on receiving these messages, the nodes change to the sleep mode.

Tree reconfiguration

Tree reconfiguration and replacement of the root node is necessary as the target moves. However, frequent reconfiguration of the convoy tree increases the overall energy consumption of the nodes, which is undesirable. Thus, a threshold distance is taken to check the condition of triggering the tree reconfiguration. The distance between the root and the predicted target location should be larger than $d_m + \alpha \times v_t$. Here, d_m is the minimum distance to enable tree reconfiguration, α is a parameter that indicates the impact of velocity, and v_t is the target velocity at time t. A new root node of the tree is selected which is near to the target location, so that the tree is constructed with less height and the energy consumption is also minimized. For multiple nodes in the option, one of them is randomly selected.

A minimum cost tree is generated based on the assumption that any node in this tree knows the locations of other nodes in the tree.

Algorithm: *MinCostTree(V, e, R)*

The notations are as follows.

- $e(i,j)$: energy consumption for transmitting from i to j
- Q: minimum cost tree
  - (1) $Q = V$
  - (2) While $Q \neq 0$
  - (3) Find $u \in Q$ such as $(\forall v \in Q)\, (e(u, R) \leq e(v, R))$
  - (4) For each v that is a neighbor of u
  - (5) If $e(v, u) + e(u, R) < e(v, R)$ then
  - (6) $e(v, R) = e(v, u) + e(u, R)$; change v's parent to be u
  - (7) $Q = Q - \{u\}$

The sequential reconfiguration scheme relies on the grid structure of the network. Root R sends an *rconf(R, l_0)* message to the nodes present in the same grid, say, g_0, and to the heads of neighboring grids, say g_1, g_2, g_3, g_4. The *rconf(R, l_0)* message contains information about the location of nodes and cost of sending data to the root. The nodes within the grid g_0 execute the *MinEngPath(R, l_0, 0)* algorithm. Any grid head g_i receiving *rconf(R, l_j)* checks whether there is any neighboring grid that is closer to R. If the condition satisfies, the *MinEngPath(R, l_i, $\cup l_j$)* algorithm is executed. Otherwise, the *rconf(R, l_i)* message is forwarded.

Algorithm: *MinEngPath(R, l_1, l_2)*

The notations are as follows.

- l_i: List contains locations of nodes in grid g_i and the packet sending cost. For a node k of g_i, $l_i = \langle i, e(i,R) \rangle$
- $d_{i,j}$: Distance between i and j
   (1) Sort items in l_1 in increasing order of $d_{i,R}$, where $\langle i, e(i,R) \rangle$ is an item in l_1
   (2) For each $\langle i, e(i,R) \rangle \in l_1$
   (3) Find $\langle j, e(j,R) \rangle \in l_2$ such that $(\langle k, e(k,R) \rangle \in l_2)(e(i,j) + e(j,R) \le e(i,k) + e(k,R))$
   (4) Node j is set to be i's parent, $e(i,R) = e(i,j) + e(j,R)$
   (5) $l_2 = l_2 + \{\langle i, e(i,R) \rangle\}$

The drawback of the sequential method is that it generates high overhead in exchanging the information about all the nodes. The localized scheme incorporates optimizations to solve this problem. In the first of the two proposed optimizations, the cost of message sending is calculated based on the distance between two nodes. This information is not included in the configuration messages further. In the second optimization, the information about the node locations is removed. This is done by reading the cache attached to the node, and assuming that the node movement is not so frequent. For high-density networks the localized scheme is better; however, the energy consumption of the nodes increases when the node density is low.

7.2.1.2 Cluster-based tracking

CODA: continuous object detection and tracking algorithm

Clusters are formed to support collaborative data processing required by the sensor nodes. One such approach is the continuous object detection and tracking algorithm or CODA [39]. There are two types of clustering approaches – either static or dynamic. Both of these approaches have some advantages and disadvantages. In static clustering, the cluster area members remain unchanged; thus, they are not suitable for tracking mobile targets. On the other hand, clusters are formed dynamically based on the detection of the event of interest, in dynamic clustering. This approach is suitable for target tracking applications. However, the communication overhead for cluster formation is quite high using this approach. CODA uses a hybrid clustering approach for continuous tracking with low message overhead.

CODA assumes a network where nodes are divided into static clusters; one cluster head (CH) is present in each cluster. Clustering may be done by any existing scheme. The CH decides the boundary nodes of the cluster, by solving the convex-hull problem [40] using the Graham scan algorithm [41, 42], after receiving the location information of all the nodes. These nodes are notified using messages from the CH. Based on the classification, the boundary sensors are named Static-cluster boundary-sensors (SBs) and the remaining nodes are named Static-cluster inner-sensors (SIs).

In the next phase, the static clustering scheme is utilized for boundary tracking of the object. The sensors send control messages to notify the CHs about the detection of an object. Two control messages are the "SENS" message used by the SIs, and the "REPORT" message used by the SBs. In the "REPORT" message, the nth bit is set against the detection of the object in the cluster n. The SB then communicates with its one-hop SBs to know the status of object detection, and sets the corresponding bits against this information. Depending on the number of clusters that detects the object, there may be different cases by which the boundary of the object with each static cluster may be identified.

The boundary sensors are then organized in a dynamic cluster. It is dynamic in the sense that the number of nodes in a cluster is changed as the target moves. CHs fuse the boundary data and send the data to the sink node. The sink node collects the information from the CHs and can determine the whole boundary of the object.

The boundary of the object changes as the object moves, and thus the membership of the boundary sensors changes. This change indicates the new set of nodes required for the detection of the object boundary.

Localized policy-based target tracking

Misra and Singh [43] designed a cluster-based tracking scheme, in which the number of active sensor nodes is minimized to achieve energy efficiency. This scheme maintains a balance between the energy efficiency of the nodes and the target tracking accuracy. With prudent usage of the sleep and wake-up mechanisms, the network lifetime can be increased. The movement of a target is modeled based on the Gauss Markov mobility model [44]. On detecting a target, the cluster head that detects it activates an optimal number of nodes within its cluster, so that these nodes start sensing the target. A Markov decision process (MDP)-based framework is designed to adaptively determine the optimal policy for selecting the nodes localized with each cluster. As the distance between the node and the target decreases, the received signal strength (RSS) increases, thereby increasing the precision of the readings of sensing the target at each node.

The sensor nodes are deployed over a two-dimensional terrain randomly and are distributed in clusters. Each node knows its location and can estimate the location from the neighboring nodes. The Gauss Markov mobility model [44] is set for the target's mobility model as it has a certain degree of randomness associated with it, which makes it more realistic. It uses a tuning parameter to bring about variation in the degree of randomness of a moving object. The target maintains constant speed, however, it changes direction by certain angle α after time δt, following the given equations [44]:

$$\theta_t = \alpha\theta_{t-\delta t} + (1 - \alpha)\bar{\theta} + \sqrt{(1 - \alpha^2)\theta_{s_{i-\delta t}}}.$$

Target velocity can be calculated using the following:

$$V_{x_t} = V_t \cos\theta$$
$$V_{y_t} = V_t \cos\theta,$$

The location after time δt is calculated as follows:

$$x_t = x_{t-\delta t} + V_{x_{t-\delta t}} \times \delta t$$
$$y_t = y_{t-\delta t} + V_{y_{t-\delta t}} \times \delta t.$$

The primary motive of this approach is minimizing the number of nodes within a cluster that should be selected and subsequently activated to sense a target. The nodes are activated for a certain time period, t_a. These nodes sense the target at sensing interval t_s. The total time taken by a CH to wake up the nodes in a cluster and for the nodes to start sensing the target is t_{as}. Assume that the time taken for the nodes to send back the sensed information to the CH is t_b. The MDP-based algorithm for node selection runs on each of the CHs. At every state, it predicts the next location of the target using a Kalman filter. Prediction takes place after the CH detects the target or receives the responsibility of tracking the target from the neighboring CH that was detecting the target before it. Subsequent to predicting the next location of the target after a time interval of δt, the CH checks whether the target lies within its range or is closer to the neighboring nodes. Once that decision is made, the CH of the cluster that was tracking the target till then, hands over the charge of tracking the target to the closest CH.

Once the CH is selected, it selects a few nodes from its child list, based on the remaining energy of the nodes. Node selection probability (NCP) is defined as the probability with which a node would be included into the list of nodes that would perform the tracking operation within a cluster. It is given by the following equation:

$$P_s = 1 - (G_t/N_j),$$

where, G_i is the number of nodes within the cluster having an NCF greater than the NCF of the ith node. The node capability function (NCF), is defined by the following equation [43]:

$$F_i = \beta \times T_i + \lambda \times 1/E_i$$

where β and λ are arbitrary constants and T_i and $1/E_i$ denote tracking efficiency and energy efficiency, respectively.

A higher NCF would imply that the node is a better candidate for being selected for the target tracking procedure. It is proved that F_i is an adequate measure to choose the nodes from the list of child nodes per CH.

7.2.1.3   Prediction-based tracking

Prediction-based energy saving scheme (PES)

This scheme addresses the issue of how object tracking in a sensor network can be done in a more energy-efficient way [45]. PES minimizes the number of nodes involved in object tracking, while putting the other nodes into the sleep mode to save energy.

The problem of object tracking discussed in [45] involves S number of sensor nodes tracking O number of moving objects. The sampling time required is X seconds and the event update rate is $1/T$. The goal is to minimize the overall energy consumption while maintaining an acceptable missing rate. Here, the missing rate denotes the ratio of sensor nodes that fail to report the detection to the total number of sensor nodes.

There may be different types of monitoring schemes for applications of object tracking. The naïve approach, which is the simplest one, however, is not energy efficient for sensor nodes. Scheduled monitoring (SM) is another approach, in which the sensor nodes do not remain active for all the time. These nodes remain active only for X seconds and sleep for the remaining $(T-X)$ seconds. The scheme is advantageous in the sense that the nodes save energy by staying in the sleep mode most of the time. However, the disadvantage of this scheme is that the required number of sensor nodes increases to ensure no missing rate. Another scheme is continuous monitoring (CM), where only a sensor node, which has detected the object, remains active. Such a sensor node predicts the next destination of the target and activates that node W seconds before the object reaches its destination. This handover happens only in the boundary area for small values of W. This scheme has the advantage that it requires only one node to be active per object.

Comparing these strategies, it can be concluded that the naïve approach requires the maximum number of active nodes and the highest sampling frequency for maintaining the zero missing rate. However, SM maintains the zero missing rate with the maximum number of active nodes and CM shows the same result with the highest sampling frequency.

In the PES scheme, the authors minimize both of these factors, the number of active nodes and the sampling frequency, to optimize the energy consumption. PES has three parts – prediction model, wake-up mechanism, and recovery mechanism. Using the prediction model, PES predicts the future movement of the target and activates only those nodes. The sensor nodes are selected to be activated based on the energy and performance in the wake-up mechanism. The recovery mechanism is used when the target is lost. The sensor nodes always try to sleep as much as they can. Prediction to determine the next active nodes uses different heuristics for selecting a target's velocity and direction such as – INSTANT, AVERAGE, and EXP_AVG. Next, a group of nodes is selected based on different wake-up heuristics – DESTINATION, ROUTE, and ALL_NBR. A recovery mechanism is proposed by the authors of this work to find the lost target. For this, all the sensor nodes are activated by the current node tracking the target.

7.2.1.4   Mobicast message-based tracking
HVE-mobicast

Hierarchical-variant-egg-based mobicast or HVE-mobicast is a mobicast routing protocol proposed for sensor networks with the goal of maintaining power efficiency [46]. Mobicast is a variant of multicast which decides the forwarding zone of a message. The message is forwarded to all sensor nodes present at zone Z at time t, and the location and shape of this message forwarding zone may be determined by a function of time interval (t_{start}, t_{end}). A spatiotemporal multicast or mobicast session is defined as $\langle m, Z[t], T_S, T \rangle$, where m denotes the multicast message, $Z[t]$ is the forwarding zone, T_S is the sending time and T is the duration of the session. The sensor nodes are assumed to have a GPS attached to them by which these nodes estimate their locations. The overall method is divided in two phases – egg estimation and distributed HVE-mobicast.

All sensor nodes estimate the variant-egg $F_{HVE}[t+1]$ at time t for the delivery zone Z_{HVE} $[t+1]$. The shape of the variant egg is calculated by $[(x)^2 + (y)^2]^2 - 2e^2[(x)^2 - (y)^2] = 0$ (the equation is reduced from [47]), where $e = \sqrt{\pi}R$, and R is the radius of the delivery zone. In the second phase, a distributed algorithm is designed to adjust the size and shape of the variant egg. The sensor nodes are divided in two groups – cluster heads and relay nodes belonging to group I, and the remaining nodes belonging to group II. To maintain low packet overhead, the nodes in group II do not forward the message. First, the message is forwarded to a hold-and-forward cluster head (CH) at time t_1. This CH again forwards this message to all neighboring CHs. This way all the nodes in group I are activated. In this scheme, the power efficiency of message delivery is improved due to the use of cluster-to-cluster delivery, instead of node-to-node delivery. Thus, HVE-mobicast predicts the forwarding zone efficiently.

7.2.1.5   Hybrid tracking method
Distributed predictive tracking (DPT)
This scheme provides a hybrid solution for target tracking applications [48]. It is a hybrid of two different schemes: the cluster-based approach is utilized for scalability, and the prediction-based approach offers a distributed and energy-efficient solution. DPT also provides robustness against node failure.

In DPT, the sensor nodes are randomly distributed over the area. These sensor nodes are of the same type and the CH knows their IDs, location and energy level. The sensing radius of these nodes may change from normal beam (r) to high beam (R). To enhance energy efficiency, the sensors remain in the sleep mode until they are instructed by the CH to perform a sensing task. The target is first detected by the boundary sensors of a cluster. A target descriptor (TD) is used to maintain the information of the target. It contains target id, present location, next predicted location, and timestamp. From the information of $TD_{i-1}(x_{i-1}, y_{i-1})$ and $TD_i(x_i, y_i)$, the speed of the target is calculated as follows [48]:

$$v = \frac{\sqrt{(x_i - x_{i-1})^2 + (y_i - y_{i-1})^2}}{t_i - t_{i-1}}.$$

The direction of movement of the target is calculated as [48]:

$$\theta = \cos^{-1} \frac{x_i - x_{i-1}}{\sqrt{(x_i - x_{i-1})^2 + (y_i - y_{i-1})^2}}.$$

Based on this information, the target's location may be predicted after time t, as follows [48]:

$$x_{i+1} = x_i + v.t.\cos(\theta),$$
$$y_{i+1} = y_i + v.t.\cos(\theta).$$

Once the next CH (CH_{i+1}) is decided, it starts selecting sensors that will perform the target sensing task. Three nodes are selected, which maintain minimum distance to the predicted location of the target. These sensors will update the TD value. However, if the appropriate sensor nodes are not found inside range r, then the algorithm searches for such nodes in range less than R. Now, these sensors will have to use high beam for sensing.

There may be failures occurring during tracking. Two types of failures are discussed – unavailability of predicted CH, and sudden change of target speed and direction. To counter this problem, three recovery mechanisms are proposed. The first level of recovery changes the sensor node's sensing radius to high beam, if it were using normal beam initially. In the second level of recovery, all the sensors present within distance r from the predicted location L_i are activated. If there is failure of both of these strategies, the sensors present inside $(2N-3)r$ distance from L_i are activated. A few energy-saving strategies are included in this algorithm – first, the nodes are at hibernation until activated, and second, the algorithm uses normal beam whenever possible.

7.2.2   Multi-target tracking

7.2.2.1   Hierarchical Markov decision process (HMDP) for target tracking (HMTT) [49]

HMTT proposes an energy saving scheme for sensor nodes based on a realistic mobility model [49]. The target tracking framework is cluster-based, and uses a two-level Markov decision process (MDP) to predict the target trajectories. Energy efficiency of the sensors is maintained by determining the optimal sleep time of the sensors. The sensors are assumed to be deployed randomly over a two-dimensional field and they are divided into few clusters. R is the detection range of these sensors, and they measure the target velocity $(\hat{v}, \hat{\theta})$, not the target location. A cluster head has three states – sensing, listening, and tracking. Target mobility is driven by the Gauss Markov (GM) mobility model.

A simple prediction model, parameter estimation, is used to predict the target trajectory. For this purpose, a Kalman filter was not used, as target mobility parameters are not known to the sensor nodes in this scenario. A maximum-likelihood estimation of $x_{t+\delta t}$ is calculated as follows [50]:

$x_{t+\delta t} = \hat{\rho}_x x_t + (1 - \hat{\rho})\hat{\mu}_x$, where $\hat{\mu}_x$ and $\hat{\rho}_x$ are the estimated mean and the autocorrelation coefficient, respectively. These values are calculated as [50]:

$$\hat{\mu}_x = \frac{1}{N}\sum_{n=1}^{N} x_{n\delta t}$$

$$\hat{\rho}_x = \frac{N}{N-1}\frac{\sum_{n=2}^{N} x_{n\delta t}x_{(n-1)\delta t}}{\sum_{n=1}^{N} x_{n\delta t}^2}.$$

Using this information, the target location P_t is predicted as [50]:

$$\hat{P}_{n\delta t} = P_{(n-1)\delta t} + \hat{v}_{n\delta t}\begin{pmatrix}\cos(\hat{\theta}_{n\delta t}) \\ \sin(\hat{\theta}_{n\delta t})\end{pmatrix}\delta t,$$

where $P_{(n-1)\delta t}$ is calculated as [50]:

$$\hat{P}_{(n-1)\delta t} = P_0 + \delta t \sum_{m=1}^{n-1} v_{j\delta t} \begin{pmatrix} \cos(\hat{\theta}_{m\delta t}) \\ \sin(\hat{\theta}_{m\delta t}) \end{pmatrix}.$$ However, P_0 cannot be calculated until the

sensor nodes are localized. To solve this problem, the authors designed the scheme to track the targets at location areas rather than exact positions. Next, a Markovian prediction model is defined as $\tilde{L}_K(t + \tau) = f_k(L, \tau_K)$, where f_k is a stochastic function and τ_K is the dwell time of a target K in location area L. For a target K detected for the nth time in interval δt, $\tau_K = n\delta t$. Values of $\hat{v}_{n\delta t}$ and $\hat{\theta}_{n\delta t}$ are calculated based on their $(n-1)$th time's value. Thus, the prediction of the target at the next location area is [50]:

$$\tilde{L}_K(t + \delta t) = f_k(v_{K,t}, \theta_{K,t}, L, \tau_K).$$

The stochastic function f_k may be changed as per the required target mobility model.

The preference of a few trajectories over the set of all trajectories creates the problem of energy drain in the nodes belonging to those areas. In this work, the authors solve this problem using prediction. The target mobility is assumed to be Markovian in nature, as mentioned earlier. Also, the targets are divided into different classes, where targets in one class have similar mobility parameters. In this work, multiple targets of only one class are considered. Based on HMDP [50], the tracking problem is formulated. The authors look to optimize the overall cost function (J) to determine the duty cycle of the sensors, and to predict target trajectories while tracking:

$$J = E\left[\sum_{i=0}^{\infty} \gamma_h^i g_i\right],$$

where g_i is the immediate cost function calculated by [50], and

$$g_i = \alpha \frac{p_i}{p_{\max}} + (1 - \alpha) \frac{\sum_{K=1}^{K_t} X_{K,i}}{\sum_{K=1}^{K_t} N_{K,i}},$$

where p_i is the used power, $p_{\max}$ is the maximum power, $X_{K,i}$ is the number of wrong predictions for target K in the interval i, and $N_{K,i}$ is the number of predictions for target K in the interval i. Note that α is a parameter required to balance p_i and $X_{K,i}$. To optimize the performance of prediction (i.e. accuracy of tracking), and energy efficiency, the overall cost function J is minimized.

The target may reach a cluster that may not have been warned before. This type of situation may happen because of incorrect or inaccurate predictions. As a result, the target may be lost. A network-wide search should be conducted to find the lost targets. However, in this scheme, such a search is not included, to maintain the low-energy consumption of the nodes. Rather, it is left to the other clusters to detect the target.

7.3   Summary

This essence of this chapter lies in the fact that localization in a WSN is highly challenging, as the intrinsic resources (sensor node) are extremely resource-constrained. Thus, the conventional global positioning system is not appropriate. The general methods of localization have been presented. Some of the standard distance estimation techniques based on received signal strength intensity, radio hop count, time difference of arrival, and angle of arrival were discussed. Localization algorithms were originally classified as centralized and distributed. The algorithms, such as MDS-MAP, and adaptive beacon placement were illustrated. Then, the distributed algorithms, viz. beacon-based, relaxation-based, coordinate system stitching-based, hybrid, interferometric ranging based and error propagation aware algorithms were studied in detail.

The later part of the chapter highlighted the application of target tracking in a wireless sensor environment. Some standard solution alternatives for tracking a single target were discussed and analyzed – Tree-based, cluster-based, prediction-based, mobicast message-based, and hybrid tracking. For multiple target tracking, a hierarchical MDP was discussed.

References

[1] W.-P. Chen, J. C. Hou and L. Sha, "Dynamic clustering for acoustic target tracking in wireless sensor networks," *IEEE Transactions on Mobile Computing*, Vol. **3**, No. 3, pp. 258–271, 2004.

[2] H. Zhang and H. Shen, "Energy-efficient beaconless geographic routing in wireless sensor networks," *IEEE Transactions on Parallel and Distributed Systems*, Vol. **21**, No. 6, pp. 881–896, 2010.

[3] L. Shu, Y. Zhang, L. T. Yang, Y. Wang and M. Hauswirth, "Geographic routing in wireless multimedia sensor networks," in *Proceedings of International Conference on Future Generation Communication and Networking*, pp. 68–73, 2008.

[4] R. N. Handcock, D. L. Swain, G. J. Bishop-Hurley, *et al.*, "Monitoring animal behaviour and environmental interactions using wireless sensor networks, GPS collars and satellite remote sensing," *Sensors*, Vol. **9**, No. 5, pp. 3586–3603, 2009.

[5] S. Kim, S. Pakzad, D. Culler, *et al.*, "Health monitoring of civil infrastructures using wireless sensor networks," in *Proceedings of Information Processing in Sensor Networks (IPSN)*, pp. 254–263, 2007.

[6] B. Son, Y.-S. Her and J. Kim, "A design and implementation of forest-fires surveillance system based on wireless sensor networks for South Korea mountains," *International Journal of Computer Science and Network Security (IJCSNS)*, Vol. **6**, No. 9, pp. 124–130, 2006.

[7] S. Bhatti and J. Xu, "Survey of target tracking protocols using wireless sensor network," in *Proceedings of International Conference on Wireless and Mobile Communications (ICWMC)*, Cannes, La Bocca, pp. 110–115, 2009.

[8] G. Mao, B. Fidan and B. D. O. Anderson, "Wireless sensor network localization techniques," *Computer Networks*, Vol. **51**, No. 10, pp. 2529–2553, 2007.

[9] A. Pal, "Localization algorithms in wireless sensor networks: current approaches and future challenges," *Network Protocols and Algorithms*, Vol. **2**, No. 1, 2010.

[10] J. Bachrach and C. Taylor, "Localization in sensor networks," *Handbook of Sensor Networks: Algorithms and Architectures*, ISBN: 978-0-471-68472-5, 2005.

[11] F. Zhao and L. J. Guibas, *Wireless Sensor Networks: An Information Processing Approach*, Morgan Kaufmann, ISBN: 9781558609143, 2004.

[12] P. Bergamo and G. Mazzini, "Localization in sensor networks with fading and mobility," in *Proceedings of International Symposium on Personal, Indoor and Mobile Radio Networks*, pp. 750–754, 2002.

[13] D. Niculescu and B. Nath, "Ad hoc positioning system (APS) using AoA," in *Proceedings of INFOCOM*, pp. 1734–1743, 2003.

[14] Y. Shang, W. Ruml, Y. Zhang and M. P. J. Fromherz, "Localization from mere connectivity," in *Proceedings of the 4th ACM International Symposium on Mobile ad hoc Networking & Computing*, Annapolis, Maryland, USA, pp. 201–212, 2003.

[15] N. Bulusu, J. Heidemann and D. Estrin, "Adaptive beacon placement," in *Proceedings of the 21st International Conference on Distributed Computing Systems*, Phoenix, Arizona, April 2001.

[16] K. Langendoen and N. Reijers, "Distributed localization in wireless sensor networks: a quantitative comparison," *Computer Networks*, Vol. **43**, pp. 499–518, 2003.

[17] N. Bulusu, V. Bychkovskiy, D. Estrin and J. Heidemann, "Scalable, ad hoc deployable RF-based localization," in *Proceedings of Grace Hopper Celebration of Women in Computing Conference*, Vancouver, British Columbia, Canada, 2002.

[18] S. Fitzpatrick and L. Meertens, "Diffusion based localization," private communication, 2004.

[19] Y. Chraibi, "Localization in wireless sensor networks," Master's Degree Project, Stockholm, Sweden, 2005.

[20] T. He, C. Huang, B. M. Blum, J. A. Stankovic and T. Abdelzaher, "Range-free localization schemes for large scale sensor networks," in *Proceedings of the 9th Annual International Conference on Mobile Computing and Networking*, San Diego, California, USA, pp. 81–95, 2003.

[21] A. Savvides, C. C. Han and M. B. Srivastava, "Dynamic fine grained localization in ad hoc networks of sensors," in *Proceedings of the 7th Annual International Conference on Mobile Computing and Networking*, Rome, Italy, pp. 166–179, 2001.

[22] Z. Yang and Y. Liu, "Quality of trilateration: confidence-based iterative localization," *IEEE Transactions on Parallel and Distributed Systems*, Vol. **21**, No. 5, pp. 631–640, 2010.

[23] T. Ojha and S. Misra, "HASL: high-speed AUV-based silent localization for underwater sensor networks," in *Proceedings of the International Conference on Heterogeneous Networking for Quality, Reliability, Security and Robustness*, LNICST 115, pp. 128–140, Greater Noida, India, 2013.

[24] T. Ojha and S. Misra, "MobiL: a 3-dimensional localization scheme for mobile underwater sensor networks," in *Proceedings of National Conference on Communications (NCC)*, pp. 1–5, New Delhi, India, 2013.

[25] A. Novikov and A. Bagtzoglou, "Hydrodynamic model of the lower Hudson River estuarine system and its application for water quality management," *Water Resources Management*, Vol. **20**, No. 2, pp. 257–276, 2006.

[26] A. Bagtzoglou and A. Novikov, "Chaotic behavior and pollution dispersion characteristics in engineered tidal embayments: a numerical investigation," *Journal of American Water Resources Association*, Vol. **43**, No. 1, pp. 207–219, 2007.

[27] N. B. Priyantha, H. Balakrishnan, E. Demaine, and S. Teller, "Anchor-free distributed localization in sensor networks," in Proceedings *of the 1st International Conference on Embedded Networked Sensor Systems*, Los Angeles, California, USA, p. 340, 2003.

[28] D. Moore, J. Leonard, D. Rus and S. Teller, "Robust distributed network localization with noisy range measurements," in *Proceedings of the 2nd International Conference on Embedded Networked Sensor Systems*, Baltimore, MD, USA, pp. 50–61, 2004.

[29] K.-Y. Cheng, K.-S. Lui and V. Tam, "Localization in sensor networks with limited number of anchors and clustered placement," in *Proceedings of Wireless Communications and Networking Conference (WCNC)*, Kowloon, pp. 4425–4429, 2007.

[30] I. Borg and P. Groenen, *Modern Multidimensional Scaling, Theory and Applications*, 2nd ed. New York: Springer, 2005.

[31] H. Lim and J. C. Hou, "Localization for anisotropic sensor networks," in *Proceedings of INFOCOM*, Vol. **1**, pp. 138–149, 2005.

[32] M. Maroti, B. Kusy, G. Balogh, *et al.*, "Radio interferometric geolocation," in *Proceedings of International Conference on Embedded Networked Sensor Systems (SenSys)*, pp. 1–12, 2005.

[33] R. Huang, G. V. Zaruba and M. Huber, "Complexity and error propagation of localization using interferometric ranging," in *Proceedings of IEEE International Conference on Communications (ICC)*, Glasgow, Scotland, pp. 3063–3069, 2007.

[34] N. A. Alsindi, K. Pahlavan and B. Alavi, "An error propagation aware algorithm for precise cooperative indoor localization," in *Proceedings of IEEE Military Communications Conference (MILCOM)*, pp. 1–7, Washington, DC, USA, 2006.

[35] S. M. Kay, *Fundamentals of Statistical Signal Processing: Estimation Theory*. Upper Saddle River, New Jersey: Prentice-Hall, 1993.

[36] S. Misra and M. Khatua, "Cross-layer techniques and applications in wireless sensor networks," in H. Rashvand and Y. Kavian (Eds.), *Using Cross-Layer Techniques for Communication Systems*, pp. 94–119, 2012, Hershey, PA: Information Science Reference. doi: 10.4018/978-1-4666-0960-0.ch004.

[37] V. V. Veeravalli and J. F. Chamberland, "Detection in sensor networks," in *Wireless Sensor Networks: Signal Processing and Communications Perspectives*, A. Swami, Q. Zhao, Y. W. Hong and L. Tong (Eds.), John Wiley & Sons, 2007, pp. 119–148.

[38] W. Zhang and G. Cao, "DCTC: dynamic convoy tree-based collaboration for target tracking in sensor networks," *IEEE Transactions on Wireless Communications*, Vol. **3**, No. 5, pp. 1689–1701, 2004.

[39] W.-R. Chang, H.-T. Lin and Z.-Z. Cheng, "CODA: a continuous object detection and tracking algorithm for wireless ad hoc sensor networks," in *Proceedings of IEEE Consumer Communications and Networking Conference*, Las Vegas, NV, USA, pp. 168–174, 2008.

[40] W. Eddy, "A new convex hull algorithm for planar sets," *ACM Transactions on Mathematical Software*, Vol. **3**, No. 4, pp. 398–403, 1977.

[41] R. L. Graham, "An efficient algorithm for determining the convex hull of a finite planar set," *Information Processing Letters*, Vol. **1**, pp. 132–133, 1972.

[42] J. O'Rourke, "Convex hulls in 2D," Chapter 3 in *Computational Geometry in C*, 2nd ed. Cambridge University Press, 1998.

[43] S. Misra and S. Singh, "Localized policy-based target tracking using wireless sensor networks," *ACM Transactions on Sensor Networks*, Vol. **8**, No. 3, pp. 27:1–27:30, 2012.

[44] T. Camp, J. Boleng and V. Davies, "A survey of mobility models for ad hoc network research," *Wireless Communication and Mobile Computing*, Vol. **2**, No. 5, pp. 483–502, 2002.

[45] Y. Xu, J. Winter and W.-C. Lee, "Prediction-based strategies for energy saving in object tracking sensor networks," in *Proceedings of the IEEE International Conference on Mobile Data Management (MDM)*, pp. 346–357, 2004.

[46] Y.-S. Chen and Y.-J. Liao, "HVE-mobicast: a hierarchical-variant-egg-based mobicast routing protocol for wireless sensornets," in *Proceedings of Wireless Communications and Networking Conference (WCNC)*, Las Vegas, NV, pp. 697–702, 2006.

[47] E. W. Weisstein, "Cassini ovals", From *MathWorld* – A Wolfram Web Resource [Online] http://mathworld.wolfram.com/CassiniOvals.html.

[48] H. Yang and B. Sikdar, "A protocol for tracking mobile targets using sensor networks," in *Proceedings of IEEE International Workshop on Sensor Network Protocols and Applications*, pp. 71–81, 2003.

[49] W.-L. Yeow, C.-K. Tham and W.-C. Wong, "Energy efficient multiple target tracking in wireless sensor networks," *IEEE Transactions on Vehicular Technology*, Vol. **56**, No. 2, pp. 918–928, 2007.

[50] W.-L. Yeow, C.-K. Tham, and W.-C. Wong, "A novel target movement model and energy efficient target tracking in sensor networks," in *Proceedings of IEEE VTC – Spring*, pp. 2825–2829, 2005.

8   Topology management and control

Wireless sensor networks (WSNs) exhibit an "autocratic" operational policy with minimal human intervention. So, such networks must be self-configurable to maintain their autonomy. The requisitions of WSN-specific applications are quenched by temporal cooperation and coordination among the sensor nodes. Naturally, these nodes are expected to perpetuate a healthy intra-network infrastructure. However, they are power constrained with a bounded communication range and low computational ability. Hence, issues related to network infrastructure should be dynamically handled with efficacy.

Topology is a vital aspect of WSNs that needs attention for both network and fault management. In this chapter, we focus on two aspects of topology: (a) topology management, and (b) topology control.

Topology management is the process of deriving a simple graph of node connectivity by determining the inter-nodal links and virtual relationships for efficient operations within a network contour. Topology management aims at conserving the energy of the nodes and consequently extending the lifetime of the network with parallel maintenance of network connectivity.

Topology control of a WSN is a measure of the degree of network coverage and internode connectivity. Topology management and control might appear analogous. However, these two aspects are distinct, and so is their categorization, which we will discuss in this chapter.

8.1   Topology management

We first discuss the taxonomy of topology management in WSNs. The two popular topological architectures employed in WSNs are as follows.

Flat topology   In the absence of proper topology management, nodes are treated impartially and data transmission follows a multi-hop route. Such a topology is flat [1, 2], since nodes are handled equally. Such a topology is also called unstructured. It is the simplest topology allowing any-to-any connectivity with minimum management overhead. However, it leads to very poor and uncertain network connectivity. It is also incapable of fault management. Moreover, routing mainly takes place through flooding of packets, which might insinuate redundancy. This kind of topology is best suited for small networks.

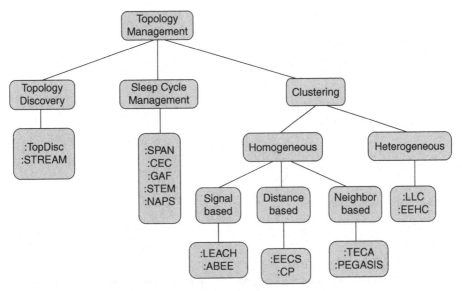

Figure 8.1   Taxonomy of topology management algorithms (adapted from [3], with minor modifications).

Hierarchical topology   In some cases, nodes are classified into groups or clusters, thereby forming a hierarchical topology. Data from nodes with a cluster are aggregated and transmitted to the base station (BS). Hierarchical topology is a sovereign approach of node organization, in which every cluster is represented and managed by a cluster head. Hierarchy improves network lifetime by introducing a "division-of-labor" concept at various tiers of network topology.

8.2   Taxonomy of topology management

The taxonomy of topology management algorithms in WSNs is (a) topology discovery, (b) sleep cycle management, and (c) clustering. Each of these categories has its own set of algorithms, as shown in Figure 8.1. We now discuss the details of each of these categories in the following sections.

8.2.1   Topology discovery

Routing in WSNs is typically executed with respect to a static base station (BS) that transmits packets to and from different sensor nodes [1]. In addition, the BS also serves as the functional unit for initiating topology diagnosis. Topology discovery involves retrieving the topological details from the nodes of the network. A base station enquires about the topological trivialities by broadcasting packets to the network. Consequently, the nodes reciprocate by sending packets to the base station itself. After receiving acknowledgements from every node, aggregation is performed. The aggregated outcome portrays the network topology. We now look at some topology discovery algorithms.

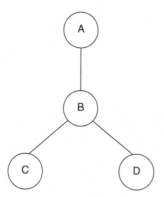

Figure 8.2   Example of topology.

8.2.1.1   TopDisc algorithm

Deb *et al.* proposed the TopDisc [4] algorithm that tries to accumulate the entire network topology from the perspective of a single node. The monitoring node sends the "topology discovery request" packet to all the active nodes. The information from these nodes is aggregated to get the entire topology. However, the sequence or divergence of response from the nodes varies as per the mode of response (Figure 8.2).

> *direct response*   (B>A), (C>B>A), (D>B>A),
> *aggregated response*   (C>B), (D>B), (B>A).

Coloring algorithms for finding the responding set (node labeling method)
TopDisc uses a coloring scheme to propagate requests to nodes and find the responding set.

The 3-coloring scheme
In the 3-coloring scheme, white, black, and gray colors are used. The significance of each of the colors is as follows.

- White: These are the nodes that are yet to be reached, i.e. they are currently undiscovered.
- Black: The cluster heads are denoted by a black color.
- Gray: These gray-colored nodes are the one-hop neighbors of the black-colored nodes.

The steps of request propagation using such a coloring scheme are listed below.

(i)   Assign white color to each node.
(ii)   Blacken the initiator node that broadcasts the topology discovery packets to its immediate neighbors.
(iii)   Color the receiver nodes gray.

(iv)   Each gray node broadcasts the received packet with a random forwarding delay heuristic that is inversely proportional to the square of the distance from the parent black node.

(v)   A white node, on receiving packets from a gray node, turns into a black node with a randomized delay. However, if it receives any packet from a gray node within the delay interval, it eventually turns gray.

(vi)   This process continues until all nodes are either black or gray colored.

The 4-coloring scheme

The modified version of the 3-coloring scheme is the 4-coloring scheme. It introduces a fourth color, dark gray. The idea behind this is to reduce the overlap between clusters. Any node that is at a distance of at least two hops from a cluster head (black node) is credited a dark gray color. After waiting for a randomized delay, as mentioned before, if it does not receive any packet, it turns black, otherwise it is gray. This scheme ensures that a particular cluster head will find another head, not more than two hops away from itself.

TopDisc responding mechanism

Having discussed the node coloring phase, we now focus on the response mechanism that TopDisc follows. Every node maintains its neighborhood and other associated information, which is as follows.

(a)   A gray node stores information about its neighboring black node.

(b)   Every node is aware of the parent black node, i.e. the sender of the topology discovery request.

After each transmission, a black node waits for responses from its children. These are aggregated and transmitted to the immediate parent node. Thus, after a series of packet transmissions, the initiator node is knowledgeable about the complete topology.

Advantages

- A step-by-step aggregation is performed.
- Energy consumption is optimized.

Disadvantages

- The communication range of nodes is not taken into consideration. So, connectivity is not ensured.
- Optimality in hop count is highly dependent on timer mechanisms and calculations.

8.2.1.2   Sensor topology retrieval at multiple resolutions (STREAM)

Before we illustrate the STREAM protocol, we need to understand the requirement for topology retrieval at multiple resolutions. The key facts for understanding STREAM are as follows [5].

- Topology resolution requirements vary according to applications. Some might require partial, whereas some others might be interested in complete topological information.
- A low-resolution topology facilitates a thorough understanding of several network properties.
- In some cases, a network spends only limited bandwidth for topology retrieval. In those cases, information retrieval might be based on an optimum resolution.

Now, we discuss the algorithmic aspects of STREAM, as proposed by Deb *et al.* [5].

(i)   The monitoring node broadcasts the topology discovery packet containing two specialized parameters – virtual range and resolution factor.
(ii) The monitoring node broadcasts a packet and turns black. It gets added to a set called the minimal virtual independent dominating set (MVIDS). MVIDS is formed for topology acquisition.
   (a)   Any node within a black node's virtual range is red color.
   (b)   Nodes within the communication range (and not within the virtual range of a black node) are blue nodes.
   (c)   White nodes are the undiscovered nodes.
   A red-colored node forwards the perceived packet, so does a blue node. However, a blue node starts a timer after forwarding. If the timer times out, it turns black. But if it receives a request in the meantime, it turns red. A white node on receiving packets from a blue/black node performs likewise, as explained in step (iv).
(iii) Black/red nodes discard packets that come to them.
(iv) This process of packet dissemination continues till all nodes are black or red.
(v)   All black nodes get added to MVIDS. These nodes are responsible for aggregation of information and subsequent transmission of the aggregated information from their children nodes.

Advantages

- Resolution-based retrieval is permitted.
- A modular aggregation is performed.
- Energy consumption is optimized.

Disadvantages

- Resolution management is a complex task.
- Too many clusters might incur delay in the process of information retrieval.

8.2.2   Sleep cycle management

These are specialized algorithms to manage and set optimal schedules of sleep and wake-up operations, thereby preserving and safeguarding the energy of the nodes. To maintain exhaustive network coverage, only a subset of nodes needs to be active at a particular instant of time. The other supplementary nodes are redundant and are scheduled to sleep. The subset of active nodes keeps changing temporally. Hence, each sensor follows a sleep–wake-up sleep cycle. The scheduling of this cycle for each sensor is entirely managed in this phase. However, this phase requires proper planning and analysis to avoid latency that might be incurred while reporting data. We explain below some of the sleep cycle management algorithms.

Sleep cycle management algorithms

8.2.2.1   **Span:** Span [6] is a distributed, randomized sleep cycle management algorithm that has its applicability in a dense wireless network. Span aims at the following.

- Each point in the network is covered by at least a single coordinator node.
- Coordinators are scheduled in a rotating fashion so that the task of maintaining coverage and connectivity is shared.
- It aims at selection of an optimal number of coordinators, thereby preventing a lossy coverage.
- Election of coordinators is locally managed.

The key functions of the SPAN algorithm (refer to Figure 8.3) are as follows.

(i)   Nodes maintain state information and proactively broadcast HELLO messages containing state information.
(ii)   A node turns on its radio after a fixed interval and decides to be a coordinator node, if two of its neighbors were unreachable.
(iii) A coordinator node, on the other hand, backs out if two of its neighbors can communicate without intervention.

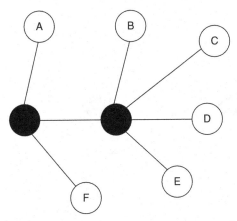

Figure 8.3   Span forwarding backbone formed by the coordinators (black nodes).

(iv) A grace period is the interval of time between withdrawal of one coordinator and replacement of the other. Each coordinator node must serve this period before going to sleep.

(v) Span rotates and distributes the role of coordinators. This leads to distribution of responsibility and reduction of energy exhaustion.

Advantages

- Network lifetime increases.
- Load distribution is done periodically.
- Energy consumption is reduced.

Disadvantages

- HELLO messages are too costly in terms of bandwidth.
- The transition from sleep to wake-up schedule involves energy expense.

8.2.2.2 Geographic adaptive fidelity (GAF)

The geographic adaptive fidelity (GAF) [7] algorithm is an energy-efficient algorithm incorporating location awareness by using the global positioning system (GPS). It was proposed by Xu *et al.* primarily for ad-hoc networks. It improves the performance of WSNs by introducing the concept of a "virtual grid."

Geographic adaptive fidelity can be analyzed by the following key features.

(i) The entire network is viewed as several square grids, where each grid is mastered by a single node, as shown in Figure 8.4

(ii) The master node is in charge of managing the grid and reporting data. The other members of a grid, the slave nodes, are redundant and put to a "sleep" state.

(iii) One of the slave nodes volunteers to be the master node. The master node, however, does not perform any aggregation.

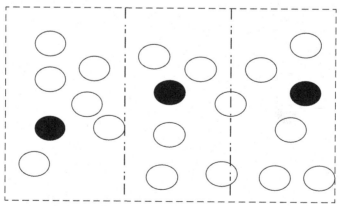

Figure 8.4   GAF virtual grids (adopted from [7], with minor modifications).

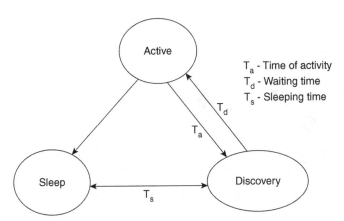

T_a - Time of activity
T_d - Waiting time
T_s - Sleeping time

Figure 8.5   GAF state transitions (adopted from [7], with minor modifications).

(iv)   The possible set of states for each node is discovery, sleeping and active.
  (a)   In the discovery state, messages are exchanged to find other nodes of that grid.
  (b)   A node moves to the active state when it is unable to determine the other nodes in the grid. A master node is in the active state.
  (c)   When another node wishes to be active, an active node moves to its sleep state.

Thus, GAF maintains network connectivity without degrading the routing fidelity. The transition diagram of GAF is shown in Figure 8.5.

Advantages

● Routing is done in a distributed manner.
● The idea of grid avails the advantages of modularity.
● Energy management is done intelligently.

Disadvantages

● Although it is energy aware, the use of GPS decelerates the performance.
● The state transitions involve energy expense.

8.2.2.3   Cluster-based energy conservation (CEC)

The cluster-based energy conservation (CEC) [8] algorithm is based on GAF, but it does not involve the use of GPS. Thus, CEC is an improvisation of GAF. In CEC, nodes are arranged into imbricated clusters.
  CEC maintains three types of nodes, as shown in Figure 8.6.

(a)   Cluster head (CH): The usual notion of a cluster head (CH) exists, which is one-hop away from the other cluster members.
(b)   Gateway nodes: Due to overlapping of clusters, some nodes may belong to more than one cluster. These nodes connect clusters and act as cluster gateways.
(c)   Redundant nodes: Nodes other than CHs and gateways are redundant nodes in sleep state.

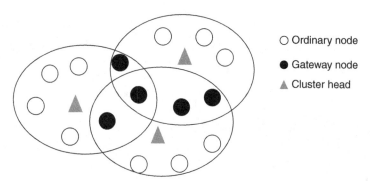

Figure 8.6   CEC cluster formation (adopted from [8], with minor modifications).

Cluster heads are periodically polled as in GAF and SPAN for energy conservation.

Advantages

- Absence of location awareness makes it highly realistic.
- It has great admissibility in networks with the heterogeneous radio range.
- Energy conservation is maintained.

Disadvantages

- It is not applicable for networks with a highly dynamic topology.

8.2.2.4   Sparse topology and energy management (STEM)

Sparse topology and energy management (STEM) [9, 10], proposed by Schurgers *et al.*, is an alternative solution to idle listening. STEM functions as an event-based collaborative sensing algorithm. STEM is a two-state algorithm composed of the following.

(a)   Monitor state: In this state, the nodes monitor and no event occurs.
(b)   Transfer state: Data transmission takes place due to event detection.

STEM considers two kinds of nodes. The initiator node is the one that begins communication and monitoring. For this purpose, the initiator node tries to establish a virtual link with a receiver node called the target node. There are two types of channels used in STEM.

(a)   Wake-up: The wake-up channel functions to check the possibility of a data transmission.
(b)   Data: This channel is activated only when there is a need for packet transmission or node communication.

STEM has two implementation versions.

STEM-B   In STEM-B, a sender node sends a beacon containing the source and target address. The target, on identifying the beacon, switches on its data channel, and starts communication.

STEM-T   In STEM-T mode, the sender transmits a continuous interrupt signal to wake up the target node. On sensing the interrupt, the target node shifts to its data channel.

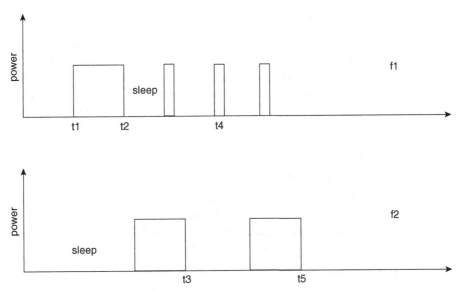

Figure 8.7   State transitions in STEM, f_1: wake-up radio, f_2: primary radio frequencies (adopted from [9], with minor modifications).

In both STEM versions, nodes other than the initiator and target are kept in the sleep mode, thereby optimizing energy. The operation of STEM is shown in Figure 8.7 [10].

Advantages

● Improves the monitoring phase of a node.
● Energy conservation is maintained.

Disadvantages

● Transition from sleep to wake-up state is energy exhaustive.
● Switching between data and wake-up channels is also an energy hungry process.
● Latency is incurred due to switching and shifting.

Initially, at t_1 instant of time, the initiator node sends a beacon with frequency f_1. At t_2 instant of time, the target node responds and shifts to data channel in frequency f_2. After completion of data transmission, the respective radios are turned off.

8.2.2.5   Naps

Naps [11] is another sleep cycle management protocol. The significance behind the name "naps" is that this protocol aims to find a subset of nodes that may turn their radio off (or take a "nap") for some period of time. Naps deals with two kinds of nodes.

● Waking: These are the nodes that operate to maintain connectivity.
● Napping: These nodes turn off their radio and remain in the sleep state.

The steps of NAPS are as follows.

(i)   The initiator node broadcasts a HELLO message and starts a timer.
(ii) For each HELLO message that the node receives, it increments a counter, that was initially set to zero.
(iii) Step (ii) repeats until the timer times out or the counter hits a threshold.
(iv) If the counter reaches the threshold, before the timer times out, the node naps till the timer stops.

Advantages

- It is very simple.
- It is flexible to dynamic topology.
- Energy conservation is highly maintained.
- It is applicable for mobile networks as well.
- Location awareness is not needed.

Disadvantages

- Rotation of sleep nodes is not performed.
- A single node may be kept in sleep node for a long period, thus exhibiting bias.
- Introduces high influx of packets.

8.2.3   Clustering

Clustering algorithms introduce hierarchy into the network. Nodes are classified into clusters governed by a cluster head. Data packets from each member node of a cluster are transmitted to the cluster head. The cluster head is responsible for aggregating the individual node data to a composite value. The reason behind clustering is a reasonable reduction in energy consumption. Clustering enables only a subset of nodes to be active, thereby indirectly contributing to sleep cycle management. The cluster head is chosen in such a way that the communication cost of other member nodes for transmitting data to the cluster head is minimal. Since communication is a major source of energy expenditure in WSN, reduction of communication expense hugely improves network lifetime. However, to avoid over-burdening of the cluster head alternative, a rotation is carried out on every member node to be the cluster head periodically. Clusters can be static or dynamic, depending on network and application requirements.

Based on cluster formation strategy, a clustering algorithm can be static or dynamic, whereas based on the nature of the network resources, these algorithms can be classified into homogeneous and heterogeneous classes.

Static clustering
In static clustering, nodes are manually grouped into fixed clusters, before network operation commences, preferably at the time of deployment. The base station is aware of the existing clusters. These algorithms are useful to conserve energy, as they do not involve dynamic grouping. However, they are less realistic.

Dynamic clustering
In dynamic clustering, nodes are added or deleted at run-time, based on application and network demands. Not only are the members dynamically chosen, so are the cluster heads. This introduces real-time flexibility and scalability.

Another possible classification of clustering is homogeneous and heterogeneous clustering.

Homogeneous clustering
From a sensor network perspective, homogeneous clusters are those that consist of similar resource types, thereby maintaining uniformity and evenness.

Heterogeneous clustering
This type of clusters consists of varied and miscellaneous resource types. Thus, it possesses a diverse nature.

Clustering algorithms

8.2.3.1   Homogeneous clustering algorithms
A homogeneous WSN consists of identical resources. Homogeneous clustering algorithms can be divided into three main categories.

(1)   Signal-based clustering algorithms.
(2)   Distance-based clustering algorithms.
(3)   Neighbor-based clustering algorithms.

We discuss each of them below.

Signal-based algorithms
Low-energy adaptive clustering hierarchy (LEACH)
LEACH [12, 13] or low-energy adaptive clustering hierarchy is one of the most popular clustering algorithms designed for microsensor networks. It works on the principle of dividing nodes into clusters governed by a cluster head (CH). Nodes inside a cluster communicate directly with the CH. The CH is responsible for data fusion and subsequent transmission to the base station (BS). However, CHs are not statically determined for the purpose of energy conservation. The load is uniformly distributed by facilitating a rotation of the CHs.

LEACH is based on the following rounds or phases.

(a)   *Setup phase* In this phase, the clusters are organized and a cluster head is determined. At the beginning of every round, each node probabilistically elects itself to be the cluster head. If a node is eventually made a cluster head, it broadcasts the information to the other nodes by an advertisement packet (ADV). These nodes, in turn, choose the cluster they wish to join, depending on the received signal strength and subsequently transmit response messages to those cluster heads by a join-request message (Join-REQ). After this, the cluster head prepares TDMA slots for each of its member nodes to avoid collisions while communicating.

(b)   *Steady-state phase*   In this phase, the nodes communicate with the heads by sending single frames in their slots. Each node strives to attain energy efficiency by turning on its radio before the time of transmission. Since a greater number of clusters increases the probability of inter-cluster interference, LEACH uses direct sequence spread spectrum (DSSS) for cluster head transmissions. Data obtained from each member node are compressed, fused, and then transmitted to the base station. LEACH uses carrier sense multiple access (CSMA) for medium access.

Advantages

●   It is a distributed algorithm.
●   It is self-configuring.
●   It is adaptive and performs data aggregation.

Disadvantages

●   DSSS introduces overhead as a result of time-synchronization aspects.
●   The broadcast messages incur energy expense.

Access-based energy-efficient cluster algorithm (ABEE)

Access-based energy-efficient cluster algorithm (ABEE) [14] is a simple clustering protocol that focuses on three crucial network parameters – network lifetime, deployment, and node correlation. Using this protocol, clusters are probabilistically arranged. ABEE is a request–response message broadcast protocol based on the following.

(a)   *Mass center merge algorithm*   This algorithm aims to maintain the energy efficacy of an entire cluster. For this purpose, every node of a cluster is given the opportunity to be the cluster head. However, the cluster head is chosen by optimizing the Euclidean metric from each member node. Hence, the node closest to the center of mass is chosen as the cluster head.

(b)   *Distance merge algorithm for rational distribution*   This algorithm involves repeated broadcasts of HELLO messages among cluster heads. Based on the message information, a cluster head analyzes the inter-cluster distances. If the distance between two clusters is sufficiently small, the clusters are merged.

The steps of ABEE are as follows.

(i)   A node broadcasts REQ_TO_JOIN message and starts a timer.
(ii)   If it receives a HELLO within the timeout period, it sends a JOIN message to be a member of the cluster.
(iii)   If the timer times out, the node itself becomes a cluster head and broadcasts HELLO messages.
(iv)   After a node joins a cluster, it periodically sends HEART_BEAT signals to keep the head aware of its current status.

(v) On receiving two HELLO messages, a node chooses to join the nearest cluster. So, a DISCONNECT signal is transmitted to the other cluster head.

(vi) Information from the member nodes is aggregated by the cluster head, before it is transmitted to the sink.

Advantages

- It is simple.
- Rotation is performed to change cluster head alternatives.
- The cluster head performs data aggregation on behalf of a cluster.

Disadvantages

- It involves location awareness or use of GPS.
- The broadcast messages incur energy expense.
- Residual energy is not taken into consideration while cluster heads are selected.

Distance-based algorithms
Energy-efficient clustering scheme (EECS)

The energy-efficient clustering scheme (EECS) [15] is a homogeneous clustering scheme proposed by Ye *et al.* EECS follows a similar clustering scheme to LEACH where clusters are headed by a cluster head that communicates with base station. The main phases of EECS are as follows.

- *Cluster head election*   This phase involves a competition of candidate nodes of a cluster to become the cluster head. Only a node that has an optimum distance and weight metric from other member nodes is picked up as a cluster head.
- *Cluster formation*   In this phase, the cluster heads broadcast special messages called HEAD_AD_MSG. Nodes obtain information about the communication range from these packets and decide which cluster to join. Data are directly transmitted by the member nodes to the cluster heads.

However, for clustering to work properly, the two phases should be correctly synchronized. For this purpose, the base station broadcasts special signals periodically.

Advantages

- Cluster formation overhead is optimized as the number of clusters to be formed is pre-decided.
- Energy consumption is reduced.
- Rotation is performed to change cluster head alternatives.
- The cluster head performs data aggregation on behalf of a cluster.

Disadvantages

- It involves synchronization overhead.
- The broadcast messages incur energy expense.

The clustering protocol (CP)

The clustering protocol (CP) [16] aims at arranging nodes into disjoint clusters. Each cluster can be viewed as approximately a circle, with the cluster head as the center and a radius of unit communication range. CP is defined as a covering problem of hexagonal packing. The key features of CP are as follows.

(i) The base station is an effective entity in terms of resource. It is also the initiator.

(ii) Every edge of a hexagon is equal to the sensing range of a sensor.

(iii) CP assumes three nodal states: (a) clustered, (b) unclustered, and (c) cluster head.

(iv) In the beginning, every node is unclustered. The base station centers itself inside a randomized hexagon, declares itself as the cluster head and broadcasts cluster head announcement (CHA) packets to a maximum of two hops.

(v) On receiving a CHA, an unclustered node becomes clustered and retransmits. However, a clustered node ignores such packets.

(vi) When an unclustered node receives packets from a clustered node, it computes the asymptotic distance of it from the center of the cluster and starts a timer. The time period is a function of the computed distance.

(vii) If no message is received within the timeout period, it declares itself as a cluster head, and if it receives any message, it sets its status to clustered.

Advantages

- It is simple and efficient.
- No two clusters overlap.
- The number of clusters does not vary with the node density.
- Communication overhead is minimized.

Disadvantages

- Residual energy of nodes is not considered while selecting cluster heads.
- Location awareness is needed in CP.
- The likelihood of a node being positioned at the desired places of a hexagon is quite small.

Neighbor-based algorithms

Topology and energy control algorithm (TECA)

Busse *et al.* proposed the topology and energy control algorithm (TECA) [17] to increase network connectivity and lifetime. TECA follows the usual clustering approach such as CP, EECS, and LEACH, and uses special nodes to maintain inter-cluster connectivity. In other words, TECA establishes a connected backbone topology. In TECA, five nodes states are defined.

(a) *Initial* A node that has its radio on is in the initial state. In this state, a node judges its neighborhood.

(b) *Sleeping* These nodes turn their communication radio component off. However, sensing and processing components are kept operational.

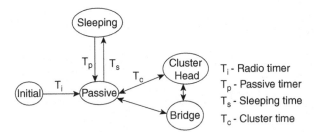

Figure 8.8   TECA node state transitions (adopted from [16], with minor modifications).

(c) *Passive* Nodes in this state are similar to the initial nodes. Passive nodes maintain the neighborhood information for certain time, after which they move to sleep.

(d) *Bridge* These nodes are the connected components between two clusters.

(e) *Cluster head* These nodes are the usual leaders of a cluster.

State transition diagram of TECA is shown in Figure 8.8. The important phases of TECA are as follows.

(i) *Cluster head selection* Initially, a node remains a cluster head while it is not assigned to a cluster. The node consults its one-hop neighborhood information and chooses a cluster head. Members from the head are one hop away. At the end of this phase, each node is either a head or a normal cluster member.

(ii) *Bridge selection* After the previous phase, nodes are selected to form a bridge. It is a difficult process to decide which nodes will be the bridging nodes. These nodes connect clusters and aim to reduce packet loss, extend network lifetime, and optimize the bridge count. Every node determines its minimum spanning tree (MST). Virtual links connecting cluster heads are considered. Such nodes that make up the local MST are the bridge nodes.

(iii) *Sleeping timeout* After the first two phases are completed, a passive node may go off to sleep. However, it must wake up periodically to check whether a bridge node is drained, or a cluster head needs to be replaced.

Advantages

●   Cluster heads are rotated.
●   Bridging ensures network connectivity.
●   Residual energy of nodes is considered during the election of cluster heads.
●   Communication overhead is minimized.

Disadvantages

●   Bridging introduces a greater count of active nodes.
●   The greater the number of active nodes, the more is the energy consumption.

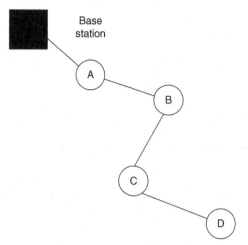

Figure 8.9   Greedy chain formation of nodes in PEGASIS (adopted from [18], with minor modifications).

Power-efficient gathering in sensor information systems (PEGASIS)

Power-efficient gathering in sensor information systems (PEGASIS) [18] is pro-posed by Lindsey and Raghavendra to optimize the number of transmissions and receptions, thereby improving the power efficiency. PEGASIS is an improvisation of LEACH. PEGASIS aims to form a chain of nodes that communicates with the nearest neighbor.

This formation of chain takes a greedy approach, starting with the farthest node from the base station. Figure 8.9 shows the formation of chain, starting with node 0. At every time instant, the closest neighbor is added to the chain.

Nodes on the chain behave in a special way. A node transmits its data to the node, immediately next to it in the chain. Each node, thus, obtains data from the previous node, aggregates the data, and transmits them to the next node. This process continues till the data reach the base station. At the end of a transmission to the base station, a new iteration begins.

At the beginning of every iteration, a new leader is elected to drive the chain. However, nodes that are too far from the base station are usually avoided from being made the leaders, because they might deplete more energy.

Advantages

- Rotation of leader nodes is performed to distribute the energy expenditure.
- Data aggregation is performed in an optimum manner.
- Communication overhead is reduced.

Disadvantages

- Residual energy is not considered while selecting leaders.
- Global position of nodes is required to form the chain.

8.2.3.2   Heterogeneous clustering algorithms

Low-energy localized clustering (LLC)

Kim *et al.* proposed the low-energy localized clustering (LLC) [19] algorithm for heterogeneous sensor networks. The key features of LLC are as follows.

(1)   It follows a two-tier architecture, in which the sensor nodes are present in the lower layer and the cluster heads are positioned in the upper layer.

(2)   It is a two-phase algorithm.

   (a)   *Initial phase*   In this phase, all cluster heads are used to compute an asymptotic equilateral triangle using Delaunay triangulation. A cluster radius decision point (CRDP) algorithm is determined, thereby obtaining the cluster radius.

   (b)   *Cluster radius control phase*   This phase follows two different types of algorithms:

   • *NLP-based approach*: A CRDP is dynamically obtained with a view to minimize cluster overlapping;

   • *VC-based approach*: It looks forward to optimality by using simple vector calculations.

The notations of determining CRDP using NLP are shown in Figure 8.10 [19].

Advantages

•   It is a low-energy scheme.
•   The CHs are given more responsibility than the sensor nodes.

Disadvantages

•   It requires location awareness.
•   The NLP approach is iterative, and, hence, complex.

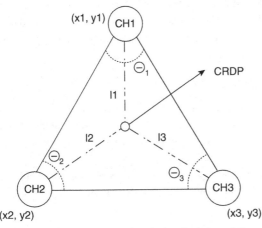

Figure 8.10   Cluster heads forming equilateral triangle (adopted from [11], with minor modifications).

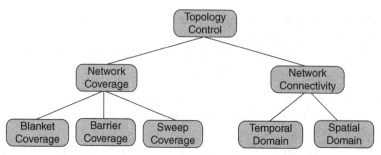

Figure 8.11   Taxonomy of existing topology control schemes (adopted from [21], with minor modifications).

Energy-efficient heterogeneous clustered scheme (EEHC)
The focus of the energy-efficient heterogeneous clustered scheme (EEHC) [20] algorithm is to find optimal cluster heads in a decentralized manner. It considers the spatial density of nodes, so that the communication cost within a cluster can be optimized. It also increases the network lifetime and performance. It classifies nodes as super, normal, and advanced, based on their health conditions. Based on a node's considerable attributes, the CHs are elected.

8.3   Topology control

Topology control strives to maximize network lifetime and optimize nodal interference during communication. As discussed earlier, topology control focuses on two important network aspects: network coverage and network connectivity. The taxonomy of topology control is shown in Figure 8.11. We discuss each of these aspects, in detail, in the following sections.

8.3.1   Network coverage

Network coverage is a measure of network performance. Network coverage can be broadly classified into three types: blanket coverage (or coverage with the highest granularity), barrier coverage (or coverage with medium granularity), and sweep coverage (coverage with lowest granularity).

8.3.1.1   Blanket coverage

In blanket coverage, each point in a network is covered or sensed by at least k sensors, where $k = 1, 2, 3$, and so on. Blanket coverage involves dense deployment of sensors. This provides maximum coverage at the expense of high deployment and maintenance cost.

Full coverage ensures that every point in the network is covered by at least one sensor. Performance of blanket coverage has been analyzed for static, hybrid, and mobile networks.

Static networks

Static networks contain statically deployed sensors that are fixed throughout their life-time. The lifetime of the network is reduced as a result of the static behavior of nodes.

The optimal geographical density control (OGDC) protocol was proposed by Zhang and Hou [22] to provide single coverage in static networks, i.e. $k = 1$. This is a distributed protocol that minimizes sensor overlap. Initially, every sensor node remains in the UNDECIDED state. DECIDED is the state of sensors that provides network coverage. A random node initially broadcasts and analyzes the degree of coverage of the sensor field. If the field is uncovered, the receiving nodes start powering on their radio until the field is fully covered.

Wang et al. [23] proposed the coverage configuration protocol (CCP) for providing multiple coverage. Each point is sensed by more than one sensor, so that the overall reliability is ensured. CCP illustrates that the connectivity of a boundary node and an internal node is expected to be equal to the degree of coverage and twice the degree of coverage, respectively. To judge the degree of coverage of each point within a network, Huang and Tseng [24] proposed algorithms that investigate the coverage issues of each point in a network.

In order to understand and analyze network coverage, sensor deployment patterns must be studied in detail. Bai et al. [25] dealt with two-dimensional deployment graphs and optimal patterns for k-connectivity, looking at pattern mutation using Voronoi diagrams.

Mobile networks

Mobile networks consist of nodes that have the ability to move. These nodes are costlier than static nodes, but they improve the network lifetime.

Wang et al. [26] studied movement-assisted sensor deployment. These techniques try to determine the existence of holes by using Voronoi diagrams. The uncovered areas are used by moving sensors around that region. Examples of such algorithms are Minimax, VEC, and VOR methods.

Howard et al. [27] proposed a special concept of virtual nodes that repel by virtual forces. The nodes can sense the presence of their neighbors and objects that might enter in their sensing region. The repelling forces depend on the distance of a node from its neighbor.

Another algorithm, called Co–Fi algorithm, was proposed by Ganeriwal et al. [28], for topology control. It is a four-phase algorithm. In the first phase, i.e. initialization, the nodes gather knowledge about their neighbors and coverage regions. The second phase is the panic request phase, in which the network topology is updated due to a dead node. In the third phase, panic reply, the neighbors of the dead node reply. The last phase is the decision phase, in which the network topology is updated.

Hybrid networks

Hybrid networks consist of both static and mobile nodes. They perform better compared to static or mobile networks, since the deficit of a static sensor can easily be covered by a mobile node.

Wang *et al.* [29] investigated maximizing sensor coverage in a simpler manner. These movement-assisted sensor deployment algorithms are based on detecting uncovered regions or holes. Based on the span of a hole, the mobile sensors are deployed to compensate the lack of coverage.

Batalin and Sukhatme [30] proposed a solution approach by the use of a mobile robot across the sensing field. With its local sensing ability, the robot moves along an uncovered route, thereby providing coverage.

8.3.1.2   Barrier coverage

Barrier coverage incurs coverage of medium granularity. It involves the intersection of crossing paths with sensing areas, and ensures the detection of intruders who cross any network path. It is similar to problems involving boundary coverage. However, the coverage is stochastic. It depends on the movement of an object and the width of the network path. The width of a path is determined by its logical boundary. Liu and Towsley [31] analysed the probability of detecting an object that crosses the path. Barrier coverage is viewed from the perspective of percolation theory.

Kumar *et al.* [32] proposed techniques to investigate k-barrier coverage of a network. The authors considered two different types of deployments. The problem of barrier coverage concerns scenarios where sensors are deterministically deployed. In the other case, coverage under random deployment of sensors is studied.

Barrier coverage can be of two types.

(a) *Weak barrier coverage*   Weak barrier coverage signifies existence of an uncovered network path. The conditions for such coverage are discussed by Kumar *et al.* [32].

(b) *Strong barrier coverage*   Such coverage ensures that no breach exists in the network paths. Strong barrier coverage issues are studied by Balister *et al.* [33] and Chen *et al.* [34]. Both distributed and localized coverage protocols are discussed.

Much work [35, 36] has been done in coverage analysis when sensors are deployed following a Poisson point process. These techniques aim to develop barriers independent of the crossing paths. For this purpose, conditions are examined for strong barrier coverage. Sensor deployment models are also conceived, so that coverage can be maximized.

8.3.1.3   Sweep coverage

This is the least granular coverage. The maintenance and deployment costs are greatly reduced compared to the other two coverage types. However, this type of coverage may give rise to breaches. Generally, in such coverage, sensor information is stored in a sensor's local memory, from where it is collected by mobile entities. Hence, some delay obviously creeps in.

Sweep coverage does not involve continuous monitoring or coverage of every point of a network. Instead, it keeps track of the frequency of regularity with which each point is covered. The active sensors are treated as points of interest (POIs).

One of the approaches to provide sweep coverage, as proposed by Wong and MacDonald [37], is to have topological maps. These maps decompose the sensing region into non-overlapping subregions. A mobile robot is also used to provide enhanced coverage. Howard and Mataric [38] used mobile robots for providing coverage. Robots add a lot of efficiency in coverage aspects of static sensor networks.

It is worth mentioning that sweep coverage has less applicability in WSNs. The main reason is the resource-constrained issue of WSNs. In [38], sweep coverage is presented for a WSN. It analyzes and computes the number of sensors that might be needed to cover all POIs. This is termed as the NP-hard problem of sensor sweep coverage. It has both centralized and distributed approaches of coverage handling. See also the work of Cheng *et al.* [39].

8.3.2   Network connectivity

Network connectivity is another important metric to be studied. Network connectivity defines the strength of inter-nodal connections. Connectivity can be studied under two domains.

(i)   *Temporal domain*   Study of connectivity under temporal domain is an optimization problem of setting up sleep and wake-up schedules, so that the selected subset of active sensors provides full network connectivity.

(ii)   *Spatial domain*   Under spatial domain, full network connectivity can be achieved by the adjustment of the communication radio level of each sensor, and, thus, obtaining continuous communication range.

8.3.2.1   Connectivity under temporal domain

These include power management techniques, which are broadly divided into three categories: synchronous, asynchronous, and hybrid protocols. In each of these categories, the radio is switched on/off, as per the requirement, to maintain complete connectivity. We discuss each of these categories of protocols in detail.

Synchronous protocols

In these protocols, the sleep and wake-up schedule of each sensor is synchronized or coordinated. Ye *et al.* [40] proposed the S-MAC (synchronized MAC) protocol that aims to conserve energy. Additionally, S-MAC attempts to reduce collision, eavesdropping, overhead due to control packets, and energy expenditure due to idle listening.

S-MAC follows a non-hierarchical topology to remain flexible to dynamic changes in a network. To achieve inter-nodal time synchronization, each node sends and receives timestamps relative to the neighbor nodes. However, one of the major lacunae of S-MAC is that it performs poorly, when the network bears a low data rate. Since the duration of schedule is not optimized, energy is misused.

With a view to solve the aforesaid issue, Dam and Langendoen proposed T-MAC [41]. T-MAC improvised the mechanism and reduced the loss of energy due to idle listening. T-MAC decomposes messages into bursts and each node sleeps in the interval between two bursts.

Asynchronous protocols

These protocols also follow having fixed schedules for sleeping and waking-up of nodes. However, the schedule of every node may not be harmonized.

SPAN [6], which we have already discussed in Section 8.2.2.1, creates a backbone for routing. In this algorithm, a node switches its radio on periodically, to decide whether to be a coordinator or not. Hence, for the rest of the time, SPAN conserves energy while some other nodes might be taking the responsibility to maintain network connectivity.

Similarly, geographic adaptive fidelity (GAF) [7] ensures that only a single node with a grid is active at a time. The rest of the nodes lie in the sleep state and the desired connectivity for routing is also established.

Tseng *et al.* [42] and Zheng *et al.* [43] proposed algorithms in which nodes wake-up in an asynchronous manner. Each node maintains a bounded wake-up schedule. This creates a dynamic network topology all the time, thereby ensuring conservation of energy to a good extent.

Another asynchronous protocol is the BMAC (Berkeley MAC) [44], which aims at collision avoidance, channel utilization and high network scalability. BMAC is an improvisation of both S-MAC and T-MAC. It is based on clear channel assessment (CCA), and low-power listening (LPL). BMAC optimizes the lifetime of a single node, thereby optimizing the network lifetime as well.

Hybrid protocols

These protocols combine the advantages of both synchronous and asynchronous protocols. An example is SCP–MAC, or the scheduled channel polling MAC [45]. A sensor that wishes to send a packet waits until the receiver senses the medium. The sender senses the carrier, and then informs the receiver about the incoming connection. After the wake-up of the receiver, the sender enters the second contention, followed by which data transmission takes place.

8.3.2.2   Connectivity under spatial domain

These are power control methods in which the sensor communication radio is adjusted to optimize connectivity. These methods have the goal to minimize energy conservation and reduce contention and interference. Connectivity under spatial domain is mainly studied under two categories – homogeneous and heterogeneous.

Homogeneous network

The main focus of power control in a homogeneous network is the critical transmission range (CTR) problem. CTR tries to find the minimum number of nodes that should be kept active to maintain connectivity.

Belonging to this category is a protocol named the common power (COMPOW), as proposed by Narayanaswamy *et al.* [46]. It is a distributed protocol that deals with the minimum transmission range that every node should have. However, a lower transmission range can be considered only if weaker connectivity standards are set. Santi and Blough [47] have experimentally demonstrated this trade-off.

Heterogeneous network

As discussed before, heterogeneous networks are composed of devices having miscellaneous resource types in terms of communication, transmission, and processing abilities. So, altering the radio range of communication for heterogeneous nodes to maintain network connectivity is a challenge. Such a problem can be viewed as the range assignment (RA) problem [48].

The RA problem was studied and unidirectional link-based routing was investigated. Despite the extreme difficulties that might arise from unidirectional links, this work [49] focuses on justifying the benefits. It shows the significance of the symmetric range assignment (SRA) problem for routing problems for symmetric links. However, SRA has also been established to be NP-hard [50], similar to RA [51], thus earning overhead.

Cardei *et al.* [52] proposed assigning a power level to each node of the network, so that the overall connectivity of the network is retained. The model is called the hitch-hiking model, which assumes that the communication power is proportional to the square of the distance between the sender and the receiver. This model computes a minimum spanning tree among all the nodes and maintains a strongly connected topology. Topology control with hitch-hiking is NP-complete. Similarly, Li *et al.* [53] proposed a distributed protocol called local MST (LMST) that generates and maintains a connected network topology.

Another approach to build and control topology was proposed by Rodoplu and Meng [54]. It aims to minimize the communication cost. Also, Lin *et al.* [55] proposed adaptive transmission power control (ATPC) to control transmission power and the received signal strength.

A different approach, called MobileGrid, proposed by Liu and Li [56], is to deviate the radio level of neighbors of a node from an optimal value. It has its applicability in mobile networks. However, since optimality cannot be defined deterministically, deviation cannot be formulated.

8.4 Summary

In this chapter, we have studied and analyzed topology management and control techniques. Each of the algorithms based on the taxonomy were separately explained under topology management and topology control, and were illustrated in detail. The advantages and disadvantages were also analyzed thoroughly. We have also shown the variants of power management and power control algorithms. Additionally, we have discussed the applicability of such algorithms in different sensor networks. However, further research should be done in these areas in future to further improve performance and applicability.

References

[1] B. Nancharaiah, G. F. Sudha and M. B. R. Murthy, "A scheme for efficient topology management of wireless ad hoc networks using the MARI algorithm," in *16th IEEE International Conference on Networks, ICON 2008*, Dec. 12–14, 2008.

[2] Q. Mamun, S. Ramakrishnan and B. Srinivasan, "Multi-chain oriented logical topology for wireless sensor networks," in *2nd International Conference, Computer Engineering and Technology (ICCET),* April 16–18, 2010.

[3] S. Misra, I. Woungang and S. C. Misra, *Guide to Wireless Sensor Networks,* London: Springer-Verlag, 2009.

[4] B. Deb, S. Bhatnagar and B. Nath, "A topology discovery algorithm for sensor networks with applications to network management," Technical Report, Rutgers University, May 2001.

[5] B. Deb, S. Bhatnagar and B. Nath, "Multi-resolution state retrieval in sensor networks," in *Proceedings of the First IEEE International Workshop on Sensor Network Protocols and Applications, 2003,* pp. 19–29,11 May 2003.

[6] B. Chen, K. Jamieson, H. Balakrishnan and R. Morris, "Span: an energy efficient coordination algorithm for topology maintenance in ad-hoc wireless networks," Mobicom, Rome, Italy, pp. 70–84, July 2001.

[7] Y. Xu, S. Bien, Y. Mori, J. Heidemann and D. Estrin, "Topology control protocols to conserve energy in wireless ad-hoc networks," Technical Report 6, University of California, Los Angeles, Center for Embedded Networked Computing, January 2003.

[8] S. Cheng, J. Li and G. Horng, "An adaptive cluster-based routing mechanism for energy conservation in mobile ad hoc networks," Personal Communication, June 2012.

[9] C. Schurgers, V. Tsiatsis and M. B. Srivastava, "STEM: topology management for energy efficient sensor networks," in *Aerospace Conference Proceedings, IEEE,* 2002.

[10] C. Schurgers, V. Tsiatsis, S. Ganeriwal and M. B. Srivastava, "Topology management for sensor networks: exploiting latency and density," *MOBIHOC'02,* Lausanne, Switzerland, ACM, June 9–11, 2002.

[11] B. P. Godfrey and D. Ratajczak, "Naps: scalable, robust topology management in wireless ad-hoc networks," *ISPN'04,* Berkeley, CA, ACM, April 26–27, 2004.

[12] W. R. Heinzelman, A. Chandrakasan and H. Balakrishnan, "Energy efficient communication protocol for wireless microsensor networks," in *Proceedings of the 33rd Annual Hawaii International Conference on System Sciences,* Jan. 4–7, 2000.

[13] W. B. Heinzelman *et al.,* "Application specific protocol architecture for wireless microsensor networks," *IEEE Transactions on Wireless Communications,* pp. 660–670, 2002.

[14] X. Hong and Q. Liang, "An access based energy efficient communication protocol for wireless microsensor networks," in *15th IEEE International Symposium on Personal, Indoor and Mobile Radio Communications (PIMRC 2004),* Vol. **2,** pp. 1022–1026, Sept. 5–8, 2004.

[15] M. Ye, C. Li, G. Chen and J. Wu, "EECS, an energy efficient clustering scheme in wireless sensor networks," in *24th IEEE International Performance, Computing and Communication Conference* IPCCC, pp. 535–554, April 7–9, 2005.

[16] A. Durresia, V. Paruchuri and L. Barolli, "Clustering protocol for sensor networks," in *20th International Conference, AINA 2006,* Vol. **2,** pp. 18–20, April 2006.

[17] M. Busse, T. Haenselmann and W. Effelsberg, "TECA: a topology and energy control algorithm for wireless sensor networks," in *International Symposium on Modeling, Analysis and Simulation of Wireless and Mobile Systems (MSWiM '06),* Torremolinos, Malaga, Spain, ACM, Oct. 2–6, 2006.

[18] S. Lindsey and C. S. Raghavendra, "PEGASIS: power-efficient gathering in sensor information systems," in *IEEE Aerospace Conference Proceedings,* 2002.

[19] J. Kim, S. Kim, D. Kim and W. Lee, "Low-energy efficient clustering protocol for ad-hoc wireless sensor network," in *15th International Symposium on Personal, Indoor and Mobile Radio Communications, 2004 (PIMRC 2004)*, Vol. **2**, pp. 1022–1026, Sept. 5–8, 2004.

[20] D. Kumar, T. C. Aseri and R. B. Patel, "EEHC: energy efficient heterogeneous clustered scheme for wireless sensor networks," *Computer Communications*, Vol. **32**, issue 4, pp. 662–667, March 2009.

[21] B. M. Li, Z. Li and A. V. Vasilakos, "A survey on topology control in wireless sensor networks: taxonomy, comparative study, and open issues," *Proceedings of the IEEE*, Vol. **25**, No. 10, pp. 2367–2380, 2013.

[22] H. Zhang and J. Hou, "Maintaining sensing coverage and connectivity in large sensor networks," *Urbana*, Vol. **1**, pp. 89–124, 2003.

[23] X. Wang, G. Xing, Y. Zhang, *et al.*, "Integrated coverage and connectivity configuration in wireless sensor networks," in *Proceedings ACM International Conference Embedded Network Sensor Systems*, pp. 28–39, 2003.

[24] C. Huang and Y. Tseng, "The coverage problem in a wireless sensor network," *Mobile Network Applications*, Vol. **10**, pp. 519–528, 2005.

[25] X. Bai, Z. Yun, D. Xuan, W. Jia and W. Zhao, "Pattern mutation in wireless sensor deployment," in *Proceedings IEEE International Conference on Computer Communications*, 2010, DOI: 10.1109/INFCOM.2010.5462076.

[26] G. Wang, G. Cao and T. La Porta, "Movement-assisted sensor deployment," *IEEE Transactions on Mobile Computing*, Vol. **5**, No. 6, pp. 640–652, June 2006.

[27] A. Howard, M. Mataric and G. Sukhatme, "Mobile sensor network deployment using potential fields: a distributed, scalable solution to the area coverage problem," *Distrib. Autonom. Robot. Syst.*, Vol. **5**, pp. 299–308, 2002.

[28] S. Ganeriwal, A. Kansal and M. Srivastava, "Self aware actuation for fault repair in sensor networks," in *Proceedings IEEE International Conference Robotics and Automation*, Vol. **5**, pp. 5244–5249, 2004.

[29] G. Wang, G. Cao and T. LaPorta, "A bidding protocol for deploying mobile sensors," in *Proceedings of the IEEE International Conference on Networks protocols*, pp. 315–324, 2003.

[30] M. Batalin and G. Sukhatme, "Coverage, exploration and deployment by a mobile robot and communication network," *Telecommunication Systems*, Vol. **26**, No. 2–4, pp. 181–196, 2004.

[31] B. Liu and D. Towsley, "A study of the coverage of large-scale sensor networks," in *Proceedings of the IEEE International Conference on Mobile Ad-Hoc Sensor Systems*, pp. 475–483, 2004.

[32] S. Kumar, T. Lai and A. Arora, "Barrier coverage with wireless sensors," in *Proceedings of the ACM International Conference on Mobile Computer Networks*, pp. 284–298, 2005.

[33] P. Balister, B. Bollobas, A. Sarkar and S. Kumar, "Reliable density estimates for coverage and connectivity in thin strips of finite length," in *Proceedings of the ACM International Conference on Mobile Computer Networks*, pp. 75–86, 2007.

[34] A. Chen, S. Kumar and T. Lai, "Designing localized algorithms for barrier coverage," in *Proceedings ACM International Conference on Mobile Computer Networks*, pp. 63–74, 2007.

[35] A. Saipulla, C. Westphal, B. Liu and J. Wang, "Barrier coverage of line-based deployed wireless sensor networks," in *Proceedings IEEE International Conference on Computer Communications*, pp. 127–135, 2009.

[36] B. Liu, O. Dousse, J. Wang and A. Saipulla, "Strong barrier coverage of wireless sensor networks," in *Proceedings ACM International Symposium on Mobile Ad Hoc Network Computers*, pp. 411–420, 2008.

[37] S. Wong and B. MacDonald, "A topological coverage algorithm for mobile robots," in *Proceedings of the IEEE/RSJ International Conference on Intelligent Robots Systems*, Vol. **2**, pp. 1685–1690, 2003.

[38] A. Howard and M. Mataric, "Cover me! A self-deployment algorithm for mobile sensor networks," in *Proceedings of the International Conference on Robotics Automation*, Washington, DC, USA, 2002, DOI: 10.1.1.16.1394.

[39] W. Cheng, M. Li, K. Liu, *et al.*, "Sweep coverage with mobile sensors," in *Proceedings of the IEEE International Symposium on Parallel Distribution Processes*, 2008, DOI: 10.1109/IPDPS. 2008.4536245.

[40] W. Ye, J. Heidemann and D. Estrin, "An energy-efficient MAC protocol for wireless sensor networks," in *Proceedings of the IEEE International Conference on Computer Communications*, Vol. **3**, pp. 1567–1576, 2002.

[41] T. Van Dam and K. Langendoen, "An adaptive energy-efficient MAC protocol for wireless sensor networks," in *Proceedings of the ACM International Conference on Embedded Network Sensor Systems*, pp. 171–180, 2003.

[42] Y. Tseng, C. Hsu and T. Hsieh, "Power-saving protocols for IEEE 802.11-based multi-hop ad hoc networks," *Computer Networks*, Vol. **43**, No. 3, pp. 317–337, 2003.

[43] R. Zheng, J. Hou and L. Sha, "Asynchronous wakeup for ad hoc networks," in *Proceedings of the 4th ACM International Symposium on Mobile Ad Hoc Network Computers*, pp. 35–45, 2003.

[44] J. Polastre, J. Hill and D. Culler, "Versatile low power media access for wireless sensor networks," in *Proceedings of the International Conference on Embedded Network Sensor Systems*, pp. 95–107, 2004.

[45] W. Ye, F. Silva and J. Heidemann, "Ultra-low duty cycle MAC with scheduled channel polling," in *Proceedings 4th International Conference on Embedded Network Sensor Systems*, pp. 321–334, 2006.

[46] S. Narayanaswamy, V. Kawadia, R. Sreenivas and P. Kumar, "Power control in ad-hoc networks: theory, architecture, algorithm and implementation of the COMPOW protocol," in *Proceedings of the European Wireless Conference*, 2002. [Online.] Available: http://citeseerx.ist.psu.edu/viewdoc/summary?doi=10.1.1.23.5186.

[47] P. Santi and D. Blough, "The critical transmitting range for connectivity in sparse wireless ad hoc networks," *IEEE Transactions Mobile Computing*, Vol. **2**, No. 1, pp. 25–39, Jan.–Mar. 2003.

[48] L. Kirousis, E. Kranakis, D. Krizanc and A. Pelc, *Power Consumption in Packet Radio Networks STACS 97*, vol. **1200**. Berlin: Springer-Verlag, 1997, pp. 363–374.

[49] M. Marina and S. Das, "Routing performance in the presence of unidirectional links in multihop wireless networks," in *Proceedings of the ACM International Symposium on Mobile Ad Hoc Network Computing*, pp. 12–23, 2002.

[50] D. Blough, M. Leoncini, G. Resta and P. Santi, "On the symmetric range assignment problem in wireless ad hoc networks," in *Proceedings IFIP 17th World Computer Congress/TC1 Stream/2nd IFIP International Conference Theor. Comput. Sci., Found. Inf. Technol. Era Netw. Mobile Comput.*, pp. 71–82, 2002.

[51] J. E. Wieselthier, J. Nguyen and A. Ephremides, "On the construction of energy-efficient broadcast and multicast trees in wireless networks," in *Proceedings of the IEEE International Conference on Computer Communications*, Vol. **2**, pp. 585–594, 2000.

[52] M. Cardei, J. Wu and S. Yang, "Topology control in ad hoc wireless networks with hitch-hiking," in *Proceedings of the IEEE Communications Society Conference on Sensory Ad Hoc Communication Networks*, pp. 480–488, 2005.

[53] N. Li, J. Hou and L. Sha, "Design and analysis of an MST-based topology control algorithm," *IEEE Transactions on Wireless Communications*, Vol. **4**, No. 3, pp. 1195–1206, May 2005.

[54] V. Rodoplu and T. Meng, "Minimum energy mobile wireless networks," *IEEE Journal on Selected Areas in Communication*, Vol. **17**, No. 8, pp. 1333–1344, Aug. 1999.

[55] S. Lin, J. Zhang, G. Zhou *et al.*, "ATPC: adaptive transmission power control for wireless sensor networks," in *Proceedings of the ACM International Conference on Embedded Network Sensor Systems*, 2006, DOI: 10.1145/1182807.1182830.

[56] J. Liu and B. Li, "Mobilegrid: capacity-aware topology control in mobile ad hoc networks," in *Proceedings of the 11th International Conference Computer Communications Networks*, pp. 570–574, 2002.

9 Performance evaluation of wireless sensor networks

Performance evaluation of Wireless Sensor Networks (WSNs), like any communications network system, can be conducted by using simulation analysis, analytic modeling, and measurement/testing techniques. Evaluation of WSN systems is needed at every stage in their life. There is no point in designing and implementing a new system that does not have competitive performance, and does not meet the objectives and performance evaluation and quality of service requirements. Performance evaluation of an existing system is also important since it helps to determine how well it is performing and whether any improvements are needed in order to enhance the performance [1].

After a system has been built and is running, its performance can be assessed by using the measurement technique. In order to evaluate the performance of a component or subsystem that cannot be measured, for example, during the design and development phases, it is necessary to use analytic and/or simulation modeling so as to predict the performance [1–46].

The objective of this chapter is to provide an up-to-date treatment of the techniques that can be used to evaluate the performance of WSN systems.

9.1   Background information

Wireless sensor networks (WSNs) are unique in certain aspects that make them different from other wireless networks. These aspects include:

(1)   small CPU with limited computation power,
(2)   small memory in the node,
(3)   limited power in the node as it is usually powered by a battery,
(4)   limited bandwidth and data rate,
(5)   weak security mechanisms,
(6)   mobility of nodes,
(7)   dynamic network topology,
(8)   communication failures,
(9)   heterogeneity of nodes,
(10)   ability to survive harsh environmental conditions, and
(11)   ease of use and unattended operation.

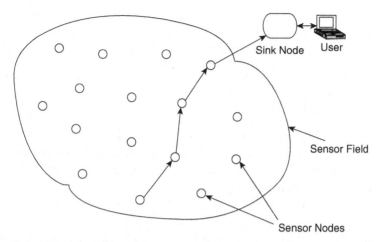

Figure 9.1 General basic structure of a wireless sensor network (WSN).

Figure 9.1 shows the basic structure of a WSN. When evaluating the performance of WSNs, such characteristics need to be taken into consideration.

In general wireless networks are made of mobile nodes and in order to transfer and communicate the information these mobile nodes communicate with one another with the help of wireless channels, whereas in WSNs the mobile nodes present in a network are usually attached to sensors in order to sense the physical environment/target and generate the data packets then send such data packets to the next or to the neighboring mobile node [10]. These mobile nodes of WSNs are of two types: (a) sensor nodes, and (b) sink nodes.

Sensor nodes are the mobile nodes that move freely in order to monitor the environment. When these nodes detect the physical target they start generating the data packets; but before generating the packets, they compare them to a threshold value which is usually set by the processor and then they send the packets.

On the other hand, sink nodes are the mobile nodes that accumulate the data from the sensor nodes. The end users gather all the collected data from the sink nodes for more analysis and decisions making [11–23].

The major reasons for engaging in modeling and simulation techniques for WSNs are listed below.

- Assessing the performance of a new application, scheme/protocol, architecture, service, or sensor device, and its effect on the performance of the WSN system.
- Constructing a prototype model in order to simulate a novel WSN design or application with the possibility to change parameters and tune the model under a simulated managed environment.
- Study the behavior of a large-scale WSN design in a managed scenario in order to find out and resolve issues or to determine the optimum design before complete setting out or in order to predict specific scenarios before implementation and investing time and money.

- Investigate using simulation analysis scaling of a deployed WSN to find out threshold indicators for capacity exhaust conditions, which can in turn be employed for life-cycle capacity planning and management of the WSN [1–15].

In order to have an effective simulation model for a WSN system, it is essential first to decide the objective of the modeling and simulation study. Questions that need to be addressed include the following [1–15].

(1)   Who will be employing the WSN and what are the main necessities?

These may include businesses, enterprises, cities in smart home environments, health and military organizations. Key requirements are privacy, security, reliability, speed of information delivery, and location accuracy.

(2)   Where will the WSN be deployed and in what environment? Is it for inside or outside; open space or with obstacles?

(a)   What present hypothesis or policies must be considered? Here, we need to consider the predetermined spectrum frequency, kinds of sensors already selected, and MAC protocols already determined.

(b)   What is the lifespan of the WSN, is there a busy time, is there an expected scaling to take place and if so over what timeline? In this context we need to differentiate between ad-hoc vs. permanent, and initial usage vs. predicted usage.

(c)   What are the projected performance aims the WSN is anticipated to accomplish? Is it Goodput, speed, cost, reliability, resiliency, response time, etc.?

(3)   What sort of WSN or WSN actions need to be investigated and evaluated, and what are the objectives of modeling and simulation?

Are these static or portable sensors; are they short power sensors; what is the life expectancy; and are they short-radius or long-radius antennas? The process needs to predict the performance and obtain accurate correlation of events from multiple sensors under different operating conditions and environments.

(4)   What laboratory environment or apparatus, if needed, will be employed?

(5)   If support such as people resources, purchasing equipment or software is required then how will that be provided?

Such questions have to be delineated in a statement of work, which can be studied, polished, and finally signed off if required in order to guarantee that outlooks are known and agreed on before work on the model commences. This step is vital to direct the right set of necessities for the simulation model design choices [1–15].

9.2   Wireless sensor networks (WSNs) modeling

There are different performance evaluation models for WSNs depending on the type of the network. WSNs have a number of challenges that have to be considered when developing these models. There are several empirical mathematical models to consider. Some of these have allocated factors based on the kind of WSN system to be analyzed and could have random number generators to simulate the conditions including attenuation over a duration of time. Models can also employ actual data provided by existing test beds, if possible, or the test data can be produced by configuring a prototype version of the network in the laboratory [39]. Below, we will look at the key models that we should consider when modeling WSNs.

Radio propagation modeling (RPM)   This model represents the radio transmission between the sender and receiver sensor nodes. Barriers in the communication route will worsen the RF signal propagating as a result of phenomena like scattering, reflection, diffraction, refraction, absorption, and polarization [40]. Issues like distance, barriers, and interference occurring in the environment and in the path between transmitter and receiver can influence the capability of completing a transmission effectively. Two performance metrics, the bit error rate (BER) and signal-to-noise ratio (SNR), are valuable in determining the radio signal performance. If the SNR is a major issue, then modeling of the sensor nodes antennas' characteristics may be considered in the model. In addition to the power of the antennas of the system, placement is also important. As an example, it is known that sensors placed on top of buildings confront lower barriers in the transmission path than sensors placed at ground level in urban settings. Various models apply for different kinds of environmental settings. The relation between received power and transmitted power can be given by:

$$P_r = P_t/(4\pi p^\lambda) \times g_t \lambda^2/4\pi,$$

where

p = distance between transmitter and receiver,
P_t = transmit power,
P_r = receive power,
g_r = receive gain (see below),
g_t = transmit gain,
λ = wavelength.

The term $4\pi p^\lambda$ represents an isotropic antenna, a hypothetical lossless antenna that transmits in all directions making a sphere around the antenna if no barriers obstruct the path. The mean radiation intensity of the transmission is equal to the overall power transmitted by the antenna divided by 4π. Actually, the overwhelming majority of antennas have some directionality known as directive gain.

The propagation loss of the RF signal is given as the ratio between the transmitted power to the received power and usually measured in decibels. This is given by:

$SNR = P_r/P_t = (4\pi p/\lambda) \ 1/(g_r \cdot g_t)$
SNR in decibels = $20 \log (4\pi p/\lambda) - \log (g_r \cdot g_t)$.

This last expression has different versions to model specific factors that affect the transmission path. Table 9.1 summarizes the popular radio propagation models for different situations [1–20].

A brief review of these models is given below.

Table 9.1 Summary of radio propagation situations

	Environment propagation	Description	Model	Density function
(a)	Line of sight	Line-of-sight example where no obstructions are faced in the transmission path	Fundamental basic transmission loss model	Constant
(b)	Ground reflection (plane earth)	Multipath case where the transmission may include two types of signals received; line of sight and ground reflection	Two-ray ground propagation	Ricean
(c)	Diffraction loss	Multipath case where topography presents obstructions in the path	Lognormal shadowing	Raleigh
(d)	Diffraction loss	Multipath case where trees introduce obstacles in the path	Lognormal shadowing	Raleigh
(e)	Multiple diffraction loss	Multipath case where topography introduces some obstructions in the transmission path	Lognormal shadowing	Raleigh

- *Basic transmission loss model*   This model considers loss mainly as a function of frequency, distance, and antenna power. It embodies perfect environmental conditions, which are seldom found. This represents entry (a) in Table 9.1.
- *Two-ray ground propagation*   The two-ray ground propagation model also represents ideal situations, with the exception that it takes into consideration a ground reflection of signals produced by the surface of the earth. This case affects an extra reflected signal in addition to the line-of-sight signal received by the receiver. Hence this is a type of multipath situation. One more point: this model presents another variable to take care of: the effect of the multipath signals received by the receiver. This represents entry (b) in Table 9.1.
- *Lognormal shadowing*   Here, the model takes into consideration the obstructions in the transmission pathway like hills, forests, and high-rise buildings. Moreover, it applies to other obstruction situations like indoor environments, which have walls or floors hindering the transmission pathway. For such outdoor model situations, there is no line-of-sight transmission received by the receiving node, but many dispersed multipath echoes of the signal sent initially. The main transmission model is revised to include an extra variable to signify an attenuated lognormal distributed variation of the signal. The multiple diffraction loss scenario described in entry (e) in Table 9.1 is hard to forecast and may lead to undependable results when simulation outcomes are compared to real measured results. The multiple obstruction situations can represent entries (c), (d), and (e) in Table 9.1.
- *Density function*   The density function recognizes the popular schemes available to account for the variables that affect each of the environmental models described above. In the line-of-sight case, a constant based on environmental surroundings, which may exist in the environment for the model, like weather, must be taken into account. As an example, humidity and temperature can impact the line-of-sight as

they change the density characteristics of the transmission path. Moreover, good analysis is essential in order to find out the correct values that apply for the different environmental situations. In the case of other models, the investigation is concentrated on the widespread Ricean and Raleigh factors in order to embody the variable for particular environmental setting. The Ricean factor is basically described as the ratio between the signal power of the line-of-sight power part over the scattered power part. Hence, this factor is most appropriate for the two-ray ground reflection case. The Raleigh factor signifies a set of multipath reflected or scattered waves received by the delivery node with no influential wave (line-of-sight signal) like the case for the Ricean factor. Thus, the Raleigh density function is most appropriate for the two diffraction loss cases, and the multiple diffraction loss case described in Table 9.1.

When modeling WSNs, it is essential to define the operating conditions of the WSNs before developing the model. Moreover, interference can influence the SNR at the receiving node. In an indoor home situation, devices like cordless phones and microwave ovens may also present interference. In order to take interference into consideration, the models described above can be adjusted [40]. More variations to these models can be found in various published papers [41–45].

Energy modeling   Unlike fixed networks, wireless networks have the challenge of dealing with energy consumption. Here, the sensor nodes are usually battery operated and, in some cases, are anticipated to operate unattended for some periods that may be of the order of months. Battery-operated wireless communication devices need regular recharging even though batteries have recently experienced substantial improvements, driven by demand for longer life. Moreover, network designers have improved their wireless communication networks to enhance energy efficiency. Manufacturers of all kinds of communication devices have improved the design of their devices and systems in order to make them energy aware. Many countries have established incentives for such manufacturers in order to encourage them to reduce power consumption and make them energy friendly and green. The trend these days is to have the power supplies for computer systems and networks at least 80% efficient. There is even an industry initiative called 80PLUS, which certifies power supplies that are at least 80% efficient [46]. The proliferation of wireless systems and applications will also place energy demands on devices and this will be an area that needs more attention, especially with the increased focus on green information, communication, and networks systems worldwide. As an example, video streaming is predicted to increase substantially and this will consume a huge amount of system bandwidth and battery resources of nodes and devices. Moreover, new types of WSNs will be used and grow as new applications are employed. WSNs have started to be used in smart homes and cities as well as for medical applications. Wireless body sensor networks are driving groundbreaking ways to manage health, especially of senior and elderly citizens.

Energy management impacts can be categorized from the point of view of the three classes of building blocks of WSNs: (a) sensor device energy efficiency, (b) network energy management, and (c) application energy management [1–40].

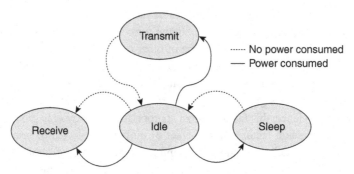

Figure 9.2   Illustration of a sensor node state transition diagram.

Here, we assume that the sensor node battery is a permanent resource. The most power-consuming job of a sensor device is the radio communications task, which deals with transmitting and receiving data. Moreover, managing battery usage by putting sensors into the sleep mode whenever possible is another important task. In this context, and in order to model a typical sensor node, we can envision a simple state-transition graph with four states as shown in Figure 9.2.

This diagram can help us develop a simplified mathematical model in order to estimate the total energy expended by a sensor node (S_n) via multiplying the power spent by a state (P_j) by the time spent in a state (t_j) plus the power spent while transitioning to a state (T_k). So, this model can then be used to find out the total energy consumed for the state changes happening over a specific amount of time [10–40]:

$$S_n = P_j t_j + T_k.$$

The next section deals with simulation models of WSNs.

9.3   Simulation models

In general, WSN simulation models can be categorized into four types:

- environment model,
- sensor node model,
- user node model, and
- communication model.

Environment model   In the environment WSN model, the nodes are used to define the physical environment. For example, in warfare applications, these nodes move wherever the enemy is moving in the war. Another example is in temperature-detection applications, where the nodes help to detect the temperature in any particular area.

Figure 9.3 shows the general architecture of the environment model. In this model, a node has an application layer, a physical layer, and a sensor layer. These environment nodes generate physical data and direct the data to a sensor channel.

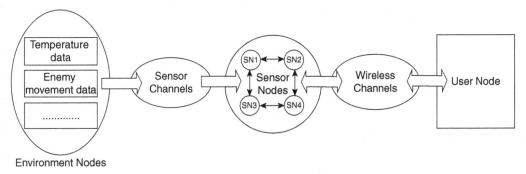

Environment Nodes

Figure 9.3   A general simulation model of a WSN.

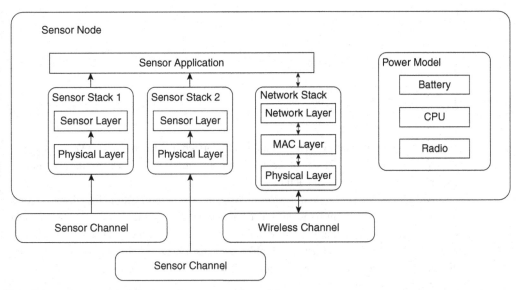

Figure 9.4   A general architecture of a sensor node model.

Sensor node model   The nodes in this model detail the conditions of nodes including communication, mobility, and routing schemes. The architecture of these nodes can be illustrated by Figure 9.4.

This model has a sensor protocol and network protocol. A sensor protocol stack is used to process and detect the data packets from previous model nodes and send them to the application layer. Then this layer will transmit them to the next model in the form of reports through a wireless channel.

User node model   This model acts like an interface between sink nodes and the user. The data packets are used by the user to analyze the targets. The architecture of this model is given in Figure 9.5. These nodes get sensor reports and transmit them to the application layer.

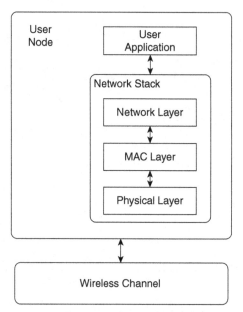

Figure 9.5   The architecture of the user node model.

Communication model   This model is categorized into three types: (1) environment–sensor communication, (2) sensor–sensor communication, and (3) sensor–user communication. A brief description of each of them is given below.

Environment–sensor communication   This is a communication broadcast of data packets from environment nodes to sensor nodes. Moreover, this communication is one-way broadcasting. The channel between these two nodes is known as the sensor channel.

Sensor–sensor communication   Unlike the sensor channel, this is two-way communication, and is between two sensor nodes. These are used for collecting signal processing of sensors in the WSN.

Sensor–user communication   This is also two-way communication, similar to sensor–sensor communication. This channel is between the sensor and the user. The nodes from the sensor to user are for transmitting sensor reports to the user and the nodes from the user to the sensor are for transmitting commands or queries to the WSN. Of course, this channel is defined as a wireless channel.

Wireless sensor networks that hold confidential or sensitive data are facing major risks. Some of these risks include modifying the data in networks, falsifying of data, and replacement or destruction of sensors. In order to counter these challenges, many applications such as sensor system replacement and conventional applications such as video monitoring have been devised. Achieving solutions for these security issues and minimizing the risks, which are caused by the weak physical protection and the openness of these wireless channels for the devices may lead to the need of deployment in large scales.

Despite the fact that state-of-the-art schemes of cryptography and other security mechanisms may address these issues to a large extent, these methods are mainly focused on the protection of high-performance applications, not for applications with tiny sensors, which have little communication bandwidth and limited computation capabilities and power. Clearly, there is a need to devise some kind of innovative "light-weight" applications in order to deal with these types of problems.

Many researches have proven recently that, even without the high-end cryptography technique, we can achieve a good level of security for wireless applications. Some of those ideas are explained next. Usage of energy in a WSN is also a main challenge since many of these networks run by using battery power. Hence, they should be energy efficient. While transmitting data from the source point to the destination point, instead of sending the data directly, we can preserve some hops in between within small distance ranges so that the energy required for transmitting the data is minimized. However, this solution has its own disadvantage; any intentional intruder can gain the information present in the network by getting access on any hop that is placed in between. Moreover, if the cryptography scheme used on the data over the network is not strong, then the intruder can easily get the information and then modify or misuse the data.

The notion which can be used here to avoid attacks to the maximum extent is to send data in double paths rather than in a single path. This way, a third party is not allowed to manipulate the data over the network. Hence, nodes share the information at each processing level, which helps to achieve confidentiality. So, if such arrangements are implemented in a WSN, then the model has to be adjusted and tuned to have proper representation of the WSN system being assessed and analyzed.

Designing an efficient topology for WSNs is also a challenge. The main objectives that are always in the mind of WSN researchers are that the sensor nodes in the system must operate for a long time and the WSN should take less processing time/power for receiving or sending data packets. At the same time, the end users want the required information quickly. Moreover, the information that is in the network should be safe and secure. By relying on current technology we cannot get all of these objectives fulfilled in a single model [13].

9.4   Modeling the behavior of sensors and sensor networks

Wireless sensor networks usually have many small sensor nodes and these nodes communicate with each other wirelessly. These nodes are used in various applications where wireless transfer of data is necessary [14]. These sensor nodes have the capability to sense, compute/process and communicate. The goal is to arrange all these given features in one single chip. Figure 9.6 illustrates the sensor node structure.

There are some main reasons for the advancement of sensor node development. These include: (a) increasing complexity of devices on the microchips, (b) advances in tools that have high performance, (c) combining sensor data skills and signal processing mechanisms, (d) availability of networking technologies (wireless) with high performance, and (e) progress in micro-electro-mechanical systems development [15–23].

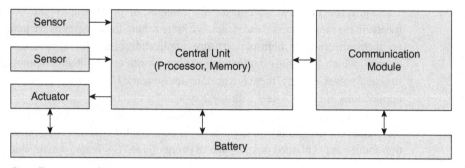

Figure 9.6   Overall structure of a sensor node.

The key requirements for the implementation and design process of WSNs are as follows [1–23].

(1)   WSNs should be designed as self-organizing.
(2)   In order to have more accurate results, more cooperative processing should be performed.
(3)   There is a need for some good security mechanisms so that WSNs can withstand any environmental conditions and attacks.
(4)   The protocols or algorithms used in a WSN should be energy aware [15].

A brief description of these requirements is given below.

9.4.1   Self-organization

Let us consider the network setup: a WSN consists of many sensor nodes; hence organizing these nodes manually is a difficult task. This means there is a need for a sophisticated automated method to do so. Software has to be embedded into all of these nodes so that the node can get the information from the neighboring nodes [24]. Therefore, the WSN can be made self-organizing and, for this sort of organization, nodes' context awareness is essential. Whilst building up the software, nodes should be designed and built in a way that they can know about the reliability and bandwidth of the network and the relations between network elements. Moreover, the main aim that needs to be taken into consideration is the update mechanism, since nodes are on the move continuously. For instance, it is essential for the node to have knowledge of the destination. However, for mobility reasons the target node often moves; hence, the update mechanism is quite important while building up software.

Software development for sensor nodes demands new programming schemes and paradigms. Most of the classical programming techniques are not adequate for these WSNs as these techniques fail to achieve the key goals such as cooperative test management in WSNs. In general, wired communication networks are based on the client server concept, where the client sends a request to the server and the latter sends its response. However, WSNs follow the event-based standard. If some limit value is attained then the particular event is activated and then transferred to the destination, which can be the data sink of the network system. The node addressing scheme for WSNs is different from wired communication networks'. In customary wired communications, the nodes are signified by using the node ID or IP address; however, that notation cannot be used in wireless networks, including WSNs, since the nodes are on the move continuously and the position of the node is not easy to identify. For these above-mentioned reasons, a new scheme is needed. This is called the attribute-based naming method, where a usable request needs a node's context and its position.

In most WSNs, data packets are usually routed from the source to the destination via intermediate nodes; all nodes must have some integrity so that the network maintains reliability and dependability. Such attributes have immediate bearing on the functionality and lifespan of the node. When designing WSN software there are several new techniques that should be followed in order to optimize the design and find a trade-off between network functionality, lifetime, and the task processing between the intermediate nodes. Of course, all of these aspects affect models of WSNs.

In the process of making a decision about routing algorithms, we have to consider important specific factors such as energy consumption, transmission delay, mobility of nodes, and WSN load [25].

9.4.2 Cooperative algorithms

While wired communication networks do not employ cooperative algorithms, WSNs have flexibility to utilize cooperative algorithms. These algorithms are mainly used in WSNs in order to decrease network traffic affected by data aggregation and preprocessing. WSNs applications do not care whether the data aggregation is performed by the node itself or the neighboring node. However, the only concern is that the communication toward the data sinks should be minimized as they are normally located in faraway distances. For instance, location determination using triangulation takes the measurements from any three dissimilar nodes and the computed positions are used for routing or addressing reasons.

9.4.3 Security mechanisms

Wireless sensor network applications and the operating conditions of the environment influence the selection of security schemes for the network. Moreover, sensor node resources such as energy, memory capacity, and processor computation power must be taken into consideration when dealing with this important issue. The minimum basic requirements for any security system include confidentiality, authenticity, availability, integrity, and non-repudiation [26]. Moreover, some of the special requirements for these sensor networks include intrusion detection, containment, message freshness, and intrusion tolerance. In order to satisfy all the above-mentioned restrictions, the security models used by the WSN should be tested and developed carefully. The software system that is built for WSNs must consider the management and selection of the security scheme, and the latter should follow the policies given by the system operator. In addition, such security schemes need to be evaluated by using modeling and simulation, and by adopting security performance metrics that are acceptable for the application at hand.

9.4.4   Energy-aware requirement

The nodes in WSNs are usually powered by batteries, and these WSNs often have a large number of small nodes, which makes it difficult to recharge or replace their batteries. Furthermore, WSNs in general are used in harsh environments or remote sites. These factors make the lifetimes of the nodes a crucial issue in the design, implementation, and launching of WSNs.

Sensor nodes typically employ microcontroller hardware that offers many schemes such as dynamic power management (DPM) for power saving. The power saving is obtained by DPM as it turns off the hardware components that are not needed. The operating system (OS) of the sensor node can help in power saving by implementing a low-energy task-scheduling technique. Scheduling algorithms may employ nonlinear battery effects for reducing the energy consumption. Moreover, choosing the right communication scheme can reduce power consumption. Emphasis on the kind of protocols makes performance evaluation models for WSNs slightly different from those for traditional communications networks.

9.5   Simulation tools for wireless sensor networks (WSNs)

Wireless sensor networks are found in many applications and scenarios. Some of these applications require monitoring of nodes in the network. The main functions that are supposed to be monitored include movement of the nodes and the data communicated between sensor nodes. In order to accommodate these functions, there is a need for a simulator that is exploited for monitoring wireless actuators. Moreover, this may be utilized to evaluate different functions, network structures, cooperation forms, and event and space scenarios, among others.

Generally, there are various routing algorithms that are employed in WSNs. Simulation analysis is one of the most effective, flexible, and convenient performance evaluation techniques for testing these routing protocols. This entails the simulation of many applications of WSNs. Among these are forest fire monitoring, chemical gas applications, and military surveillance. The output of this simulation can be given either in the form of text or by using a graphical interface, depending on the type of application. In addition, for the applications of network topology even the physical environment can be used as the background. This means that we can have hardware in the loop simulation, which has the potential of providing better accuracy to the simulation modeling process.

A simulator is the computer program that can simulate certain characteristics, operation, or some conditions of WSNs for research and development purposes. This provides a means of adding events or removing some events, emptying event queues, transforming the units of time, generating random numbers and random variates, among others. A simulator can run a simulation for any given time interval or until the events queue is empty. This may use binary heap data structure for managing events according to the time at which they are started [27].

In simulation of WSNs, we may need to define the following terms depending on how the simulation model is implemented. In general, we will need the following categories.

The event category has the details about all the events, such as the time any particular event should start at. Events can also provide some methods for comparing events with respect to their fire times, the basic method for firing any particular event, and also for determining whether two events are equal or not. Events have subclasses, which can provide an interface to the listeners of an event. Moreover, they have members that act like parameters for another event [1–45].

The medium-category models wireless transmission medium, permits all the nodes of the WSN to broadcast signals, and is also responsible for giving nodes the information about the signals that affect them. In order to perform these functions, the medium should have some inputs that are updated such as the presence of all nodes in the network, the changes that occur in the position of nodes, and the changes that occur with respect to the radio properties like receiver sensitivity and transmitter power. It has the characteristics of the medium being modeled, such as the wavelength and bandwidth. The propagation model of the medium gives strength to the receiver from a transmitted signal by the given transmitter. Moreover, it has reference to the propagation model, which is given to that medium during the construction time.

The propagation model is used to define the strength of the signal at any receiver from any transmitter at a specific period of time. The propagation model interfaces provide methods to get the maximum signal strength to any given receiver and transmitter pair. The output of this interface is used to find the interface relationships. Signal propagation has many practical models as well as theoretical models. Some of these signal propagation models are two-ray ground reflection, Rayleigh and Ricean fading, free space propagation, and RIM (radio interface model). In addition to these above-mentioned models, there are many other models for signal propagation, and there is active research in this area [1–46].

The environment module is similar to the medium module. Implementation of the environment module includes features that are related to the physical phenomenon, and this is the key variation between environment model and the medium module.

The propagation model of the environment module is similar to that of physical phenomena module. The main physical phenomena that are included in the sensor networks include temperature, humidity, light, magnetic field, chemical presence, and sound, among others [28].

There, "node" means an individual sensor node in a WSN. This acts like a container for each component in hardware and software. Some of these components are transceivers, actuators, processors, sensors, energy sources like battery, and applications and network protocols. Every node of the WSN has important features like identification and location.

The processor of a sensor node is modeled in simulation by the processor module. It models states of the processor like energy consumption and power modes at each state. In this module the processor employs the events for generating events that can report the power consumption changes.

Below is a review of some of the major network simulation packages that are used to simulate WSNs [1–5, 20–45].

NS-2   This simulation tool is a discrete event simulator and it was developed in 1989. It is an open source network simulation tool. This is a popular network simulator, which is built using OTcl and C++ object-oriented programming language. This simulator is pretty much limited to the IP networks because of the low-level assumptions [29]. Although this tool was originally designed for LAN protocols, it has been revised to be used for mobile ad-hoc networks and wireless networks. There are two MAC protocols that are supported by the NS-2 simulator: the IEEE 802.11 and TDMA protocols. NS-2 now can support wireless sensor networks. However, the main disadvantage is that the accuracy of simulator is not appreciable [2, 30].

GloMoSim   This simulator project started in 1998, initially for the mobile wireless networks. This simulator is programmed using PARSEC, which is a C-based simulation language, developed by the Parallel Computing Laboratory at University of California-Los Angeles, for sequential and parallel execution of discrete event simulation models. PARSEC can also be used as a parallel programming language. This simulator has many different choices for propagation of radio signals, 802.11 CSMA MAC protocols, implementations of TCP and UDP, mobile routing protocols, among others.

The GloMoSim tool simulates mobile IP networks efficiently. It is also limited to IP networks due to the assumptions taken at the low-level design phase. Hence, this simulator also has same drawbacks as those of NS-2 simulation tool. This simulation package cannot give support to sensors, physical phenomena, and environmental conditions or actuators. In general, it does not provide very accurate simulation results [2, 29].

J-Sim   This simulation package was initiated in 1999 and it was meant as a generic component, compositional simulation framework. This tool is also based on two languages, like the NS-2 simulator: JAVA and Jacl (which is JAVA version of Tcl). J-Sim simulator framework has a general packet-switching network, which permits various Internet protocols. This tool now supports sensor networks and mobile networks. It can also support physical phenomena and sensors [29]. However, it is not very accurate.

SENSE   This simulation tool is one of the most recently developed packages. The features of SENSE are almost similar to these of J-Sim. SENSE is component based, but it is built using C++, so that the apparent inefficiency of JAVA is prevented. SENSE simulator supports the energy model and only some protocols, which include NullMAC. The NullMAC messages arrive instantaneously leading to the avoidance of collisions or errors. Simulations using this tool are also not very accurate [2–29]. Newer versions are expected to support physical phenomena, sensors, and actuators only for the sound model.

Visual Sense   The Visual Sense simulator offers an accurate radio model. This radio model is primarily based on the energy propagation model, which can also be reused for the physical phenomena. Moreover, this tool affords a sound model with the help of the propagation model, which is considered accurate for the localization. Visual Sense has a very good framework, but it does not provide any other protocols over wireless medium and it does not provide any physical or sensor phenomena except the sound model [29].

Prowler and JProwler   These are two different wireless sensor network simulators. Prowler was developed using Matlab, whereas JProwler was built using JAVA. These simulation tools provide very accurate radio models. They do not support physical phenomena or the sensors. Prowler and JProwler support one MAC protocol, which is a default of TinyOS.

Sidh   This simulation package was developed especially for WSNs. Sidh is very efficient; it can simulate networks with thousands of nodes much faster compared to real-time on a normal desktop. This simulation tool has a component-based framework and can be reconfigured easily. Moreover, it can be adapted for different environmental circumstances, protocols, communications transmission media, actuators and sensors, and in various simulation granularity levels and accuracies.

This simulation tool was built using JAVA, and has many different sets of modules. These modules are all well-defined in the JAVA interface. Some of the modules are accessed by using method calls. Other modules communicate with each other by using a simulator that makes use of the events. Modules of the simulator are not directly reliant on each other with the help of the events. Furthermore, the modules can be categorized as simulator, medium, events, transceivers, protocols, and environment, among others [29–45].

Optimized network engineering tool 10.5 (OPNET)   The optimized network engineering tool (OPNET) is meant to study the performance of all communication networks. The performance is projected by modeling the network using discrete event simulation [1, 2].

OPNET models the network at the packet level. Although it was designed initially to simulate fixed networks, it is now able to analyze the performance of wireless networks.

OPNET has been successfully employed to analyze the performance of ZigBee-based WSNs by offering components of a ZigBee network such as ZigBee coordinator, ZigBee router, and ZigBee end device, which can be fixed or mobile. Examples on the use of OPNET to simulate WSN can be found, for example in [33–35].

The simulation package is improved every few years and a new version of the package is then released. The major aspects of OPNET are as follows [1–2, 6–7, 17–28, 36].

- Simulation and modeling cycle: it has effective tools that aid in model building, simulation running, and analysis of the simulation outputs.
- OPNET supports ordered organization of modeling.
- It includes an abundant set of library modules that aid protocols and network-related topologies.
- It has a decent troubleshooting capability and the model can be easily compiled and executed.

There are three types of editors in OPNET for modeling three types of networks [1].

- Network topology models are modeled by the network editor.
- Data flow models are created via the node editor.
- Control flow models are conveyed through the process editor.

Moreover, the simulation program can be run by using a simulation tool or debugging tool. A simulation tool is employed for running the model in a customary manner; however, if we want to collaborate while executing the simulation program then it is better to execute it utilizing the debugging tool [1, 6–7].

There are a number of available tools that can be used for analyzing the simulation results. The probe editor is used for collecting the data. Statistical results are acquired by employing the analysis tool. Processing of data is performed with the aid of the filter tool. The dynamic actions of the model can be observed by using the animation viewer.

The probe editor in OPNET encompasses different types of probes for gathering output data. Statistic probe is employed to collect statistics like bit error rate, latency, and throughput performance metrics. To generate the animation, sequences in the simulation automatic animation probe are utilized. A custom animation probe may be employed to gather the animation features for process and link models. Statistic probe collects the data; nevertheless, it does not generate the output data. In order to generate the statistical output data, a statistic probe is used. The analysis tool used in OPNET is employed to report the information in terms of graphs. The analysis panels are used to show these graphs [1, 37].

TOSSIM   This is specially designed for TinyOS and simulates entire TinyOS applications. This was intended to facilitate the development of sensor network applications. One good aspect of this tool is that it scales to thousands of nodes, and compiles directly from the TinyOS code. TOSSIM simulates the TinyOS network stack at the bit level, permitting experimentation with low-level protocols in addition to top-level application systems. Here, users can connect to TOSSIM and cooperate with it using the same tools as one would for real-world networking, making the transition between the two as easy as possible [38].

This package has a GUI tool, and TinyViz, which allows visualization and interaction with running simulations. By utilizing a simple plug-in model, users can develop new visualizations and interfaces with TinyViz [38].

This tool simulates a TinyOS network stack allowing all low-level network protocols. This simulation helps in monitoring the packet traffic, and packets can also be dynamically injected into the network. This tool is less flexible when compared to other simulators and also does not address energy consumption.

9.6   Performance metrics

Wireless sensor networks are different from traditional networks, and therefore different performance metrics/measures may need to be used in order to assess their performance. The key performance metrics for WSNs are reviewed below [1–35].

(1)   Energy efficiency   The energy efficiency refers to the number of data packets that can be sent successfully by using the unit of the energy. Tasks and operations like routing overhead, packet collision at the MAC layer, packet loss, and data packet retransmitting reduce the energy efficiency and decrease the life of the battery.

(2)   Lifetime of the WSN system   This may be defined by using: (a) the period of the time until the required quality-of-service (QoS) of the application cannot be provided, (b) the period of the time until some nodes consume all the energy/power, or (c) the period of time until the WSN becomes separated.

(3)   Reliability   In the WSNs, event steadiness is employed as a measure to show how consistently the sensed incident could be sent to the sink node. For the applications that can endure packet losses, reliability could be explained as the ratio of the effectively reached packets to the total number of data packets that are sent.

(4)   Coverage   The full coverage range of a WSN refers to the overall space that can be observed by using sensor node devices. When some sensor nodes are not working because of energy weariness, there is some definite quantity of the space that can no longer be perceived. Here, coverage is defined as the ratio of monitored gap to the entire space.

(5)   Connectivity   In the case of multi-hop WSNs, it is easily possible that the WSN becomes disordered due to some faulty nodes. This performance metric can be used to assess the degree of the interconnectivity of the WSN or how many nodes have been separated.

(6)   QoS metrics   There are some WSN applications that require a specific QoS metric, such as constant bit rate/real-time applications. Such applications require low delay and low delay variations (jitter). Other applications such as e-mail traffic or file transfer require low error rate/low loss rate.

9.7   Fundamental models

9.7.1   Traffic model

Traffic characteristics in WSNs are different from those of traditional communication networks. For example, whereas the broadly utilized applications for Internet are file transfer, e-mail, Web services, and the peer-to-peer services, WSNs have rather different properties. Hence, traffic and the data-sending models are also dissimilar. Currently, there are four traffic designs utilized in WSNs: (a) continuous delivery, (b) event-based delivery, (c) hybrid delivery, and (d) query-based delivery [32]. Traffic models actually influence protocol designs and affect the performance. These four models and associated performance features are explained next.

9.7.1.1   Continuous delivery

Data gathered by the sensor devices have to be updated frequently, maybe constantly, or from time to time. For example, a WSN can be utilized to monitor the propagation behavior of a specific breed of bird, such as the parrot on an island. In this case, the sensor node devices are stationed inside tunnels and on the surface in order to evaluate pressure, humidity, temperature, and ambient light level. Each sensor device reports the sample's values to its sink node.

9.7.1.2   Event-based delivery

In this case, the sensor node monitors the occurrences of the events constantly. If one event occurs, the sensor nodes begin to report events. Every time it delivers an event to a sink, the routing protocol is invoked to find the path to the sink. This kind of procedure is often called an on-demand routing scheme. Whenever an event appears repeatedly, at one node or a group of the nodes, the routing function will be invoked repetitively, which can result in extra energy consumption. Alternative schemes should be set up in advance in order to utilize the frequent path. Therefore, routing efficiency of the delivery design is greatly dependent on the frequency of occurrence of the events. The adaptive routing protocols are needed to establish the path by its own in advance whenever an event occurs frequently; or else, the path will be set up on demand [32].

9.7.1.3   Hybrid delivery

In some WSNs, the kind of sensor devices and data they sense may be exceptionally varied. For example, the data could be updated continuously by using some of the nodes, and the sink may need to give query information from the other sensor device nodes.

9.7.1.4   Query-based delivery

Sometimes, sinks may show interest in particular part of the information that has previously collected in sensor nodes. The sink will transmit a query message to the sensor nodes in order to find out the updated values of the information of interest. The query message may also acquire some commands from the sink to sensor nodes about the reporting frequency and the other metrics of interest to the sink. In this scheme, the sink transmits query messages. The path is established automatically whenever a query reaches the sensor node. Then the sensor node conveys the results based on the requests in the query messages.

9.7.2   Energy models

There are ways by which we can reduce the consumption of power by sensor communications. The main scheme is to plan so that it can save energy naturally, for instance, by turning off the transceiver for a period of time. The other scheme is to reduce the amount of the communications in the network. This approach requires functions such as data aggregation and data compression, while the computation process usually needs lower energy than performing the communication process.

9.8   Summary

In this chapter, we have reviewed the basics of performance evaluation of wireless sensor networks, especially the modeling and simulation aspects. We presented the features and characteristics that make WSNs different from other wireless networks. We then tackled the modeling techniques of WSNs and identified the key models that we should consider when modeling WSNs, including radio propagation modeling (RPM), basic transmission loss model, two-ray ground propagation model, lognormal shadowing model, and density function models. Then we investigated the WSN simulation models, which can be categorized into four types: (a) environment models, (b) sensor node models, (c) user node models, and (d) communication models. The software simulation tools that can be used to simulate WSNs have been reviewed and analyzed; those include NS-2, GloMoSim, J-Sim, SENSE, Visual Sense, Prowler and JProwler, Sidh, OPNET, and TOSSIM. We have also identified the major performance metrics that can be used to assess the performance evaluation of WSNs, including energy efficiency, lifetime of the WSN, reliability, coverage, connectivity, and the QoS metric.

References

[1] M. S. Obaidat, *Fundamentals of Performance Evaluation of Computer and Telecommunication Systems*, Wiley, 2010.

[2] M. S. Obaidat and G. I. Papadimitriou, Eds., *Applied System Simulation: Methodologies and Applications*, Springer, 2003.

[3] R. Jain, *The Art of Computer Systems Performance Analysis*, New York: Wiley, 1991.

[4] K. Kant, *Introduction to Computer System Performance Evaluation*, New York: McGraw-Hill, 1992.

[5] D. J. Lilja, *Measuring Computer Performance*, Cambridge: Cambridge University Press, 2000.

[6] M. S. Obaidat, "Advances in performance evaluation of computer and telecommunications networking," *Computer Communication Journal, Elsevier*, Vol. 25, Nos. 11–12, pp. 993–996, 2002.

[7] M. S. Obaidat, "ATM systems and networks: basics issues, and performance modeling and simulation," *Simulation: Transactions of the Society for Modeling and Simulation International, SCS*, Vol. 78, No. 3, pp. 127–138, 2003.

[8] M. S. Obaidat, "Performance evaluation of telecommunication systems: models issues and applications," *Computer Communications Journal, Elsevier*, Vol. 34, No. 9, pp. 753–756, 2003.

[9] M. C. Ghanbari, J. Hughes, M. C. Sinclair and J. P. Eade, *Principles of Performance Engineering for Telecommunication and Information Systems*, Hertfordshire, United Kingdom: IEE, 1997.

[10] F. Bao, R. Hu, R. Li and L. Cheng, "Principles and implementation challenges of cooperative transmission for realistic WSNs," *Proceedings of the Second International Conference on Signal Processing Systems (ICSPS)*, Vol. 2, pp.V2-334–V2-338, 5–7 July 2010.

[11] E. Altman, T. Basar, T. Jimenez and N. Shimkin, "Competitive routing in networks with polynomial costs," *IEEE Transactions on Automatic Control*, Vol. 47, No. 1, pp. 92–96, 2002.

[12] W. K. Seah, Z. A. Eu and H-P. Tan, "Wireless sensor networks powered by ambient energy harvesting (WSN-HEAP) – survey and challenges," in *Proceedings of the 1st International*

Conference on Wireless Communication, Vehicular Technology, Information Theory and Aerospace & Electronics Systems Technology, Wireless VITAE 2009*, pp. 1–5, May 2009.

[13] C. Enz, N. Scolari and U. Yodprasit, "Ultra low-power radio design for wireless sensor networks," in *Proceedings of the 2005 IEEE International Workshop on Radio-Frequency Integration Technology: Integrated Circuits for Wideband Communication and Wireless Sensor Networks*, pp. 1–17, 30 Nov./Dec. 2005.

[14] Q. Wang and W. Yang, "Energy consumption model for power management in wireless sensor networks," in *Proceedings of the 4th Annual IEEE Communications Society Conference on Sensor, Mesh and Ad Hoc Communications and Networks, SECON '2007*, pp. 142–151, June 2007.

[15] S. Wielens, M. Galetzka and P. Schneider, "Design support for wireless sensor networks based on the IEEE 802.15.4 standard," in *Proceedings of the 2008 IEEE International Symposium on Personal, Indoor and Mobile Radio Communications, PIMRC 2008*, pp. 1–5, Sept. 2008.

[16] M. Conti, R. Di Pietro, L. V. Mancini and A. Mei, "Requirements and open issues in distributed detection of node identity replicas in WSN," in *Proceedings of the 2006 IEEE International Conference on Systems, Man and Cybernetics*, pp. 1468–1473, Oct. 2006.

[17] M. Gupta, M. S. Obaidat and S. Dhurandher, "Energy-efficient wireless sensor networks," in *Green Information and Communication Systems*, M. S. Obaidat, A. Alagan and I. Woungang, Eds. Elsevier, 2013.

[18] C. Ramachandran, M. S. Obaidat, S. Misra and F. Pena-Mora, "A secure, and energy-efficient scheme for group-based routing in heterogeneous ad-hoc sensor networks and its simulation analysis," *Simulation: Transactions of the Society for Modeling and Simulation International*, Vol. **84**, No. 2–3, pp. 131–146, Feb./March 2008.

[19] S. Dhurandher, S. Misra, M. S. Obaidat and S. Khairwal, "UWSim: a simulator for underwater wireless sensor networks," *Simulation: Transactions of the Society for Modeling and Simulation International, SCS*, Vol. **84**, No. 7, pp. 327–338, July 2008.

[20] S. Misra, V. Tiwari and M. S. Obaidat, "LACAS: learning automata-based congestion avoidance scheme for healthcare wireless sensor networks," *IEEE Journal on Selected Area on Communications (JSAC)*, Vol. **27**, No. 4, pp. 466–479, May 2009.

[21] S. K. Dhurandher, M. S. Obaidat and M. Gupta, "An efficient technique for geocast region holes in underwater sensor networks and its performance evaluation," *Simulation: Modeling Practice and Theory, Elsevier*, Vol. **19**, No. 9, pp. 2102–2116, Sep. 2011.

[22] S. Misra and M. S. Obaidat, "Fire monitoring and alarm system for underground coal mines bord-and-pillar panel using wireless sensor networks," *Journal of Systems and Software*, Vol. **85**, No. 3, pp. 571–581, 2012.

[23] S. Dhurandher, M. S. Obaidat, S. Misra and S. Khairwal, "Efficient data acquisition in underwater wireless sensor ad-hoc networks," *IEEE Wireless Communications*, Vol. **16**, No. 6, pp. 70–78, Dec. 2009.

[24] Z. Li, W. Dehaene and G. Gielen, "A 3-tier UWB-based indoor localization system for ultra-low-power sensor networks," *IEEE Transactions on Wireless Communications*, Vol. **8**, No. 6, pp. 2813–2818, June 2009.

[25] R. Frank, *Understanding Smart Sensors*, 2nd ed. Norwood, MA: Artech House, 2000.

[26] M. S. Obaidat and N. Boudriga, *Security of e-Systems and Computer Networks*, Cambridge University Press, 2007.

[27] M. S. Obaidat and N. Boudriga, *Fundamentals of Performance Evaluation of Computer and Telecommunication Systems*, Wiley, 2010.

[28] I. F. Akyildiz, W. Su, Y. Sankarasubramaniam and E. Cayirci, "A survey on sensor networks," *IEEE Communications Magazine*, Vol. **40**, No. 8, pp. 102–114, Aug. 2002.

[29] P. Rentala, R. Musunuri, S. Gandham and U. Saxena, "Survey on sensor networks," in *Proceedings of International Conference on Mobile Computing and Networking*, 2001.

[30] L. Zhou and Z. J. Haas, "Securing ad hoc networks," *IEEE Network Magazine*, Vol. **13**, No. 6, pp. 24–30, Nov./Dec. 1999.

[31] S. Sundresh, K. Wooyoung and G. Agha, "SENS: a sensor environment and network simulator", in *Proceedings of 37th Annual Simulation Symposium*, pp. 221–228, Apr. 2004.

[32] W. Wang, "Traffic analysis, modeling and their applications in energy-constrained wireless sensor networks – on network optimization and anomaly detection," PhD Thesis, No. 78, Dept of Information Technology and Media, Mid Sweden University, Sundsvall, Sweden, 2010.

[33] N. Krishnamurthi and S. J. Yang, "Feasibility and performance analysis of sensor modeling in OPNET," in *Online Proceedings of OPNETWORK 2007*, Washington, DC, Aug. 2007.

[34] A. Timm-Giel, K. Murray, M. Becker, *et al.*, "Comparative simulations of WSN," *ICT-MobileSummit*, 2008.

[35] S. Leung, W. Gomez and J. J. Kim, "Zigbee mesh network simulation using Opnet and study of routing selection," ENSC 427: Communication Networks, final project, 2009.

[36] C. Marghescu, M. Pantazica, A. Brodeala and P. Svasta, "Simulation of a wireless sensor network using OPNET," in *Proceedings of the 2011 IEEE International Symposium for Design and Technology in Electronic Packaging (SIITME 2011)*, pp. 249–252, 2011.

[37] X. Chang, "Network simulations with OPNET," in *Proceedings of the 1999 Winter Simulation Conference*, pp. 307–314, 1999.

[38] H. Sundani, H. Li, V. Devabhaktuni, M. Alam and P. Bhattacharya, "A survey and comparisons on wireless sensor network simulators," *International Journal of Computer Networks (IJCN)*, Vol. **2**, No. 5, pp. 249–261, 2010.

[39] http://db.csail.mit.edu/labdata/labdata.html, 29 Feb. 2012.

[40] M. Ikeda, E. Kulla, L. Barolli and M. Takizawa, "Wireless ad-hoc networks performance evaluation using NS-2 and NS-3 network simulators," in *Proceedings of the IEEE 2011 International Conference on Complex, Intelligent, and Software Intensive Systems*, pp. 40–45, 2011.

[41] T. Yang, G. Mino, E. Spaho, *et al.*, "A simulation system for multi mobile events in wireless sensor networks," in *Proceedings of the IEEE 2011 International Conference on Advanced Information Networking and Applications*, pp. 411–418, 2011.

[42] H. K. Kalita and A. Kar, "Simulator based performance analysis of wireless sensor network – a new approach," in *First IEEE International Workshop on Wireless Communication and Networking Technologies for Rural Enrichment*, pp. 461–465, 2011.

[43] A. Al-Dhalaan and I. Lambadaris, "Semidefinite programming for wireless sensor localization with lognormal shadowing," in *Proceedings of the 2009 IEEE International Conference on Sensor Technologies and Applications*, pp. 187–193, 2009.

[44] N. Sah, N. Prakash, A. Kumar, Da Kumar and De Kumar, "Optimizing the path loss of wireless indoor propagation models using CSP algorithms," in *Second International Conference on Computer and Network Technology*, pp. 324–328, 2010.

[45] M. Awad, K. Wong and Z. Li, "An integrated overview of the open literature's empirical data on the indoor radiowave channel's delay properties," *Proceedings of the IEEE Transactions on Antennas and Propagation*, Vol. **56**, No. 5, May 2008.

[46] M. S. Obaidat, A. Alagan and I. Woungang, Eds., *Handbook of Green Information and Communications Systems*, Elsevier, 2013.

10 Security issues in wireless sensor networks

Providing security to wireless sensor networks is very challenging, as they include protection against damages, losses, attacks, and dangers. Moreover, a wireless sensor node has limited computation power, limited memory, and limited I/O resources. The classic security issues that are usually considered in wireless sensor networks are upholding the secrecy and avoiding intrusion. Securing access to wireless networks in general is a difficult task when compared to fixed/wired networks because wireless networks use wireless transmission medium. Securing access to WSNs is more challenging than for traditional wireless networks. This is mainly due to the limited resources of WSNs and to the harsh working environments of these systems in most cases.

In this chapter, we present key issues, challenges, vulnerabilities, attacks, existing solutions, and comparison of major security techniques related to WSNs.

10.1   Background

In general, WSNs are heterogeneous systems. They contain general-purpose computing elements with actuators and tiny sensors. Moreover, these computing elements have limited computational power, limited power, limited bandwidth, and limited peripherals. These aspects of WSNs make it difficult and challenging to design a secure WSN system [1–71], as secured schemes require computational power, large memory, and more power consumption, among other resources.

Moreover, providing security in WSNs is not an easy task because of the resource limitation on sensor nodes, high risk of physical attacks, density and size of networks, unknown topology prior to deployment, and also due to the nature and characteristics of wireless communication channels. The goal to provide security for WSNs is to support confidentiality and integrity for all the messages and data using means such as authentication and encryption.

When the information is passing between nodes it may be modified because of the uncontrolled wireless channels and node environments as well as because of attacks from intruders. Owing to the similarities between wireless sensor networks and wireless ad-hoc networks, the requirements of security are similar for both [1–11].

The chief security requirements that wireless sensor networks should maintain include confidentiality, availability, and integrity. These main requirements can be realized in a distributed manner by efficient key connectivity, scalability, and resilience.

The likelihood of storing the identical key or the keying material for two or more sensor nodes is the key connectivity. On the other hand, scalability is the capability to maintain a vast number of sensor nodes with proportional improvement in performance.

To provide flexibility, increase the size of the WSN, and support large number of WSNs, a key distribution technique can be employed. The restrictions of sensor nodes' computation power capabilities and their limited memory capacity make it difficult to have resilience against any sort of node capture. Hence, reducing the number of compromised links will show a greater resilience.

In general, major attacks on wireless sensor networks can be passive or active attacks, and these can be cryptographic primitive attacks, mote class, or laptop class attacks. In addition, these attacks can be on the WSN availability [1, 11].

The chief characteristics of wireless networks, including WSNs, are as follows [1].

(1) High bit error rate (BER), which is due to: (a) atmospheric noise, (b) multipath propagation, and (c) interference.
(2) Dynamic topologies – hidden terminal problems.
(3) Energy limitations.
(4) More prone to security threats than wired networks due to unguided transmission medium.

Regardless of the size of the WSN, there is always the likelihood of observing the different sensors and taking any action needed; however, in the case of large-scale WSNs, it is difficult to observe each distinct sensor. The key types of threats that WSNs are most likely to receive are from nearby networks and transportation systems, as well as attacks on physical structure.

The threats can be categorized into two main kinds: (a) at the sensor level, and (b) at the laptop level. From the viewpoint of causing harm to the WSNs, a laptop-level adversary can do a lot more harm than a sensor node since a laptop level can hold high communication abilities, has larger computation power, and also has good power supply when compared to sensor nodes.

Moreover, threats can be classified based on whether the attack is internal or external. Detecting and defending against the internal attackers are very difficult. As far as the external attacker is concerned, he cannot access cryptographic means that the inside attacker can possess [3–15].

Table 10.1 shown below summarizes the main types of threats that sensor networks may face.

There exist three main kinds of keying paradigms that can be considered when comparing different relations among WSNs' security and operational requirements. These are: (a) network keying, (b) pair-wise keying, and (c) group keying.

Network keying scheme

The network keying scheme has an advantage over the other two paradigms as it is easier to use and employs a small amount of resources. Moreover, it requires much less management effort. In general, this scheme needs only one key in the entire WSN

Table 10.1 Summary of traditional threats in wireless sensor networks

Threat type	Layer	Protection scheme to realize
Tampering	Physical	Tamper-proofing, efficient key management techniques
Jamming	Physical	Spread spectrum scheme, reduction of duty cycle
Exhausting	Data link	Rate restriction
Collision	Data link	Error control code scheme
Routing information control	Network	Authentication, and encryption
Selective forwarding	Network	Redundancy, probing
Sybil attack	Network	Authentication
Sinkhole	Network	Authentication, checking, redundancy
Hello flood	Network	Two-way authentication, two-way authentication, and three-way handshake
Wormhole	Network	Elastic routing, and observing
Flooding	Transport	Restricting number of connections
Clone/replica attack	Application	Distinct pair-wise keys

through which it efficiently manages and handles the work. The major disadvantage of this approach is that it is not robust.

Pair-wise keying scheme

This scheme of keying uses $(N-1)$ keys in each node, where N represents the number of nodes in the WSN. Such a keying scheme will not compromise any sensor node. Hence, it is robust; however, it does not have good scalability characteristics.

Group keying model

This scheme contains the features of both pair-wise keying and network keying schemes. In this technique, as in the network keying scheme, group keying uses a single shared key in order to carry out communications. A cluster is created by using various types of nodes. In this context, the connection to pair-wise keying scheme comes from employing various keys in between the members of the communication session. Accessibility can be attained since there is some data aggregation. The technique has shown some robustness; however, there is a need for setup, which takes time, and also it is an application-dependent process in creating the groups [1–71].

The answer for a security system in this regard will rely on the strength of the network without controlled environment, and on the effectiveness of the key distribution system used [4–25].

If the capacity of storage means, ability of data processing, and endurance to temperature are all taken into account, then base stations are crucial when dealing with transmission rate. Base stations (BSs) make an important frame structure for WSNs, where a sensor node has the capability to be deployed to either a sole or several hop neighborhoods. Hence, the BS acts like a main distribution center. With the use of the state-of-the-art antenna technologies such as directional antennas and multiple input multiple output (MIMO) systems, there is a higher chance for heterogeneity in sensor nodes. Base stations in WSNs act as distribution centers, and inside any cluster they are able to share the pair-wise master key with any node; this will be used to form other keys. Hence, for a communication session between a base station and a sensor node, a pair-wise key is needed and this is realized by having the base stations share a pair-wise master key with the sensor node.

10.1.1 Software updating in WSNs

Software updating is considered a key issue that needs to be looked at carefully while setting up communication networks including WSNs. The reasons that WSNs' software should be updated include technology refreshment and supplement, and some other minor releases.

In general, a lot of effort is required after preservation of the software for the embedded WSNs. Since several releases of the software used in WSNs are expected, careful updating is needed. Insertion of software is believed to play a key role in this context because of the progress of new algorithms and protocols realized as well as because of the progress in research in WSNs [8–20].

Carrying out software updates can be a hard task when accessing the sensor nodes at the same time. The main issues that may be faced when doing remote updating include: (a) interference, (b) cost, and (c) the consequences of having a fault occur during the process of upgrading [1–71].

The majority of attacks on the physical layer of WSNs are jamming, interference, Sybil, and tampering. Channel hopping and black listing can be used to protect against jamming and interference, while physical protection of devices can be used to protect against Sybil. In order to protect against tampering, physical protection and key changing can be used [1–48].

As far as attacks on the data link layer, we can identify collisions, unfairness, and attacks of exhaustion. Collision is also considered as a type of jamming at the link layer. The frame can become corrupted by the attacker, and ends up in having a mismatch in the checksum. Unfairness is a weak type of denial of service (DoS) that occurs due to abuse of the schemes of priority in MAC sub-layer. This kind of attack can lead to the loss of the deadlines in real-time, which leads to QoS degradation. Exhaustion of the supply of the battery arises with the realization in the layer of a local link attempting retransmission that may recur after the collisions [49]. To deal with collision, cyclic redundancy code (CRC) can be employed. Protection of the WSN ID and other information needed to join services is used to deal with exhaustion. In order to handle spoofing at this layer, different routes can be used for re-sending the messages. Changing the keys on regular basis is used to deal with Sybil. In order to tackle eavesdropping, special keys can be employed to protect the protocol data units (PDUs) of the data link layer.

Attacks in the network layer include wormhole, traffic analysis, eavesdropping, selective forwarding, denial of service, and Sybil attacks [16–22, 34].

In wormhole attacks, tunnels of the attacker's messages are received within a part of a network over a link to another segment in the network in which replaying of messages will be performed with a small delay. The easiest sort of this attack is to have a node that is malicious by dispatching data between two sensor nodes that are genuine. These can persuade sensor nodes far away that those were neighbors, which may direct to the fast end of energy resources. If this kind of attack is combined with the forwarding selective and the Sybil attack it becomes hard to detect. To deal with this attack, physical monitoring of devices as well as regular monitoring of networks using source routing can be employed as a countermeasure. A traffic analysis scheme is considered a sort of passive attack, where the intruder tries to listen to the channel in order to guess communication patterns. The attacker could discover the position and identity of communicating hosts and could notice the frequency and length of encrypted messages being exchanged. Such information could be used by the intruder/attacker as, when analyzed, may disclose information that helps to guess the nature of information being exchanged. In order to deal with this attack, dummy packets may be sent in light load hours as well as by regular monitoring of the WSN [1–48].

Eavesdropping is also considered a passive attack. It is an attack on the confidentiality of the WSN. Here, the attacker tries to access information that is being transmitted. To tackle this, special session keys that can protect network PDU are often employed.

In selective forwarding, the attackers include themselves in the path of the data flow of interest. Then the intruder can have the chance not to forward some specific selected packets and drop them from the network [9–13, 22, 49]. In general, these kinds of attacks are harder to detect when comparing them with traditional black-hole attacks. As a countermeasure against these attacks, regular network monitoring using source routing is employed.

The denial-of-service (DoS) attack attempts to prevent genuine users from using the service. It has several variants, including flooding the networks in order to prevent legitimate network traffic, disrupting connections between two sensor nodes, and preventing some nodes from using some services, among others. To address this attack, it is recommended to protect the WSN-specific data such as the ID as well as using physical and inspection means.

In the Sybil attack, which is a distinct attack in WSNs, a malicious node acts as if it were a greater number of nodes, by mimicking other nodes or just declaring incorrect identities. In the cruelest situations, an intruder can produce a random number of extra node identities by employing only one physical gadget [50]. Sybil attacks perplex the protocols during the routing process. To protect from this attack, devices are relocated and session keys are changed.

As for the attacks in the transport layer, we can identify desynchronization and flooding. The chief aim of the flooding attack is to exhaust the WSN memory resources. As in the TCP SYN attacks, the intruder sends the request several times in order to establish the connection, and force the attacked systems to allot memory space in order to maintain a state for each connection. In the desynchronization attack, the messages are faked by an intruder between the communicating nodes [15]. The intruder may avoid those communicating nodes from exchanging messages as retransmission would be requested constantly, which may lead to wasting power resources.

Examples of some of the general models of threats to WSNs include the following. The intruder may get access only to several nodes that are compromised. Also, attackers may get access to devices that are more powerful, such as laptops. The latter are called "attackers of laptops." This type of attacker has CPUs that have powerful battery and radio transmitters, and antennas that are sensitive. Here, you may get some nodes that are capable of jamming some links of the radio, and the laptop has the ability of jamming the whole WSN.

Attacks on a WSN can be from outside or inside the network system. In the former case, the attackers do not have special access to the WSN. In the latter case, attackers are considered participants that have authorization to use the network system. These sorts of attacks may be from the WSN nodes that are compromised or from laptops that use stolen data (keys of cryptography and code) from nodes that are valid.

10.2   Limitations in WSNs

In general, a WSN is composed of a large number of sensor nodes that are essentially resource-constrained. In addition, these nodes have small physical size, restricted processing power, limited energy, small memory, and limited I/O capabilities and communication bandwidth. Because of these limitations, it is not easy directly to utilize traditional security schemes used in typical wired and wireless networks in WSN systems. Thus, it is important to keep in mind these limitations when devising a security scheme for WSNs. The chief limitations of WSNs are reviewed briefly below [1–71].

Memory constraints
A sensor node in a WSN is a very small device with limited memory and storage space. It can be composed of a RAM and a flash memory. Usually, there is not enough space to execute large programs after loading the operating system (OS) and application program.

Power constraints
Power is the most critical constraint that faces WSNs. The power consumed in wireless sensor nodes can be of three types: (a) power consumed by nodes to perform the needed computation, (b) power consumed by the sensing transducers, and (c) power consumed due to inter-processor communication. Keep in mind that if a high level of security is needed in WSNs, then there will be high power consumption by the nodes [51–71].

Unpredictable communication
The unstable communication link in WSNs is considered a major challenge to WSN security. Typically, routing in WSNs uses a connectionless service, hence the process is

essentially undependable. Moreover, since these are wireless networks, the received signal suffers from fading, reflection, and scattering, which may lead to high bit error rate and high loss rate [1].

High delay in communication

Owing to network congestion in intermediary nodes in WSNs, these networks suffer from great latency in packet transmission, which complicates the synchronization process. In some cases, it may be critical in the context of security since some security schemes can depend on crucial event reports and cryptographic key dissemination [52].

Remote sites and unattended setup of WSNs

In many cases, WSNs are deployed in remote or harsh locations and environments. This means that the probability that a sensor node confronts physical attacks is very high. Furthermore, in remote environments, it is hard to detect physical tampering in WSNs.

10.3   Security requirements in WSNs

The security services in WSNs are similar to these of any wireless networks with a few differences. The most important security requirements in WSNs are summarized below.

Data confidentiality

This service means the protection of data being carried by the network from attacks. Another aspect of confidentiality is the protection of traffic from a hacker who attempts to eavesdrop and analyze it. In other words, there must be some measures that deny the hackers from observing the frequency and length of use, as well as content [11, 8–15].

This requirement entails that no information in the WSN should be understood except by the intended receiver. In this context, secrecy must deal with the following requirements [52–54]: (a) the key distribution scheme must be very strong, (b) a node must not permit its readings to be read by its nearby nodes without its approval, and (c) public keys of the sensor nodes must be encrypted in critical situations in order to safeguard against traffic analysis attacks.

Confidentiality is crucial to the proper operation of WSNs. Upholding confidentiality for WSNs is related to the following [11–54].

(1)   In order to safeguard WSNs from attacks by traffic analysis, strong and robust encryption schemes for public sensor information should be realized.

(2)   Key distribution is a crucial process that is utilized by many applications in relaying highly sensitive data. Hence, designing a secured channel is very important for proper and reliable WSN systems.

(3)   Sensor measurements and readings must not be disclosed to nearby networks and systems.

(4)   Data that are available in sensor nodes should be treated as confidential and sensitive, especially if they are related to military applications.

Data integrity

This ensures that the messages are sent properly without duplication, modification, reordering, or replay. This requirement ensures that the message would not experience any change when traveling from the source to the destination.

Data inside a packet may be modified or some bits of data may be inserted by malicious nodes. Hence, this altered packet may be transmitted to the recipient who originally produced the data. From time to time, because of some harsh communication environment, damage or loss of the data may occur. So establishing a data integrity procedure can help in making sure that data are not altered when they are in transit.

Self-organization

Wireless sensor networks are one type of wireless ad-hoc network. Here, each sensor node is autonomous as well as flexible. Based on various settings, the nodes should be self-healing as well as self-organizing [50–71]. Such an attribute presents a difficulty to the security of the WSN. It is not easy and not practical to deploy any pre-installed shared key schemes between the nodes and the base station of WSNs due to their dynamic behavior [50–56]. There are several key pre-distribution techniques that have been devised in the literature, which address symmetric encryption such as the ones reported in [12–20, 48–71]. Nevertheless, we need a capable methodology to distribute keys if public key cryptography is to be employed.

Data newness

This means that the data are updated and guarantees that no attacker can replay old data. In shared key environments, this is a critical requirement. To alleviate this issue, a nonce or time-specific counter can be augmented to each packet in order to find out if the data are updated or not. Maintaining data confidentiality and data integration is important; however, this is not enough to ensure the security of WSNs as they have to be checked for data freshness. Guaranteeing data freshness means that only updated data are kept. If there is any existence of shared keys, then they need to be updated over time [57].

Authentication

In general, the authentication service is used to ensure that the message is from an authentic source. In other words, it assures that each communicating party is the entity that it claims to be. Also, it ensures that the connection is not interfered with in a way that a third party impersonates one of the authorized parties [11]. Hence, it is vital for a receiver to have a scheme to make sure that the delivered packets are from the intended sending node. Data authentication between two communicating nodes can be obtained by using a message authentication code (MAC) that is calculated by using the shared symmetric secret key between these nodes.

Time synchronization

Synchronization is a process needed for the overwhelming majority of WSNs applications that necessitate critical timing. Hence, the associated security scheme must also be well-timed/synchronized. The sensor node radios may be kept offline in order to save the life of battery. The authors in [58] devised several schemes for secured synchronization.

Secure localization

Localizing sensor nodes in WSNs in a secured manner is essential in many environments and applications. If the information on location is not safeguarded properly, an intruder may manipulate and deliver fabricated location information by broadcasting inaccurate signal strength, or replaying messages. There are several techniques that can be used to accurately find the location of a node. Capkun *et al.* [56] have described a scheme known as verifiable multilateration (VM), where the position of a device is precisely estimated from a sequence of identified reference points. They employed authenticated ranging and distance bounding to guarantee the precise position of a node. Owing to the utilization of distance bounding, an affecting node can only intensify its requested distance from a reference location. In order to guarantee the position steadiness, the intruder needs to attest that his location from another location reference point is smaller. Since this is very difficult to achieve, then it becomes feasible to discover the intruder [60].

Availability

This term is also known by its opposite, which is denial of service (DoS). An authorized individual should not be prevented or denied access to objects to which he has legitimate right to access. This requirement guarantees that a WSN should always be accessible to its legitimate users even under attack situations. Various techniques have been devised in the literature to address this issue, including the use of extra communication between nodes, and use of a principal access control system to guarantee effective provision of all messages to the target destination. This access applies to both service and data. Some attacks may result in loss or reduction of availability of the system, or denial of service (DoS). Automated schemes can resolve some of these issues while others require some type of physical procedures [11, 57].

To abridge the algorithm some of the schemes may oblige stringent restrictions on accessing data or offer some central point type of improper structures. Such a situation may lessen the sensor node accessibility. In this case, extra energy will be spent due to the required additional manipulations and calculations. Hence, data may not be available if there is no extra power. Moreover, employing a central point of control leads to a single point of failure, which impacts the availability of the WSN.

Denial-of-sleep attacks

This attack is particular to wireless sensor networks because of their limited computational and power resources. Owing to the fact that it is hard to reload the energy, it is essential to make sure that the nodes do not operate and consume power and efforts on unwanted or bad packets [54, 58]. These attacks include continuously transmitting request-to-send (RTS) packets, which avert the nodes from switching to the low power "sleep states," wearing the computational power by having sensor nodes performing unwanted expensive computations and regularly transmitting forged or repetitive packets.

10.4   Vulnerabilities and attacks specific to wireless sensor networks (WSNs)

Wireless sensor networks are threatened by several types of attacks. These attacks can take different forms, such as denial-of-service attacks, attacks against privacy, traffic analysis, physical attacks, and many more. WSNs are susceptible to security attacks due to the nature of their broadcast type of transmission. They are vulnerable to various types of attacks including: [62] (a) attacks on network availability, (b) attacks on the secrecy and authentication, and (c) stealthy attacks against service integrity.

The attack on availability is basically a denial-of-service (DoS) attack. In the attacks on privacy and authentication, the attacker attempts to eavesdrop, replay packets, or modify/spoof packets. In stealthy attacks, the aim of the intruder is to cause the network to accept an incorrect data value. For instance, the attacker compromises a sensor node and inserts a false value via that attacked node.

Proper operation of wireless sensor networks requires data integrity, authentication and confidentiality, as well as trust. Trust is vital as the sink nodes of WSNs are typically launched in harsh or remote sites that are not usually secured or friendly.

Intruders can easily distort or tamper with the sensor nodes. Among the main attacks in this context are [11, 63–64]: (a) routing attacks, (b) node replication attacks, (c) attacks on information in transit, and (d) denial-of-service attacks.

Routing attacks
There are some types of network attacks that are aimed at the routing schemes of WSNs. Typically these are internal attacks from within the network. Among these we can identify the ones listed below.

- **Hello flood attack**   Here, a high-power transmitter is employed by the attackers by sending numerous hello packets. Basically, they send by the publicizing links/routes of the gateway. The intruder may be seen as the neighboring node that has the shortest path to the gateway. As the route vanishes, the sensor nodes will become perplexed.
- **Spoofing**   This is also referred to as message altering. Here, the message is altered, which leads to delivering to the recipient inaccurate information that leads to confusion. Because of these modified messages, wrong routing actions will be taken, which can lead the data to travel long distances. This results in increases in the overall latency of the network and waste in the power of the nodes of the WSN. One way to tackle these types of attacks is to employ integrity checks like message authentication code (MAC) [11, 57, 65].
- **Replay attack**   In this kind of attack, the same messages are repeated again and again over the WSN, and this will degrade the throughput/bandwidth of the network. This is a major limitation in WSNs. To address this attack, encryption cannot be effective. Hence, a nonce or timestamp is employed. The timestamps are often preferred in WSNs environments as they need a lower number of messages compared to the use of nonce.

- **Sybil attack**   This is a serious and persistent problem in many systems, especially in WSNs. Trust-based schemes are often targeted by such attacks. This kind
of attack relies on the capability of falsifying the nodes or impersonating nodes'
identifications in order to generate a great number of analogous IDs, which may
end up influencing the trust schemes. Using identical IDs can produce a very large
number of fake sent messages. Hence, the hacker tries to reduce the trust of nodes
that are actually harmless [66–71]. One effective way to deal with this attack is to
use a key registration system.

- **Wormhole attack**   This kind of attack is typically executed by two attackers who
can correspond in any way of communication other than the customary method of
communication such as any two computers that are deployed at two far ends of a
WSN with these nodes communicating at different data rates. The intruders may
publicize a new route that may appear shorter than other routes, which leads all the
nodes in the network to employ this fake route publicized by the intruders as their
transitional hop.

- **Selective forwarding**   This type of attack may be realized if the attacking node is
positioned in the routing track of any other node by itself. The intruder may select
the packets that have to be thrown. A sinkhole attack is the basic selective
forwarding attack. In this latter attack, all the packets that arrive merely go to
the "sinkhole" [64].

- **Acknowledgement spoofing**   This is considered a major issue in routing protocols. Here, the invader attacks the acknowledgement packets. The aim of this
attack is to induce the nodes in the WSN by indicating that a dead node is alive,
which will lead to confusion and misleading in the routing process.

- **Compromised nodes attacks**   This is unique for WSNs. If the WSN is launched
in the battlefield to detect the enemy's movements, then the intruders may reach
the sensor nodes physically and compromise some of these nodes, if not all of
them. If any of these sensor nodes becomes compromised, then an intruder can get
access to confidential information, including security keys.

   Trust protocols can help in distinguishing compromised nodes from uncompro-
mised nodes. These schemes observe the neighboring nodes for any improper
behavior that may occur [2–20, 57–59].

- **Node replication attacks**   In this kind of attack, the attacker intrudes on a node
in the WSN that is at present replicated from the formerly standing nodes. This is
simply sustained by the hardware of the nodes. The latest node that is injected into
the WSN may behave in a similar manner to the former one that has been cloned.
The cloned node has additional attributes such as sending data that look sensitive
to the hacker. In the worst-case scenario, which is difficult, if the hacker was also
able to clone the base station, then this will have serious consequences that makes
this kind of attack much more severe.

- **Attacks on information in transit**   This kind of attack occurs while the data are
passing in transit from one node to another. It is considered one of the further-
most common attacks on wireless networks. The data passing in the WSN are
typically susceptible to injection, modification, and eavesdropping attacks.

These attacks can be dealt with by employing effective integrity, privacy and confidentiality, replay, encryption, and authentication protocols on the WSN. Traffic analysis is considered a serious challenge in WSNs [65, 66].

- **Denial-of-service (DoS) attack** This is an attack on the availability of the WSN system. Here, the attacker's aim is to make the WSN system resources unattainable for the legitimate user. A denial of service prevents or reduces the normal use, operation, or management of communication facilities. Such attacks attempt to block or disable access to resources at the victim system. These resources can be network bandwidth, computing power, buffers, or system data structures [11, 34, 54, 58]. In Section 10.8, we will elaborate on this type of attack.

10.5   Physical attacks on WSNs

Wireless sensor networks are subject to all kinds of physical attacks [1, 11]. Because of the nature of WSNs, especially, limited computational capabilities and limited energy resources, unreliable transmission media, harsh deployment environments, the majority of conventional computer network security schemes are inadequate for WSNs [1–71]. Thus, security schemes including physical security for WSNs should be designed carefully in order to address these constraints.

Note that WSNs have some weakness when compared to other wireless networks. They also have some unique characteristics such as [1–71]: (a) the sensor nodes can be constant or mobile, (b) harsh or hostile environment, (c) unreliable transmission medium, (d) dynamic/unpredictable WSNs' topology, (e) they are ad-hoc-based wireless networks, (f) non-central management, autonomously and infrastructure-less, (g) high density, and (h) sensor nodes have limited resources.

Wireless sensor networks are susceptible to all types of attacks mainly as a result of the following: (a) restricted resources and abilities, (b) inadequate traditional security schemes for them due to WSNs' nature and limited capabilities, (c) unsystematic placement, (d) remote locations and unattended operation, (e) high possibilities for inside attack, (f) usually deployed in an ad-hoc manner or randomly, (g) challenging and unreliable communication channel, (h) usually no central supervisory point, and (i) greater number of potential attacks and higher vulnerability [1–71].

Physical attacks on WSNs can be classified based on damage or access level. In this regard, we can identify the following main classes.

Active attacks
These include inserting faulty files into the WSN system, masquerading, packet alteration, illegal access, monitoring, eavesdropping and modification of resources and data flows, overloading the network, and creating holes in the security schemes [20]. These attacks cause the following effects: (a) disruption of the functionality of the WSN, (b) isolating/cutting off specific sensor nodes from the rest of the WSN nodes, (c) modification of data, (d) denial of service, (e) degradation of the overall performance, and (f) destruction of the sensor nodes [11–71].

Passive attacker

These are inherently eavesdropping or snooping on the communication transmission of wireless systems. The attacker tries to access information that is being transmitted. There are two subclasses [11]: (a) release of message contents, and (b) traffic analysis. In the first type, the attacker reaches the e-mail messages or a file being transferred. In traffic analysis, the intruder could learn the location and identity of communicating hosts and could monitor the frequency and length of encrypted messages being communicated. Such information could be valuable to the attacker as, when analyzed further, can disclose valuable information in predicting the nature of information being communicated. In general, passive attacks are difficult to detect; however, there are ways that can be employed to avoid them. On the other hand, it is not easy to prevent active attacks [11, 26–30].

In general, passive intruders/attackers may perform the following actions [11]: (a) the attacker behaves like a regular sensor node in the WSN and can collect data from the WSN, (b) observing and eavesdropping [1–71] on the communication links by the intruder, and (c) invasion of privacy.

The aims and effects of this kind of intruders are [11]: (a) eavesdropping, gathering, and stealing information, (b) compromised privacy and confidentiality requirements, (c) degrading the WSN functionality, and (d) partitioning the WSN into non-cooperating entities.

Attacks based on attacker location

The intruder may be located inside or outside the WSN. If the attacker is within the network's sensing range, then he is called an insider/internal attacker, while if he is located outside the sensing range of the network, then he is called outsider/external attacker. Hence, WSNs' physical layer attacks can be categorized into two classes based on the attacker's location [34, 57, 61].

External attacks

These are carried out by outside intruders. The main consequences of these attacks include [34, 57, 61]: (a) jamming the entire communication of the WSN, (b) consumption of the resources of the WSN, especially its battery, memory, and computational power, and (c) initiating DoS attacks, which are serious attacks with severe consequences.

Internal attacks

These are carried out by inside intruders. These are critical threats to WSNs as they come from inside the system. They come from inside the WSN due to: (a) an authorized node in the WSN that becomes compromised, (b) an execution or use of a malicious code/data, and (c) an authenticated node that compromises a number of the WSN's nodes.

The major consequences of these attacks include [34, 57, 61]: (a) entry to cryptography keys, (b) disclosing of classified keys, (c) a high risk to the operational efficiency of the entire WSN, (d) degradation of the performance, and (f) disruption of the operation [1–69].

In addition, attacks on WSNs can be categorized based on type of operation. In this context, the physical layer attacks in WSNs may be classified into three types, based on their main functionality: (a) secrecy, (b) availability, and (c) stealthy.

Moreover, physical attacks on WSNs can be classified into different categories. The two main categories that are used to classify physical attacks are: (a) degree of control over the sensor node that the attacker gets, and (b) time duration during which regular operation of a node is disturbed. As an example, we briefly describe in-the-field attacks and countermeasures.

Many applications store all the important data on the external EEPROM such as flash memory. One simple form of attack in such a situation is an eavesdropping attack, which is a sort of passive attack. The attack can be performed on the conductor wires that connect the external memory to the CPU. The attacker can read all the data with the aid of a microcontroller logic analyzer. Here, an intruder reading the data would not change the normal operation of the node or WSN. Hence, this attack cannot be found for a long period of time. If the attack is serious then it can break the connection and make the microcontroller and flash connect with the intruder. Then, the attacker counterfeits the external memory.

10.6   Recent security issues in WSNs

There are three main security issues that lead to attacks in WSNs. These are reviewed below [34, 57, 61].

(1)   **Context and design implication**   The environment that correlates to the set of features describing the WSN must be taken into consideration in the security design of the system. We have here two main issues [11, 34, 57, 61]: (a) hacker motives and (b) vulnerabilities.

These matters influence the design of WSN systems and, hence, may increase the cost and energy efficiency. Of course, they can influence performance. There should be a compromise between security and performance from the software, protocols, and service aspects [11–39]. In some situations, the protocol may expose the position of the BS which increases the possibility of attacks. Nevertheless, these may be alleviated by using strong encryption schemes at the expense of computational power and resources [2–13].

(2)   **Middleware applications**   Provision for WSN application development is augmented via making custom-tailored cases for object platforms. The idea is identified by using the extensible markup language (XML), with the XML document being amalgamated into each data type delineation. The structure is realized as a plug-in for a basis where a user may employ a graphical user interface to input parameters. The application designer can realize his system by using security mechanisms and key material that are encompassed in the developed middleware [2–24].

(3)   **Integrity, authenticity, and confidentiality**   The major aims of the security of any communications networks including WSNs are: integrity, authenticity, and

confidentiality. Confidentiality of service means the protection of data being carried by the network from passive attacks. The broadcast service should protect data sent by users. Other forms of this service include the protection of a single message or a specific field of a message. Another aspect of confidentiality is the protection of traffic from hackers who attempt to analyze it. In other words, there must be some measures that deny the hackers from observing the frequency and length of use, as well as other traffic characteristics in the network. Integrity also refers to examining if data were modified between source and destination. Authenticity means data coming from a permitted user along with information ratify the ID of the sender. In general, the authentication service must ensure that the message is from an authentic source. In other words, it guarantees that each communicating party is the entity that it claims to be. Also, this service must ensure that the connection is not interfered with in a way that a third party impersonates one of the approved users [11, 16–30]. Integrity means that operations such as substitution, insertion, or deletion of data can be performed only by authorized users using authorized ways. Three aspects of integrity are commonly recognized: authorized actions, protection of resources, and error detection and correction [11].

As an example, let us look at the self-originating wireless sensor networks (SOWSN) [2–45]. The security needs include: prohibiting replay of sent alerts, prohibiting DoS attacks carried out by malicious nodes, prohibiting masquerade attacks, and ensuring integrity and confidentiality of warnings transmitted by nodes. Hence, the chief duties of the base station in this context are to [1–11]: (a) verify the integrity of the messages gathered depending on the signature delivered and (b) validate or authenticate requests for route creation before the creation of any path is approved.

10.7   Secure protocols for wireless sensor networks

This section sheds some light on some examples of secure protocols for WSNs. Basically, we will review popular schemes such as SPINS, TinySec and LEAP.

10.7.1   SPINS

The secure protocol in wireless sensor networks (SPIN) is a group of security protocols that handle the data integrity and confidentiality issues in WSNs. This group of security protocols was proposed in [51].

SPINS consists of two major modules SNEP and μTESLA [51, 70]. The first offers different security attributes to the WSN. For pair-wise links, it offers data confidentiality, privacy, and integrity, as well as newness. Newness (or freshness) is used in WSNs to mean that the same packets must not be permitted to recur, which can form mix-ups and lead to waste of the power of the battery. Weak freshness implies that there is no delay assurance despite the fact that recurrence of packets is not permitted. SNEP employs a pair wise link key, k, for each pair.

The encryption of data in the cipher block chaining method guarantees that the succeeding ciphered blocks do not generate identical sequences. The initialization vector is not transmitted with data; it is obtained from a common counter. This counter ensures that an old packet is not replicated, so listing is retained.

As for authentication, this may be performed by using a function that considers two arguments [54]: (a) the pair-wise key, and (b) the concatenation of the ciphered message and the counter. Since this information is already obtainable by the recipient and the sender, then this task helps as a useful tool for verification.

The outcome generated is basically an 8 byte digital signature that is verified by the recipient.

SNEP is an effective security scheme; nevertheless, it only can be employed for pair-wise connections and not broadcast services because, if the key is permitted to be distributed between various users the opponent may endanger a node, which can result in compromising the entire WSN.

Authentication in the broadcast method is offered by µTESLA [51, 70]. The latter employs a one-way function and in the broadcast mode of operation it offers authentication [70]. Basically, it produces a sequence of keys to be employed in the message authentication code (MAC) scheme.

10.7.2   TinySec

This is a data link layer security scheme that comes integrated with the formal version of TinyOS. TinySec is basically a light-weight and a common package that may be incorporated with the network. Tiny security offers various security features including [70–71]: (a) privacy of messages using cryptography, (b) authentication of communications and reliability through the MAC scheme, and (c) shield from replay attacks using the initialization vector.

Moreover, TinySec maintains two forms of security: (a) the only-authentication approach, and (b) the authentication and encryption approach. In the first scheme, TinySec verifies the whole packet; however, encryption is not offered for the payload portion of the packet. In the second scheme, the data payload is encrypted and the whole packet is authenticated.

10.7.3   LEAP

The localized encryption and authentication protocol (LEAP) is a key management protocol that is developed to hold in network management by the nodes as well as limiting the security effect of a compromised node on the WSN neighborhood. As an alternative to permitting only one key, this scheme permits a node in the WSN to require four key types [70–71].

- A collection of keys: This is usually shared by all nodes in the WSN and utilized for communicating a non-confidential broadcast of knowledge.
- Arranged in clusters: This is employed for connecting with a set of nodes in a cluster.
- Arranged in pairs: Here, the key is employed among nodes that create a pair-wise link.
- Single key: This is usually shared with the BS.

10.8   Denial of service (DoS) in WSNs and related defenses

Although we briefly discussed these kinds of attacks earlier, we elaborate more on this subject in this section because of their importance and serious consequences on WSNs.

A denial-of-service attack exhausts the network resources available to the victim's node by transmitting needless packets and hence denies the legal users from getting system services and resources that they are entitled to. There are various kinds of DoS attacks that can be carried out at diverse levels within a WSN [11, 34, 64]: (a) at the physical layer the DoS attacks could be jamming and tampering, (b) at the data link control layer, this can be collision, exhaustion, and unfairness, (c) at the network layer, it can be neglect and greed, homing, misdirection, and black holes, and (d) at the transport layer this attack can be carried out by mischievous flooding and resynchronization.

By capturing a node we may know its information including disclosure of cryptographic keys and thus compromise the whole sensor network.

Most of the attacks would result in a kind of denial of the service; however, here we address the kinds of attacks that waste the resources and interrupt the service in the WSN in a way that would far exceed the efforts of the attacker [23, 34, 40, 42, 57, 61].

The significance of a DoS attack is very well anticipated when the wireless technology is involved in more crucial applications, where precision plays a vital role and the entire application relies on the proper timing of the messages [34, 57, 61]. One good example is the fire alarm, which explains the seriousness of the denial-of-service attack. In such a scenario, if a building with many residents in it is on fire, the denial-of-service attack may postpone the sending of an alarm to the fire department, which will reduce the likelihood of saving lives of residents of the building and of reducing the number of injuries and damages.

The DoS attacks are more serious than the other WSNs attacks. There are several kinds of DoS attacks. Jamming is considered a typical type of DoS on WSNs. Here, the attacking node diverts the radio frequency of other nodes. This will postpone the information to be delivered between the sensor nodes.

Jamming is considered the major physical layer attack in WSNs. Usually, spread spectrum technology is employed to guard against jamming in WSNs. Regrettably, WSN nodes have limited energy supply and computational resources. This means that sensor nodes cannot employ spread spectrum technology to combat jamming. If a sensor node recognizes a jamming attack, a logical shield is to set sensor nodes into the sleep mode and have them awakened cyclically to check the channel for sustained jamming [13, 34, 57]. While this cannot avoid a DoS attack, it may considerably enhance the lifespan of the nodes by decreasing the amount of power consumption.

Xu *et al*. [42] devised a scheme to identify jamming attacks in WSNs. They categorized them into the following classes: constant, intermittent, deceptive, random, or reactive.

In constant jamming, the transmission between the pair of nodes is entirely jammed and no message is sent/received between them. Here, jamming is similar to a wedge and the data flow from one side cannot pass to the other end. This may inhibit the entire WSN from doing its normal work. Hence, a constant jamming attack corrupts packets as they are sent between WSN nodes. Nevertheless, it necessitates a large amount of power and hence may not be achievable if the invader is under the same power restrictions as the intended WSN.

In intermittent jamming, the WSN is jammed sporadically, not continuously. Therefore, the messages are transmitted for a specific duration and are then obstructed. So, the entire function of the WSN is disrupted, which will lead to unexpected results [13, 34].

In the deceptive jamming case, the jammer transmits a continuous flow of bytes into the WSN in order to create the illusion of authentic traffic. For instance, in the TinyOS, if the sensor node obtains a constant flow stream of preamble bytes, all nodes within the transmission limit will stay in the receive mode only.

In random jamming, the jammer swings between napping and jamming. During the jamming phase, the attacker can act as a constant jammer or a deceptive jammer. The key difference between this mode and the modes described above is that this scheme attempts to take power saving into account [13, 42].

The reactive jammer remains silent when the channel is not active. However, it begins to send a signal immediately after it detects traffic activity. Consequently, a reactive jammer aims at the delivery of a message. In this mode, the reactive jammer does not essentially save power as the jammer's radio should be switched on all the time so as to detect the channel. The key benefit of the reactive jammer is that it is not easy to discover. It is worth noting that detecting reactive jamming is not easy since it may behave and look like customary packet collisions [42].

The schemes that are often used to detect jamming occurrences include [42]: (a) statistically examining the received signal strength indicator (RSSI) values, (b) packet delivery ratio (PDR), and (c) mean time needed to detect an idle channel.

In order to have accurate detection outcomes of jamming, it is recommended to use more than one of the above schemes. Nevertheless, algorithms that integrate these schemes can consistently detect the four jamming types. For example, there is a scheme that initially detects the weak link utility using packet delivery rate (PDR) investigation, then employs received signal strength indicator (RSSI) analysis to validate the decision made by the PDR scheme. Of course, this will provide a reliability test to find out whether jamming is triggering the weak WSN performance.

A different approach to protect from a jamming attack is to let sensor nodes collectively detect the jammed zone and, based on this, re-pass traffic around it. In the multi-hop WSNs environments, this approach is considered unnecessary in the case of constant jamming as it is expected that the routing scheme will certainly avoid jammed regions.

As for the intermittent jamming scenario, paths that go via jammed segments of the WSN are typically erratic. Thus, WSN routing schemes like TinyOS destination-sequenced distance-vector routing, which relates a connection quality-of-service metric with every path in order to create paths employing bidirectional connections with good quality, would pass around these segments of the WSN [13, 42].

The data link-layer risks contain collisions, reply, and interrogation. The first attack, collision, is the same as the reactive jamming attack we discussed earlier. Error control codes may be employed to alleviate the effect of collisions. Nevertheless, using error control codes means adding transmission overhead, and this requires extra power.

In the interrogation intrusion, the attacker exploits the two-way request-to-send/clear-to-send (RTS/CTS) mechanisms that are employed to alleviate the hidden-terminal or hidden-node challenge in wireless networks including WSNs, and that could degrade the overall throughput of the network and consume the power of the battery of the nodes [1].

The RTS/CTS messages sent by the attacker can drain the sensor node's computational and energy resources. Although efficient authentication schemes at the link layer can help alleviate such attacks, an attacked sensor node getting the deceiving RTS messages can still drain the node's power and other resources such as bandwidth, I/O, and computational resources [1].

Denial of sleep is another attack on WSNs that basically denies the radio from switching to the sleep mode of operation. It affects the power resource of the node as well as its other resources. An intruder may perform a denial-of-sleep attack over a modest jamming-based attack in order to reduce the period's attack. It is worth noting that a denial-of-sleep attack can drain the node's battery much quicker than a jamming attack [30, 34, 42, 61].

The overwhelming majority of denial-of-sleep attacks do not necessitate a persistent signal, which makes them hard to detect and classify as malign as well as to locate the illegitimate and hostile node. The data link layer in WSNs determines at what time the transmitter must send frames, attend to the channel in order to get information, and nap to save power. All WSN MAC protocols are devised to be energy aware to save power by employing various techniques.

In order to start a denial-of-sleep attack, the attacker usually finds out the specific MAC layer protocol used in the WSN by examining and investigating the WSN traffic carefully. The information learned from the traffic is sufficient to launch a serious denial-of-sleep attack [42].

Raymond *et al.* [58] have modeled a variety of attacks to show the vulnerability of WSN MAC protocols. They found that if an intruder is familiar with the protocol, then he can launch a denial-of-sleep attack without digging into the data link-layer encryption. If the intruder succeeds in infiltrating into the encryption, then further active attacks are feasible, which can lead to shortening the WSN's lifespan. They also devised a structure for alleviating these denial-of-sleep risks, which includes robust data link-layer authentication mechanism, jamming detection and mitigation, and anti-replay guard, among others.

As for the network layer, routing-disruption attacks may head to DoS attacks in the case of multi-hop WSNs. Major attack types on the routing protocols comprise modification of traffic, spoofing, and replaying. Of course, we can avoid these attacks by using authentication and anti-replay techniques. A mischievous sensor node that undermines the WSN's routing scheme may launch a DoS attack by letting itself be a member of several links and then sinking all packets. Moreover, it may choose to forward packets in order to minimize the possibility of uncovering. One way to fight black holes and selective forwarding is to employ implicit acknowledgements that guarantee that packets are forwarded as they were transmitted. Another scheme is to use the multipath routing that transmits the same data across multiple paths in order to offer more chances to reach the target. Nevertheless, both schemes are unappealing for WSN environments. In general, implicit acknowledgements necessitate that the node's radio must be working, which means an increase in the energy spending. Moreover, these are not stable if bidirectional connections are not certain. Routing using multipath spends energy on unnecessary links and eats extra network bandwidth [26, 51, 58].

Hello flooding is a threat that does not need the intruder to deal with the encryption. Several routing schemes contain some nodes that send hello packets to notify sensor nodes that are one hop away of their existence. An intruder can launch a hello stream by falsifying hello packets, transmitting them from a node sending at a high transmission power. Such replayed hello packets arrive at the sensor nodes that the initiating node is not capable of communicating with right away. A sensor node that utilizes the initiating node as the succeeding hop in a route, but that is not in the node's transmission range cannot dependably advance the packets. The two schemes described above are types of DoS attacks [23, 26–36, 58].

At the data link layer, the attacker can breach the regulations of the protocol and keep sending messages to produce collisions between the packets. If the packets collide, then the retransmission of one packet is the common inclination of the protocol and, as normal, the packet is re-sent. The intruder employs this to gain admission over the various sensor nodes and produces risks to the power supply (usually battery) of the sensor nodes [1, 34, 54, 58]. The network layer is mainly in charge of the routing of packets using routing protocols. The invaders have smart ways to breach network security. The attacker performs actions that reduce the performance of the WSN [11–71]. While the message moves through such a node in regular routing process, the node basically declines the message that degrades the performance of the WSN. The rejection of the messages may be done irregularly or continuously. In both cases, harm is caused to the WSN whether we get regular or temporary denial of service.

In the transport layer, the intruder sends continuous "n" number of requests to a directed node that needs further resources to manage. Such resources may exist; however, the system will be able to manage only a specific number of them. After a particular bound the resources become drained, which can lead to the node being ineffective [30–34, 54, 58]. For instance, in a TCP SYN flood attack, an intruder transmits several linking requests without finishing the connection, hence overfilling the target's half-open connection memory. Connectionless transport schemes are invulnerable to this; however, they may not offer the proper transport-layer operation to requests. The key shield to deal with this is to use SYN cookies that code data from the client's TCP SYN message and then send them to the client in order to evade keeping the situation at the server. The high message and computational overhead make it detrimental for the WSN environment. In the case of a desynchronization attack, an intruder disturbs a working link between two sensor nodes by sending fake packets with phony sequence numbers that desynchronize endpoints so that they can transmit data. Of course, authenticating the packet's header or even the entire packet has the potential to fail these kinds of attacks [30–34, 54, 58].

In general, we can say that jamming is the major physical layer attack on WSNs. One popular technique that is used to combat jamming is the spread spectrum (SS). Regrettably, sensor nodes are limited in their power, computational, and storage resources. Hence, the SS technique cannot be employed in WSNs, mainly because of the computational resources needed. If the sensor nodes are able to detect a jamming intrusion, a sound defense is to place the nodes in a long-lasting sleeping course and then let them awaken cyclically in order to check if the jamming is still there. Despite the fact that this may not avoid a DoS attack, it may considerably enhance the lifespan of the sensor nodes by saving the energy spending.

Therefore, the intruder must jam for a significantly extended duration, which may mean that it may run out of energy before the affected sensor nodes do [30–34, 54, 58].

10.9   Summary

Security for WSNs is a major concern for developers, researchers, and users of these fascinating systems. The limited resources of the WSNs, which include constraints on the energy, computational power, storage space, bandwidth, and speed, make it hard to design and operate an efficient security and authentication protocol.

The chapter has focused mainly on the security issues, challenges, techniques, vulnerabilities, attacks, and existing solutions. As the applications of WSNs have become more and more widespread, security issues in these systems have grown at an impressive rate. We have addressed the general security issues that are commonly seen in WSNs and stressed the need to have in place working security systems with proper confidentiality, authentication, cryptography, and energy-aware security schemes, among others. We have also investigated the current WSN security protocols and their major aspects and characteristics, and compared and contrasted them.

In addition to all the above-mentioned constraints, security of WSNs is dependent on various layers of the OSI reference model as it has to meet limitations at different levels. We have reviewed the key security issues faced by WSNs as well as the major attacks such as the routing attack, node replication attack, attacks on information in transit, denial-of-service attack, denial-of-sleep attack, hello flood attack, and wormhole attack. Denial of service is one of the major attacks that seriously affects WSNs and that is why we focused on it in a separate section. Finally, we investigated techniques that can be used to avoid and prevent DoS attacks.

References

[1] P. Nicopolitidis, M. S. Obaidat and G. I. Papadimitriou, *Wireless Networks*, John Wiley & Sons, 2003.

[2] X. Du and H.-H. Chen, "Security in wireless sensor networks," *IEEE Wireless Communications*, Vol. 15, No. 4, pp. 60–66, Aug. 2008.

[3] J. C. Lee *et al.*, "Key management issues in wireless sensor networks: current proposals and future developments," *IEEE Wireless Communications*, Vol. 14, No. 5, pp. 76–84, Oct. 2007.

[4] K. Lu *et al.*, "A framework for a distributed key management scheme in heterogeneous wireless sensor networks," *IEEE Transactions on Wireless Communications*, Vol. 7, No. 2, pp. 639–647, Feb. 2008.

[5] Z. S. Bojkovic, B. M. Bakmaz and M. R. Bakmaz, "Security issues in wireless sensor networks," *International Journal of Communications*, Issue 1, Vol. 2, pp. 108–110, 2008.

[6] R. Dobrescu *et al.*, "Embedding wireless sensors in UPnP services networks," *NAUN International Journal of Communications*, Vol. 1, No. 2, pp. 62–67, 2007.

[7] S. Bravn and C. J. Screenen, "A new model for updating software in wireless sensor networks," *IEEE Network*, Vol. 20, No. 6, pp. 42–47, Nov./Dec. 2006.

[8] S. Misra, K. Abraham, M. S. Obaidat and P. Krishna, "LAID: a learning automata based scheme for intrusion detection in wireless sensor networks," *Security and Communications Networks, Wiley*, Vol. 2, No. 2, pp. 105–115, March/April, 2009.

[9] S. Dhurandher, S. Misra, M. S. Obaidat and N. Gupta, "An ant colony optimization approach for reputation and quality-of-service-based security in wireless sensor networks," *Security and Communications Networks, Wiley*, Vol. 2, No. 2, pp. 215–224, March/April, 2009.

[10] M. S. Obaidat, P. Nicopolitidis and J.-S. Li, "Security in wireless sensor networks," *Security and Communications Networks, Wiley*, Vol. 2, No. 2, pp. 101–103, March/April, 2009.

[11] M. S. Obaidat and N. Boudriga, *Security of e-Systems and Computer Networks*, Cambridge University Press, 2007.

[12] G. Wu, X. Chen and M. S. Obaidat, "A high efficient node capture attack algorithm in wireless sensor network based on route minimum key set," *Security and Communication Networks, Wiley*, Vol. 6, No. 2, pp. 230 238, 2013.

[13] S. Misra, S. Dash, M. Khatua, A. V. Vasilakos and M. S. Obaidat, "Jamming in underwater sensor networks: detection and mitigation," *IET Communications*, Vol. 6, No. 14, pp. 2178–2188, 2012.

[14] I. F. Akyildiz, W. Su, Y. Sankarasubramaniam and E. Cayirci, "A survey on sensor networks," *IEEE Communication Magazine*, Vol. 40, No. 8, pp. 102–114, 2002.

[15] V. Rathod and M. Mehta, "Security in wireless sensor network: a survey," *GANPAT University Journal of Engineering & Technology*, Vol. 1, No. 1, pp. 24–43, 2011.

[16] K. Sharma and M. K. Ghose, "Wireless sensor networks: an overview on its security threats," *International Journal of Computer Applications (IJCS), Special Issue on Mobile Ad-hoc Networks*, pp. 42–45, 2010.

[17] Y. Zhou, Y. Fang and Y. Zhang, "Security wireless sensor networks: a survey," *IEEE Communication Surveys*, Vol. 10, No. 3, 3rd Quarter 2008.

[18] H. Vogt, "Exploring message authentication in sensor networks," in *Security in Ad-hoc and Sensor Networks (ESAS), First European Workshop, Vol. 3313 of Lecture Notes in Computer Science*, pp. 19–30. Springer, 2004.

[19] N. Boudriga and M. S. Obaidat, "Mobility and security issues in wireless ad-hoc sensor networks," in *Proceedings of IEEE Globcom 2005 Conference*, 2005.

[20] Y. Wang, G. Attebury and B. Ramamurthy, "A survey of security issues in wireless sensor networks," *IEEE Communication Surveys*, Vol. **8**, No. 2, pp. 2–23. 2006.

[21] J. Yick, B. Mukherjee and D. Ghosal, "Wireless sensor network survey," *Elsevier's Computer Networks Journal*, Vol. **52**, No. 12, pp. 2292–2330, 2008.

[22] G. Padmavathi and D. Shanmugapriya, "A survey of attacks, security mechanisms and challenges in wireless sensor networks," *International Journal of Computer Science and Information Security (IJCSIS)*, Vol. **4**, Nos. 1 & 2, pp. 1–9, 2009.

[23] C. Karlof and D. Wagner, "Secure routing in wireless sensor networks: attacks and counter-measures," *AdHoc Networks (Elsevier)*, Vol. **1**, pp. 299–302, 2003.

[24] C. Karlof, N. Sastry and D. Wagner, "Tinysec: a link layer security architecture for wireless sensor networks," in *Second ACM Conference on Embedded Networked Sensor Systems (SensSys 2004)*, Nov. 2004.

[25] D. Juneja, N. Arora and S. Bansal, "An ant-based routing algorithm for detecting attacks in wireless sensor networks," *International Journal of Computational Intelligence Research*, Vol. **6**, No. 2, pp. 311–330, 2010.

[26] S. Dhurandher, M. S. Obaidat, G. Jain, I. Mani Ganesh and V. Shashidhar, "An efficient and secure routing protocol for wireless sensor networks using multicasting," in *Proceedings of the IEEE/ACM International Conference on Green Computing and Communications, Green Com 2010*, pp. 374–379, Hangzhou, China, Dec. 2010.

[27] S. Dhurandher, M. S. Obaidat, D. Gupta, N. Gupta and A. Asthana, "Network layer based secure routing protocol for wireless ad hoc sensor networks in urban environments," in *Proceedings of the IEEE ICETE 2010-International Conference on Wireless Information Networks and Systems, WINSYS 2010*, pp. 23–30, Athens, Greece, 2010.

[28] S. Misra, A. Ghosh, A. Sagar and M. S. Obaidat, "Detection of identity-based attacks in wireless sensor networks using signalprints," in *2010 IEEE GlobCom 2010 Workshop on Web and Pervasive Security (WPS)*, Miami, FL, Dec. 2010.

[29] S. Misra, M. S. Obaidat, S. Sanchita and D. Mohanta, "An energy-efficient, and secured routing protocol for wireless sensor networks," in *Proceedings of the 2009 SCS/IEEE International Symposium on Performance Evaluation of Computer and Telecommunication Systems, SPECTS 2009*, pp. 185–192, Istanbul, Turkey, July 2009.

[30] S. Misra, K. I. Abraham, M. S. Obaidat and P. V. Krishna, "Intrusion detection in wireless sensor networks: the S-model learning automata approach," in *Proceedings of the 4th IEEE International Conference on Wireless and Mobile Computing, Networking and Communications: The First International Workshop in Wireless and Mobile Computing, Networking and Communications (IEEE SecPriWiMob'08)*, pp. 603–607, Avignon, France, Oct. 12–14, 2008.

[31] A. D. Wood and J. A. Stankovic, "Denial of service attacks in sensor networks," *IEEE Computer Magazine*, pp. 54–62, 2002.

[32] T. Kavitha and D. Sridharan, "Security vulnerabilities in wireless sensor networks: a survey," *Journal of Information Assurance and Security*, Vol. **5**, No. 1, pp. 31–44, 2010.

[33] C. Karlof and D. Wagner, "Secure routing in wireless sensor networks: attacks and counter-measures," *Elsevier's AdHoc Networks Journal, Special Issue on Sensor Network Applications and Protocols*, in *First IEEE International Workshop on Sensor Network Protocols and Applications*, University of California at Berkeley, Berkeley, USA, 2003.

[34] A. Dimitrievski, V. Pejovska and D. Davcev, "Security issues and approaches in WSN," *International Journal of Peer to Peer Networks (IJP2P)*, Vol. **2**, No. 2, pp. 24–42, April 2011.

[35] S. Zhu, S. Setia and S. Jajodia, "Leap: efficient security mechanisms for large-scale distributed sensor networks," in *Proceedings of the 10th ACM Conference on Computer and Communication Security*, pp. 62–72, 2003.

[36] L. Eschenauer and V. D. Gligor, "A key management scheme for distributed sensor networks," in *Proceedings of the 9th ACM Conference on Computer and Communications Security*, pp. 41–47, 2002.

[37] K. Xing, S. Sundhar, R. Srinivasan, *et al.*, "Attacks and countermeasures in sensor networks: a survey," in *Network Security*, S. Huang, D. MacCallum and D.-Z. Du, Ed. Springer, 2005, pp. 1–28.

[38] J. Yick, B. Mukherjee and D. Ghosal, "Wireless sensor network survey," *Elsevier's Computer Networks Journal*, Vol. **52**, No. 12, pp. 2292–2330, Aug. 2008.

[39] G. Padmavathi and D. Shanmugapriya, "A survey of attacks, security mechanisms and challenges in wireless sensor networks," *International Journal of Computer Science and Information Security*, Vol. **4**, Nos. 1 & 2, 2009.

[40] E. Shi and A. Perrig, "Designing secure sensor networks," *IEEE Communications Magazine*, Vol. **44**, No. 4, 2004.

[41] A. Perrig, J. Stankovic and D. Wagner, "Security in wireless sensor networks," *Communications of the ACM*, Vol. **47**, No. 6, pp. 53–57, June 2004.

[42] W. Xu, K. Ma, W. Trappe and Y. Zhang, "Jamming sensor networks: attack and defense strategies," *IEEE Network*, Vol. **20**, No. 3, pp. 41–47, Spring, 2006.

[43] J. Deng, R. Han and S. Mishra, "Defending against path-based DoS attacks in wireless sensor networks," in *Proceedings of the 3rd ACM Workshop on Security of Ad Hoc and Sensor Networks*, 2005.

[44] C. Kraub, M. Schneider and C. Eckert, "Defending against false endorsement-based DoS attacks in wireless sensor networks," in *Proceedings of the 1st ACM Conference on Wireless Network Security*, 2008.

[45] C. Kraub, M. Schneider and C. Eckert, "An enhanced scheme to defend against false-endorsement-based DoS attacks in WSNs," in *Proceedings of the IEEE International Conference on Wireless & Mobile Computing, Networking & Communication*, 2008.

[46] K. Paul, R. R. Choudhuri and S. Bandyopadhyay, "Survivability analysis of ad hoc wireless network architecture," in *Proceedings of the IFIP-TC6/European Commission International Workshop on Mobile and Wireless Communication Networks*, Springer, Vol. **1818**, pp. 31–46, 2000.

[47] A. Perrig, "Secure routing in sensor networks." http://www.cylab.cmu.edu/default.aspx?id=1985.

[48] S. Slijepcevic, M. Potkonjak, V. Tsiatsis, S. Zimbeck and M. B. Srivastava, "On communication security in wireless ad-hoc sensor networks," in *Proceedings of 11th IEEE International Workshop on Enabling Technologies: Infrastructure for Collaborative Enterprises, WETICE2002*, June 2002.

[49] A. Perrig, J. Stankovic and D. Wagner, "Security in wireless sensor networks," *Communications of the ACM*, Vol. **47**, No. 6, pp. 53–57, June 2004.

[50] D. W. Carman, P. S. Kruus and B. J. Matt, "Constraints and approaches for distributed sensor network security," NAI Labs Tech. Rep. No. 00–010, 2000.

[51] A. Perrig, R. Szewczyk, J. D. Tygar, V. Wen and D. E. Culler, "SPINS: security protocols for sensor networks," *Wireless Networks Journal*, Vol. **8**, pp. 521–534, 2002.

[52] Y.-C. Hu, A. Perrig and D. B. Johnson, "Packet leashes: a defense against wormhole attacks in wireless networks," in *Proceedings of IEEE Infocom*, 2003.

[53] L. Eschenauer and V. D. Giligor, "A key management scheme for distributed sensor networks," in *Proceedings of the Adaptive Random Key Distribution Schemes for Wireless Sensor Networks*, 2002.

[54] D. R. Raymond and S. F. Midkiff, "Denial-of-service in wireless sensor networks: attacks and defenses," *IEEE Pervasive Computing*, Vol. 7, pp. 74–81, March 2008.

[55] S. Ganeriwal, S. Capkun, C. Han and M. Srivastava, "Secure time synchronization service for sensor networks," in *Proceedings of the ACM Workshop on Wireless Security (WSNA '05)*, 2005.

[56] S. Capkun, M. Cagalj and M. Srivastava, "Secure localization with hidden and mobile base stations," in *Proceedings of the 2006 IEEE INFOCOM*, 2006.

[57] L. Lazos, S. Capkun and R. Poovendran, "ROPE: robust position estimation in wireless sensor network," in *Proceedings of the Fourth International Conference on Information Processing in Sensor Networks (IPSN' 05)*, 2005.

[58] D. Raymond *et al.*, "Effects of denial of sleep attacks on wireless sensor network MAC protocols," in *Proceedings of the 7th Annual IEEE Systems, Man, and Cybernetics (SMC) Information Assurance Workshop (IAW)*, IEEE Press, pp. 297–304, 2006.

[59] E. Shi and A. Perrig, "Designing secure sensor networks," *IEEE Wireless Communications Magazine*, pp. 38–43, Dec. 2004.

[60] J. Granjal, R. Silva and J. Silva, "Security in wireless sensor networks", *CISUC UC*, June 2008.

[61] V. Kannan, and S. Ahmed, "A resource perspective to wireless sensor network security," in *Proceedings of the 2011 Fifth International Conference on Innovative Mobile and Internet Services in Ubiquitous Computing*, pp. 94–99, 2011.

[62] D. E. Burgner and L. A. Wahsheh, "Security of wireless sensor networks," in *Proceedings of the 2011 Eighth International Conference on Information Technology: New Generations*, pp. 315–320, 2011.

[63] S. Chen, G. Yang and S. Chen, "A security routing mechanism against Sybil attack for wireless sensor networks," in *Proceedings of the 2010 International Conference on Communications and Mobile Computing*, Vol. 1, pp. 142–146, 2010.

[64] L. Teng and Y. Zhang, "SeRA: a secure routing algorithm against sinkhole attacks for mobile wireless sensor networks," in *Proceedings of the 2010 Second International Conference on Computer Modeling and Simulation*, Vol. 4, pp. 79–82, 2010.

[65] D. O. Awduche, "MPLS and traffic engineering in IP networks," *IEEE Communications Magazine*, Vol. 37, No. 12, pp. 42–47, Dec. 1999.

[66] J. Deng, R. Han and S. Mishra, "Decorrelating wireless sensor network traffic to inhibit traffic analysis attacks," *Pervasive and Mobile Computing Journal, Elsevier*, Vol. 2, No. 2, pp. 159–186, 2006.

[67] Intrusion detection in wireless networks – Micheal Krishnan. http://walrandpc.eecs.berkeley. edu/228S06/Projects/KrishnanProject.pdf.

[68] T. Stathopoulos, J. Heidemann and D. Estrin, "A remote code update mechanism for wireless sensor networks," Tech. Rep. CENS-TR-30, University of California, Los Angeles, Center for Embedded Networked Computing, Nov. 2003. http://www.isi.edu/~johnh/PAPERS/ Stathopoulos03b.pdf.

[69] J. Hui and D. Culler, "The dynamic behavior of a data dissemination protocol for network programming at scale," in *Proceedings of the 2nd International Conference on Embedded Networked Sensor Systems (SenSys)*, pp. 81–94, 2004.

[70] W. Carman, P. S. Kruus and B. J. Matt, "Constraints and approaches for distributed sensor network security," Tech. Rep. 00–010, NAI Labs, Network Associates, Inc., Glenwood, MD, 2000.

[71] R. Chandrasekar, M. S. Obaidat, S. Misra and F. Peña-Mora, "A secure and energy-efficient scheme for group-based routing in heterogeneous ad-hoc sensor networks and its simulation analysis," *SIMULATION: Transactions of the Society for Modeling and Simulation International*, Vol. **84**, No. 2/3, pp. 131–146, Feb. 2008.

11 Wireless mobile sensor networks

As we discussed in the previous chapters, a wireless sensor network (WSN) consists of tiny nodes with limited resources. The nodes sense their surroundings and communicate with the sink(s) through multi-hop mechanisms. Nodes are powered by batteries. Often the batteries of a node are not of the rechargeable type or the replacement of the batteries, after exhaustion of battery power, is not a suitable option. Generally, static or stationary WSNs are used. If all nodes of a WSN are stationary, then the WSN is referred to as the stationary wireless sensor network (SWSN), as discussed in [1]. Some limitations of a SWSN can be overcome by using mobile nodes. As an example, if the battery power of a node becomes exhausted during operation, another node can move to the position of the former node and provide services offered by that exhausted node. If all nodes, or at least some, of a WSN are capable of moving, then this type of WSN is called a mobile wireless sensor network (MWSN). Obviously, a mobile node must have the capabilities of communication, computation, and locomotion.

Some of the drawbacks of SWSNs are discussed here. In WSNs, the sensor nodes communicate with the sink or base station through multi-hop mechanism. Moreover, the communication pattern is many-to-one. As a consequence, nodes nearer to the sink forward their own data, as well as data from other more-distant nodes. An example of an SWSN, where sensor nodes are deployed randomly, is shown in Figure 11.1. In the presence of static base station(s), a WSN encounters the funneling/bottleneck effect and the hotspot problem [2].

Funneling/bottleneck effect As already mentioned, distant or boundary sensor nodes in a multi-hop communication scenario communicate with the sink through that sink's one-hop sensor nodes. If several nodes that are not one-hop neighbors of the sink try to communicate with it simultaneously, owing to the sudden increase of data packets, there is a high probability of the occurrence of congestion at its one-hop neighbors. The effect of this congestion is packet dropping and/or retransmission of packets, and the consequence is performance deterioration of the network.

Hotspot problem The one-hop neighbors of a sink transmit more data to it than the other nodes. Also, the transmission of data in a WSN is a costly affair with respect to energy consumption. So, the battery power of a sink's one-hop neighbors is depleted more rapidly than the rest of the network. The hotspot problem generates two sub-problems – (i) sink isolation and (ii) network partitioning [2].

The second issue is related to node deployment. Nodes are deployed in the region of interest (ROI) in a deterministic or random manner. In random node deployment, it is

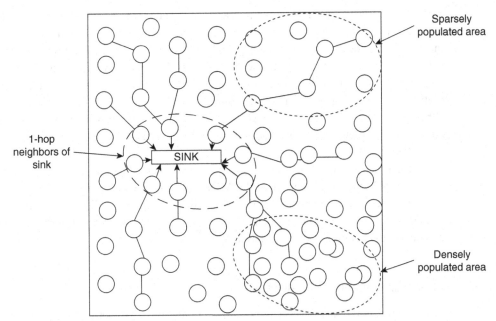

Figure 11.1 An example of multi-hop communication in a single sink SWSN (adopted from [2], with minor modifications).

very difficult to distribute the nodes equally throughout the whole region. Some portions of a ROI are overpopulated by the nodes, whereas some other portions are sparsely populated. Owing to this uneven sensor node distribution, the WSN may not provide the required coverage, thereby resulting in reduction in the effectiveness of the WSN.

11.1 Coverage and mobile sensors

Applications of WSNs can be broadly categorized into two types: monitoring and surveillance. In the coverage problem, the number of sensor nodes needs to be optimized, so that the nodes can monitor the sensed area as per the requirement of applications. Area coverage, target coverage, and barrier coverage are three types of coverage [3] commonly observed in these networks. The problem of area coverage concerns how to cover or monitor an area with the deployed sensor nodes. Target coverage is also known as point coverage. In target coverage, only some points of interest are covered by the deployed sensor nodes. A sensor barrier is created by the deployed nodes along the boundary of the area of interest to detect any intruder into that region. Coverage holes can be created in the region of interest due to the lack of appropriate number of sensor nodes, limited sensing range of the individual nodes, and the random deployment of the nodes. Stationary sensor nodes are deployed either deterministically or randomly in the area of interest. Deterministic node deployment is applicable if the area of interest is small and "friendly." On the other hand, in a hostile environment, and/or in a large sensed area, the sensor nodes are deployed randomly, for example by air-dropping from an airplane. The general solution for reaching the desired level of coverage for stationary nodes is to deploy the nodes densely. According to [3, 4], there exists a threshold value for node density. Complete coverage is possible if the node density is greater than this threshold value. But highly dense WSNs have their own problems. To be energy efficient, the nodes need to be coordinated and follow a schedule. Otherwise, precious battery power will be wasted due to redundancy, without improving the efficiency of the network. A coverage hole may even be created due to the effect of natural phenomena such as landslides and earthquakes.

Wireless sensor networks with mobile nodes can be classified into two broad categories – hybrid WSNs and mobile WSNs. Hybrid WSNs consist of both mobile and stationary nodes, whereas mobile WSNs are built from mobile nodes only. The strategies for mobile node deployment depend on the principles of virtual force, computational geometry, and grid-based approaches.

11.1.1   Voronoi diagram-based approaches

The Voronoi diagram [5] is an important computational geometric structure. It has several important applications in physics, astronomy, robotics, and many more fields. It provides proximity information about a set of geometric nodes or points. The Euclidean distance of two nodes p, q is denoted by $ed(p, q)$. Let us consider a set of sensor nodes S deployed in a field. The position of the ith sensor node is denoted by p_i. The Voronoi diagram of S is the partitioning of the field into n cells; each cell corresponds to one sensor. A point q is in the cell corresponding to the ith sensor, if $ed(q, p_i) < ed(q, p_j)$, for all nodes $\in$S and $i \neq j$. In the left-hand part of Figure 11.2, a Voronoi diagram for three points or three sensor nodes is shown. An example of Voronoi polygon is shown in the right-hand part of Figure 11.2. The vertices of the polygon are v1, v2, v3, v4, and v5. The procedure to estimate the coordinate position of vertices of a Voronoi polygon is given in [6]. When sensor node s0 of Figure 11.2 tries to calculate the position of v1, it

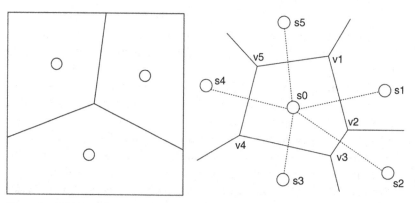

Figure 11.2   Example of Voronoi diagram (left) and Voronoi polygon (right) (adopted from [7], with minor modifications).

requires the location information of two neighbors, s1 and s5. Let us assume that the location of sensor node si is (x_i, y_i), and the line which is perpendicular to the line connecting si, sj is L_{ij}. The gradient of L_{05} is $-((x_0 - x_5)/(y_0 - y_5))$ and the coordinate of the middle point of the line connecting $s0$ and $s5$ is $((x_0 + x_5)/2, (y_0 + y_5)/2)$. From the above information, the equation of L_{05} can be estimated and the equation is represented by equation (11.1). From the equations of L_{05} and L_{01}, $s0$ is able to estimate the position of point $v1$:

$$y - \frac{(y_0 + y_5)}{2} = -\left(\frac{x_0 - x_5}{y_0 - y_5}\right)\left(x - \frac{(x_0 + x_5)}{2}\right). \tag{11.1}$$

In [7], a Voronoi diagram-based node-deployment protocol is illustrated to optimize network coverage. It is assumed [7] that all sensor nodes are mobile. The protocol runs iteratively. Initially, all sensor nodes are deployed into the region of interest and the sensors broadcast their locations. Each sensor node calculates its Voronoi polygon from the received neighborhood information. In the next step, the sensor nodes check whether a coverage hole exists in their respective Voronoi polygon or not. If it does, the sensor nodes estimate their next locations to reduce or to omit the coverage hole.

11.1.1.1 The Voronoi-based algorithm (VOR)

In VOR, a sensor node first estimates the existence of a coverage hole within its Voronoi polygon [7]. If the furthest Voronoi vertex is not covered by the sensor node, then the node assumes the existence of a coverage hole within its Voronoi polygon. When such a hole exists, it then moves toward the furthest uncovered Voronoi vertex to cover that vertex. A sensor node S and its Voronoi polygon are shown in Figure 11.3. B is the furthest uncovered Voronoi vertex of sensor node S. So, S moves towards B to cover B. Sensor node S

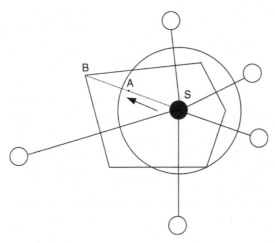

Figure 11.3   Example of sensor movement in VOR (adopted from [7], with minor modifications).

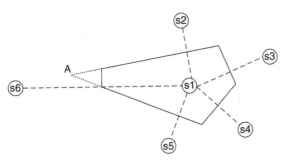

Figure 11.4   Inaccurate Voronoi polygon due to incomplete neighborhood information (adopted from [7], with minor modifications).

estimates its new location along the path SB and it moves to A to cover B. The Euclidean distance between A and B is equal to the sensing range of a sensor node.

The maximum moving distance for a sensor node is the difference between half of the communication range and the sensing range of a node. A typical scenario is shown in Figure 11.4. Sensor node s_6 is outside of the communication range of node s_1, and as a result s_1 estimates its Voronoi polygon incorrectly, the dotted one. Node s_1 estimates point A as the furthest Voronoi vertex and it tries to cover A by moving forward towards it. However, in reality, A is nearer to node s_6 than node s_1. So, when s_1 tries to cover point A, it covers an area that is already covered by sensor node s_6. The maximum moving distance of a node is limited so that a node can move toward a coverage hole in multiple steps. At each step, a node gathers more information about its neighbors and corrects its previous estimation. As an example, node s_1 moves towards point A in steps and, in each step, it collects information about its neighbors. When node s_1 becomes aware of the presence of node s_6, node s_1 recalculates its Voronoi polygon. Another problem associated with VOR is the oscillation of a node between its new and previous locations. Whenever a node moves to a new location, a new coverage hole may be created in the previous location of that node. A mechanism for oscillation control is taken to prevent the immediate backward movement of a node. Before movement, a node checks its movement direction, and if this direction is the reverse of the previous round, then the node abandons its movement for that step.

11.1.1.2   Minimax algorithm

The Minimax algorithm [7] is also based on the Voronoi diagram, and is similar to VOR. Here a node moves toward the furthest Voronoi vertex (V_{fur}), but the node estimates its target location to minimize its distance to V_{fur}. In the Mimimax algorithm, the target position, known as the Minimax point and denoted by p_m, is chosen in a way that it reduces the variance of distances to the Voronoi vertices. This way, the nodes avoid the changing of a close vertex into the furthest one. The circumcircle of three vertices V_a, V_b, V_c is represented by $C(V_a, V_b, V_c)$. The algorithm to estimate the Minimax point of a given Voronoi polygon is given below.

(1)   Initialize $n = |V_p|$, where V_p denotes the vertices of given Voronoi polygon.
(2)   For a = 1, 2, ..., $n-2$
       for b = a+1, a+2, ..., $n-1$
          for c = b+1, b+2, ..., n
             $C(V_a, V_b, V_c)$ is calculated.
             If V is inside of $C(V_a, V_b, V_c)$ for all $V \in V_p$ then
             the circle is recorded.
(3)   For a = 1, 2, ..., $n-1$
          for b = a+1, a+2, ..., n
             $C(V_a, V_b,)$ is calculated.
                If V is inside of $C(V_a, V_b,)$ for all $V \in V_p$ then
                the circle is recorded.
(4)   The circle with minimum radius is selected and the target location is the center of
       the selected circle.
(5)   The sensor node moves to newly selected location.

The VOR and Minimax algorithms terminate when there is no improvement of local coverage. If the requirement of coverage is not so high, the algorithms need to terminate earlier. In that case, a threshold value is used and, if the improvement of coverage does not exceed that threshold value, a node will not change its location. If, initially, the sensor nodes create clusters, the node deployment procedures take a prolonged time. To avoid such a situation, in the very beginning of deployment, an explosion technique [7] is used. Each node tries to find whether it is within a cluster or not. If a node finds itself within a cluster, it selects a random location within the area where the node will have fewer neighbor nodes.

11.1.1.3   Centroid and dual-centroid schemes

The centroid and dual-centroid schemes are based on the centroid of a polygon. According to the authors of [8], performance of centroid and dual-centroid schemes is better than VOR or Minimax. In the centroid scheme, a sensor node, at the beginning of each round, calculates its Voronoi polygon from its neighbors' location information. If there is no coverage hole within the polygon, the node skips the coverage enhancement procedures for that round. If the node finds any coverage hole, then it estimates the centroid of the polygon by using equations (11.2) and (11.3). The location of centroid of a polygon with n vertices is denoted by (C_x, C_y) and area of the polygon is shown by A:

$$C_x = \frac{1}{6A}\sum_{i=0}^{n-1}(x_i + x_{i+1})(x_i y_{i+1} - x_{i+1} y_i),$$

$$C_y = \frac{1}{6A}\sum_{i=0}^{n-1}(y_i + y_{i+1})(x_i y_{i+1} - x_{i+1} y_i),$$

(11.2)

$$A = \frac{1}{2}\sum_{i=0}^{n-1}(x_i y_{i+1} - x_{i+1} y_i).$$

(11.3)

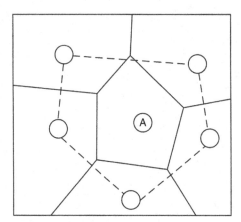

Figure 11.5 Voronoi polygon and Voronoi neighbor polygon of a node (adopted from [8], with minor modifications).

After calculating the centroid of the Voronoi polygon and choosing the centroid as its new location, the sensor node estimates the improvement of local coverage at that new location. The node moves to the centroid, if there is some improvement of local coverage at that location, otherwise it does not move.

In the dual-centroid approach, the next location of a sensor node depends on two polygons; the first is the Voronoi polygon of the sensor node, and the second is the polygon constructed by the neighbor sensor nodes of the node under consideration and called Voronoi neighbor polygon by the authors of [8]. The vertices of the Voronoi neighbor polygon are represented by the neighboring sensor nodes of the current node. The Voronoi polygons of sensor nodes are shown in solid line and Voronoi neighbor polygon of node A is shown by dashed line in Figure 11.5. To estimate its new location, a sensor node calculates the centroid of its Voronoi polygon as well as the centroid of its Voronoi neighbor polygon. The next location L is calculated by a node by using equation (11.4):

$$L = \alpha C_{Vp} + (1 - \alpha)\, C_{Vnp} \quad \text{where} \quad 0 \le \alpha \le 1. \tag{11.4}$$

C_{Vp} and C_{Vnp} represent the centroid of Voronoi polygon and the centroid of Voronoi neighbor polygon of a node.

11.1.1.4 Bidding protocols

Two distributed bidding protocols for healing coverage holes by mobile sensor nodes were proposed by Wang *et al.* [9]. The network is considered to be composed of static sensor nodes as well as mobile sensor nodes. After initial deployment of nodes, the static nodes detect coverage holes, and estimate the area of the holes and the possible location of a mobile node within the coverage hole. Mobile nodes, treated as servers in [9], provide their services to heal coverage holes. Whenever a mobile node moves to a new location, a new coverage hole may or may not be created at its previous location. Each mobile node has a base price, which is proportionate to the area of the coverage hole generated due to the movement of the mobile node. A mobile node moves only if the node is able to heal a hole that is greater than its base price.

Each round of bidding protocol has three phases: service advertisement, bidding, and serving. In the service advertisement phase, each mobile node advertises its location and base price. Initially, the base price of all mobile nodes is zero. At the end of the service advertisement phase, all the static nodes have information on their neighboring mobile nodes. Static nodes detect coverage holes by scrutinizing their respective Voronoi polygons in the bidding phase. If such a hole exists within its Voronoi cell, a static node then estimates the area of the hole, or it can be said that the node calculates the bid. Following that, the node finds a suitable location for the mobile node. The node selects the furthest uncovered Voronoi vertex as the probable location for the mobile nodes.

The next task of a static node is searching for a mobile node whose base price is lower than the bid and which sends a bidding message to that node. In the serving phase, a mobile node selects the highest bid among the received bids and moves to that area. After moving, a mobile node updates its base price and its location. The protocol terminates when all the static nodes are unable to provide higher bids than the base price of any mobile node.

11.1.2   Virtual force-based approaches

Virtual force-based approaches, proposed in [7, 10–12], are used to optimize the coverage area by the sensor network. Here nodes are assumed as virtual particles or electrostatic particles. Movement of a node is influenced by virtual forces. Owing to the effect of repulsive force, nodes move away from one another and the obstacles. This force is inversely proportional to the nodes' distance. Moreover, the repulsive force between two nodes is zero if the distance is greater than the communication range of the nodes. There is also an attractive force, called the viscous friction force [10], which helps the nodes to reach a static equilibrium state. The basic principle of virtual force-based approaches is that, after initial deployment, nodes spread from a densely populated area to all over the monitored area. Each node repels its neighboring sensor nodes and is repelled by the local obstacles. The network reaches a static equilibrium state when all nodes stop due to the viscous force. It is assumed that, nodes are capable of finding nearby sensor nodes as well as obstacles through communication or sensing.

Potential fields are used to deploy nodes by the authors of [10]. A virtual force F is applied on every node. The force F is a gradient of a scalar potential field P and the relationship between F and P is shown in equation (11.5) as

$$F = -\nabla P. \tag{11.5}$$

The potential field has two components: the field P_O due to the presence of obstacles and the field P_N due to neighbor nodes. These forces generate repulsive forces F_O and F_N. Thus, F and P can be expressed by equation (11.6) as

$$P = P_O + P_N \quad \text{and} \quad F = F_O + F_N. \tag{11.6}$$

It may be assumed that each node or obstacle holds some amount of electrostatic charge and is repelled by nearby nodes and objects. The resultant electrostatic potential, due to objects, on a node, is expressed by equation (11.7):

$$P_O = c_O \sum_i \frac{1}{ed_i},$$

(11.7)

where c_O is a constant expressing the strength of a field and ed_i is the Euclidean distance between the node and the ith object. The position of the node is denoted by x and position of obstacle i is denoted by x_i and $ed_i = |x_i - x|$. So, the force F_O is computed by using equation (11.5):

$$F_O = -\frac{dP_O}{dx} = -\sum \frac{dP_O}{ded_i} \cdot \frac{ded_i}{dx}.$$

(11.8)

The appropriate value of F_O is obtained by solving equation (11.8):

$$F_O = -c_O \sum_i \frac{1}{ed_i^2} \cdot \frac{rp_i}{ed_i},$$

(11.9)

where rp_i is the relative position of obstacles and can be shown as $rp_i = x_i - x$. The potential field P_N and F_N is calculated in a similar way:

$$P_N = c_N \sum_i \frac{1}{ed_i} \quad \text{and} \quad F_N = -c_N \sum_i \frac{1}{ed_i^2} \cdot \frac{rp_i}{ed_i}.$$

(11.10)

The trajectory of a node is estimated by using equation (11.11):

$$ac = (F - \eta v)/m.$$

(11.11)

Acceleration of a node is denoted by ac, η is the viscosity coefficient, v is velocity, and m is the mass of the particle. The expression ηv represents the viscous friction [10] which makes a node stand still in the absence of any external force. If v is the velocity of a node at time t and Δv is the change of velocity between time t and $t + \Delta t$, then the updated velocity is computed by using equation (11.12):

$$\Delta v = \frac{F - \eta v}{m} \cdot \Delta t \quad \text{and} \quad v = v + \Delta v.$$

(11.12)

Control laws are defined to estimate the effects of a virtual force on a node. Each node has potential energy and kinetic energy. Potential energy results from the interactions among neighbor nodes and obstacles, and a node gains kinetic energy due to that node's movement. The total energy of the system is estimated by summing up the energy of all individual nodes. In the presence of viscous friction, mentioned in equation (11.11), the system reaches static equilibrium after dissipating energy. Ultimately the system dissipates its total kinetic energy and reaches static equilibrium. An important assumption is made here – the environment of the system itself is a static system. By "static" it is implied that the space and the energy, in which the latter is expressed in terms of nodes and obstacles, are static or constant during the node deployment phase. In a dynamic environment, the system does not reach the equilibrium phase, but if the system changes periodically, ultimately it reaches such a state.

11.1.3   Grid-based approach

Scan-based movement-assisted sensor deployment (SMART) [13] is applied to unevenly distributed sensor networks to balance the sensor distribution. Here, all the nodes are assumed to be mobile [13]. The monitoring area is partitioned into $n \times n$ grids or mesh of clusters. The number of nodes in each grid cell is assumed to be the load of that cell. In [13], the sensor deployment problem is viewed as a load balancing problem. Initially, sensor nodes are randomly distributed and the loads of the cells are not equal. SMART helps to deploy the sensor nodes evenly and, hence, balances the load of the cells. The basic principle of SMART comes from a two-dimensional scan-based approach. Generally, a scan operation consists of a binary operator * and an ordered set $[c_1, c_2, \ldots, c_n]$. In [13], c_i represents the number of sensors of the ith grid cell. The scan returns the ordered set $[c_1, (c_1*c_2), \ldots, (c_1*c_2*, \ldots, *c_n)]$. In [13], integer addition and Boolean AND are used as operators. Sensor nodes of each grid cell make a cluster and select one of them as the cluster head. Every cluster head communicates with the adjacent grid cells and knows the position, i, in the grid and the number of sensors or the load of the cell, c_i.

The balancing is performed by a two-round scan; the first one balances the rows, and the second balances the columns. An example of a two-round scan is shown in Figure 11.6. In Figure 11.6b, the scan process balances all the rows and in Figure 11.6c, all the columns are balanced, and, hence, the total area is balanced.

Scan process   To illustrate the scan process, a one-dimensional array of cluster or grid of n cells is considered. The total number of sensors from cluster 1 to cluster j is expressed as a prefix sum and denoted as s_j. The value of s_j is expressed as:

$$s_j = \sum_{k=1}^{j} c_k.$$

(11.13)

The total number of sensors in that one-dimensional array is

$$s_n = \sum_{k=1}^{n} c_k.$$

(11.14)

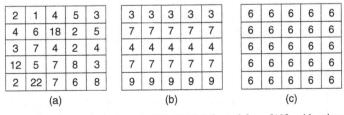

2	1	4	5	3		3	3	3	3	3		6	6	6	6	6
4	6	18	2	5		7	7	7	7	7		6	6	6	6	6
3	7	4	2	4		4	4	4	4	4		6	6	6	6	6
12	5	7	8	3		7	7	7	7	7		6	6	6	6	6
2	22	7	6	8		9	9	9	9	9		6	6	6	6	6

(a)   (b)   (c)

Figure 11.6   Example of a two-round scan of SMART (adopted from [13], with minor modifications).

Table 11.1 Scan process of a one-dimensional array

j	1	2	3	4	5	6
c_j	7	15	9	18	3	8
s_j	7	22	31	49	52	60
s_j'	10	20	30	40	50	60

The average number of sensors in each cell in a balanced state is $l = (s_n/n)$. The prefix sum in a balanced state is:

$$s_j' = i \times l.$$

In the balanced state, for any two clusters j, k, $|c_j - c_k| \leq 1$.

The scan process has two phases. In the first phase, the scan process starts from one end of the array and moves towards the other end. After reaching the other end, the scan process moves backward and returns to the starting end. In the first phase, each cluster head, j, calculates its prefix sum by $s_j = s_{j-1} + c_j$ and forwards the value of s_j to the cluster head of the $(j + 1)$th cluster. The cluster head of the last cluster estimates s_n and l. Thereafter, the last cluster head begins the second phase of the scan process by forwarding the value of l to the cluster head of the $(n - 1)$th cluster. During the second phase of the scan, each cluster head calculates the prefix sum in the balanced state. An example of the scan process is shown in Table 11.1.

Each cluster head is able to calculate whether its corresponding cluster is overloaded or not by getting the value of the load in the balanced state. If $c_j > l$, then the jth cluster is overloaded and is in the "give" state. If $c_j < l$, then the jth cluster is underloaded and is in the "take" state. If $c_j = l$, the cluster is in the "neutral" state. Referring to Table 11.1, it can be stated that the fourth cluster is overloaded, where the fifth cluster is underloaded.

Each cluster head, if in the give state, determines the number of sensor nodes required to send to each direction: c_{jr}, "give right," the number of sensor nodes required to send to the right direction; and c_{jl}, "give left," the number of sensor nodes required to send to the left. The values of c_{jr} and c_{jl} are deduced as

$$c_{jr} = \min((c_j - l), \max((s_j - s_j'), 0)), \tag{11.15}$$

$$c_{jl} = (c_j - l) - c_{jr}. \tag{11.16}$$

In Table 11.1, cluster 4 is in the given state. So, the cluster head of cluster 4 can estimate the value of c_{4r} by equation (11.15) and the value is $\min((18 - 10), \max((49 - 40), 0)) = 8$.

In the two-dimensional scan, first a row scan, and then a column scan are performed.

11.1.4 Event coverage

Mobile sensors, with collaboration of stationary sensors, are used in [14] for reliable detection and location estimation of events. The mobile sensors perform two different tasks. Either the mobile sensors monitor the area sporadically covered by static sensors or, when static sensors inform the nearby mobile sensors about a suspected event, they move to the suspected area for closer inspection. The mobile nodes autonomously plan their paths on the basis of local information such as their own measurements and information collected from the neighboring static and mobile nodes. The objectives of path planning are to reach the target area as fast as possible and to improve the area coverage.

Static sensors are deployed randomly into the monitored area. The sensing range of a static sensor is divided into two parts: the detection range and the suspicion range. If an event occurs within the detection range of a sensor, that sensor is able to detect the event reliably and report to the sink about that event. But if an event occurs within the suspicion range of a static sensor, it transmits its suspicion, by a suspicion message, either to the sink or to the nearby mobile sensors. An example is shown in Figure 11.7.

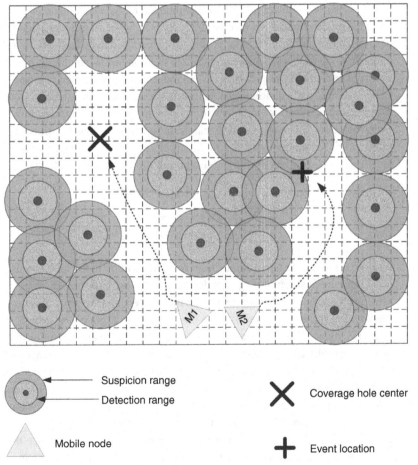

Suspicion range
Detection range

✕ Coverage hole center

△ Mobile node

✚ Event location

Figure 11.7 Collaboration of stationary nodes and mobile nodes (adopted from [14], with minor modifications).

The whole area is divided into a grid of size $R \times C$. A grid cell is covered if it is within the detection range of a stationary sensor node. The grid is associated with an $R \times C$ matrix M_k, $k = 0, 1, 2 \ldots$ The value of $M_k(i, j) = 1$ for all k, if the (i, j)th cell of the grid is covered. Otherwise, at $k = 0$, the initial value of $M_k(i, j) = 0$ and the changes of the value depend on the presence and behavior of nearby mobile nodes. The value of $M_k(i, j)$ increases, if a mobile node approaches the (i, j)th cell and provides coverage to that cell. When the mobile node moves to another location, the value of $M_k(i, j)$ gradually decreases with time. Each mobile node uses a coverage cognitive map, an $R \times C$ matrix, to hold the state of the field, for its path planning. During the kth step, to select its next location, a mobile node first chooses several candidate locations. These candidate locations are dependent on the current location of the mobile node, its heading direction, and its current speed. The mobile nodes evaluate each candidate location by a cost function. This cost function depends on the neighboring stationary and mobile nodes and the target location. The neighboring nodes push the mobile node by a repulsive force, while the target location attracts the mobile node. The mobile node selects the location that minimizes the cost function as its destination location.

A mobile node is either in the coverage mode or in the search mode. A mobile node is in coverage mode if it does not receive any suspicion messages from its neighbors. The target destination of a mobile node in the coverage mode is the nearby largest coverage hole. When a mobile node receives a suspicion message from one of its neighbors, it transits from the coverage mode to the suspicion mode and moves toward the suspected event location.

Each mobile node continuously senses its surroundings and gathers information about location and sensing measurements from its neighbors while approaching the suspected event location. By using a nonlinear least squares estimation process and received information from neighbors, the mobile node estimates the target location. The target location can be estimated from the solution of minimization problem:

$$P = \sum_{i \in \varphi(k)} \left(s_i - \frac{V}{\left[(x_t - x_i)^2 + (y_t - y_i)^2 \right]^{a/2}} \right)^2, \qquad (11.17)$$

where s_i is the sensed value of the ith sensor. The position of the ith sensor is expressed by (x_i, y_i) and the position of the target destination is represented by (x_t, y_t), $\varphi(k)$ is the set of sensing measurements from neighbors at the kth step, and the sensing value of the mobile sensor up to the kth step, and V is the initial signal strength of the event.

In [15], hybrid WSN is used for event coverage by Wang *et al.* Initially, the static sensors detect an event and suitable mobile sensors equipped with more sensing and computation power are dispatched to the event location for better coverage and further actions. A mobile sensor consumes more energy for movement than sensing and computation. An energy-efficient dispatch mechanism for mobile sensors is proposed by Wang *et al.* [15]. They propose a centralized dispatch mechanism CentralSD (Centralized sensor dispatching). The CentralSD has two objectives: (i) reducing the total distance moved by each sensor, and (ii) balancing the moving distance of sensors, so that their energy consumptions to visit event locations are almost the same. For reducing the number of transmitted messages and computation complexity, a distributed mechanism GridSd (Grid-based architecture for sensor dispatching) is also proposed in [15].

CentralSD (Centralized sensor dispatching) algorithm.
    By CentralSD, a set of mobile sensors $M = \{m_1, m_2, \ldots, m_n\}$ is dispatched to a set of event locations $E = \{e_1, e_2, \ldots, e_l\}$. The sensor dispatching problem is formulated as a maximum-matching problem of a weighted bipartite graph.

Case 1: $|M| \geq |E|$
A weighted bipartite graph G is constructed from M and E. The vertex set of G contains the mobile sensors and event locations and each mobile sensor is connected to every event location by a weighted edge (m_i, e_j), where $m_i \in M$ and $e_j \in E$. The weight w_{ij} of an edge (m_i, e_j) is proportional to the distance between sensor m_i and event location e_j. The objective is to find a maximum-matching P with (i) minimum total edge weight, and (ii) minimum standard deviation of edge weights. To estimate P, a preference list $PL(m_i)$ is associated with each mobile sensor $m_i \in M$. The preference list of node m_i, $PL(m_i)$, contains all the event locations sorted in an ascending order on the basis of the weight w_{ij} for each $e_j \in E$. Similarly, each event is associated with a preference list $PL(e_j)$ containing all the mobile sensors in an ascending order on the basis of weight. A bound B_{ej} is used to limit the standard deviation of edge weights in P. A mobile sensor m_i is matched with an event location e_j, if $w_{ij} \leq B_{ej}$. The initial value of B_{ej} is the average of the minimum weight of each $PL(e_j)$, for all $e_j \in E$. For each $e_j \in E$, a mobile sensor m_i is matched with e_j, such that w_{ij} is minimized and $w_{ij} \leq B_{ej}$. If no match is found, the value of B_{ej} is expanded by ΔB and the matching process is resumed. The increment value ΔB is calculated as:

$$\Delta B = \frac{\alpha \times \left(\sum_{j=1}^{l} \max_{\forall (i,j)} w_{ij} - \sum_{j=1}^{l} \min_{\forall (i,j)} w_{ij} \right)}{l \times n}, \tag{11.18}$$

where α is adjustment factor.
    To find a match for e_j, $PL(e_j)$ is searched until $w_{ij} \leq B_{ej}$. If the first unvisited candidate m_i is also unmatched, then the pair (m_i, e_j) is added to P. If the candidate m_i is already matched with another location e_k, a competition is launched between e_j and e_k for m_i. In such cases, m_i is matched with e_j if:

(i) $B_{ej} > B_{ek}$, to reduce the standard deviation of edge weights in P;
(ii) $B_{ej} = B_{ek}$ and e_j precedes e_k in $PL(m_i)$, to decrease the total weight of P;
(iii) $B_{ej} = B_{ek}$ and m_i is the last candidate of $PL(e_j)$ but not in $PL(e_k)$, as e_j does not have any other option in this case.

    An example is illustrated in Tables 11.2 and 11.3. Here α is assumed to be 2, the initial value of bound is $(105 + 80 + 95)/3 = 93.3 \approx 93$ and $\Delta B = 2 \times ((218 + 235 + 230) - (105 + 80 + 95))/(3 \times 4) = 67.1 \approx 67$. There is no mobile sensor in $PL(e_1)$ with initial bound. After increasing the initial bound of e_1, B_{e1}, by ΔB, B_{e1} becomes $(93 + 67) = 160$. The mobile sensor m_1 is selected among the three candidates, m_1, m_2, m_3, as m_1 is the first unvisited

Table 11.2 Weight matrix of mobile nodes and event locations

Cost	e_1	e_2	e_3
m_1	105	145	95
m_2	155	235	175
m_3	130	80	230
m_4	218	105	180

Table 11.3 Preference lists of mobile nodes and event locations

Preference lists	
$PL(m_1) = \{e_3, e_1, e_2\}$	$PL(e_1) = \{m_1, m_3, m_2, m_4\}$
$PL(m_2) = \{e_1, e_3, e_2\}$	$PL(e_2) = \{m_3, m_4, m_1, m_2\}$
$PL(m_3) = \{e_2, e_1, e_3\}$	$PL(e_3) = \{m_1, m_2, m_4, m_3\}$
$PL(m_4) = \{e_2, e_3, e_1\}$	

node among the candidates. The pair (m_1, e_1) is added in the matching P. Similarly, the pair (m_3, e_2) is also added to P. But in the case of e_3, its best match is m_1, which is already matched with e_1. So, a competition is performed between e_1 and e_3 for m_1. For both e_1 and e_3, the bound values are the same ($B_{e1} = B_{e2} = 160$). But in the preference list of m_1, $PL(m_1)$, e_3 is ahead of e_1. So, (m_1, e_3) replaces the pair (m_1, e_1) in P. Similarly, (m_3, e_2) is replaced by (m_3, e_1). Finally, e_2 is matched with m_4 and (m_4, e_2) is added to P.

Case 2: $|M| < |E|$

In this case, $|E|$ events are clustered into $|M|$ clusters by the K-means clustering algorithm. Using the previous matching scheme, each mobile sensor is dispatched to each cluster. After reaching the closest event location of the assigned cluster, a mobile sensor visits all the locations of that cluster by using the traveling-salesman approximation algorithm.

GridSD   Grid-based distributive sensor dispatching algorithm.

In GridSD, the whole monitored area is divided into grids. A sensor node is selected as the grid head for each grid. A grid head collects information about all the mobile sensors and the event locations with the grid area. When static sensors detect and report events, a grid head dispatches the mobile node by using the CentralSD. If no mobile sensor is available within the grid, the grid head searches other grids for suitable mobile sensors.

A grid-quorum based approach is proposed in [15] to save the energy of grid heads during searching of mobile nodes by reducing the number of transmitted messages. Each grid head advertises the number of mobile nodes within its grid to the other grids of the same column by sending the advertisement (ADV) messages. Whenever a grid head needs to search the mobile nodes in other grids, it sends a request (REQ) message to other grid heads located in the same row. When a grid head receives both the ADV and REQ messages, it informs the sender of REQ message about the available grid location of mobile sensors. An example of

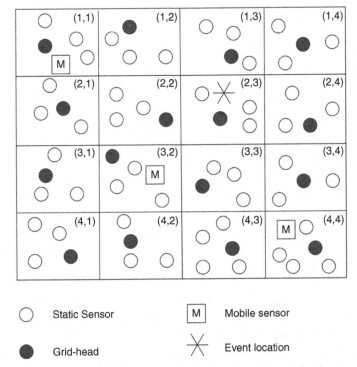

	Static Sensor	M	Mobile sensor
	Grid-head		Event location

Figure 11.8   An example of GridSD (adopted from [15], with minor modifications).

GridSD is given in Figure 11.8. The grid head of grid (1,1), (3,2) and (4,4) informs all the grid heads of column 1,2, and 4, respectively, about the number of available mobile nodes in those grids by transmission of ADV messages. When the grid head of grid (2, 3) needs to dispatch a mobile sensor to an event location, it sends a REQ message to all the grids of row 2. After receiving the REQ message, the grid head of grid (2, 2) sends the location of mobile sensor, the grid (3, 2), to the grid head of (2, 3).

11.2   Network lifetime improvement

11.2.1   Predictable and controllable mobile sink

Luo and Hubaux [16] proposed a mechanism that considers the mobility of sink(s) and routing strategy of sensor nodes, jointly, to increase the lifetime of a WSN. It is assumed that the sensor nodes are densely deployed by a Poisson process within a circular area of radius r. Luo and Hubaux [16] transposed the network lifetime maximization problem into a load balancing problem. The load of the ith node, $load_i$, represents the power consumed by node i during transmission and reception of data. Higher load implies shorter lifetime of a node. The network lifetime is, roughly, inversely proportional with the "network load," $load_N$. The load balancing problem is formulated as a min-max problem in terms of the average load of sensor nodes:

$$\text{Minimize } load_N \equiv \max_{i \in N} load_i'(R, M)$$

$$\text{Constraints}: \quad R \text{ constraints, } M \text{ constraints.}$$

The average load, $load_i'$, of the ith sensor node depends on the routing strategies, R, taken by the nodes and the mobility strategies, M, taken by the sink. Luo and Hubaux [16] showed that the average load of a sensor node decreases with the increase in the distance between the node and the sink. The mobile sink changes the location of hotspots over time, and, thus, the mobile sink prolongs the overall network lifetime. Finding the optimum joint mobility and routing strategies is a two-phase process. In phase 1, the optimum mobility strategy of a sink is estimated by fixing the routing of sensor nodes to the shortest path routing. After the estimation of the optimum mobility strategy, a better routing strategy than the shortest path routing is sought. Only the periodic mobility strategies with finite period are considered at the time of the estimation of optimum mobility strategy. It is shown in [16] that any non-symmetric trajectory can be replaced by an at-least equally efficient trajectory that is symmetric around the center of the network. The optimum symmetric trajectory is a circular trajectory around the center of the network. The load of the network is minimized when the radius of the circular trajectory is equivalent to the radius of the network. However, if the sink moves in a circular trajectory, the nodes near to the center of the trajectory have higher load than the nodes near to the boundary of the network. The routing strategy taken by nodes helps to reduce the load of central nodes while increasing the load on the peripheral sensor nodes. To distribute the load on the nodes, the network is partitioned into two parts, as shown in Figure 11.9.

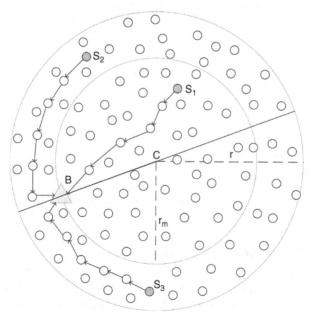

Figure 11.9 Example of routing by nodes using joint mobility and routing (adopted from [16], with minor modifications).

The sink moves in a circular path of radius r_m around the center of the network, where $r_m < r$. This path divides the network into two parts: the circular area of radius r_m and the annulus between the boundary of network and the circular path of sink. A sensor node uses shortest path routing if it is within the circular area enclosed by sink's trajectory. When node S_1 tries to communicate with base station B, it uses the shortest path routing. The nodes in the annulus transmit packets using a two-step routing process called "round routing." A packet is transmitted along a circular path around the center of the network C until it reaches CB. Then the packet is forwarded using shortest path routing. Nodes S_2 and S_3 use round routing to communicate with B. The value of r_m is determined from simulation to be roughly $0.9 \times r$.

11.2.2   Predictable but uncontrollable mobile sink

Chakrabarti *et al.* [17] proposed a framework for saving power of individual sensor nodes of WSNs in the presence of a predictable, but uncontrollable, mobile sink or "observer." Good examples of such mobile sinks are public transportation vehicles such as buses and trains. The public transportation vehicles follow the same route repeatedly and periodically.

In the proposed framework of [17], the sensor nodes are distributed over the area of interest A. Two different kinds of distribution are considered here. In the first type, the sensor nodes are distributed randomly and uniformly over A while, in the second, the minimum distance between any two sensor nodes is d. The sink S, with speed v, follows the same path repeatedly. All the sensor nodes are identical and their communication range is R_C. Let us assume that each sensor node requires t_{data} time for transferring its sensed data to sink. A sensor node fails to communicate successfully with a sink if the sink does not stay within the node's communication range for at least t_{data} time. An unsuccessful communication is called an "outage" by the authors of [17]. See Figure 11.10.

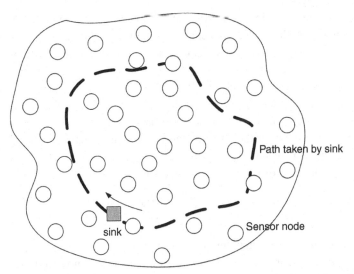

Figure 11.10   A WSN and path of a predictable but uncontrollable sink (adopted from [17], with minor modifications).

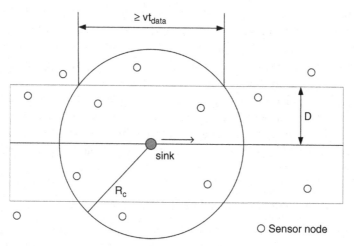

Figure 11.11   Relationship between R_C and D (adopted from [17], with minor modifications).

Single-hop communication is used between a node and the sink. The sink must come within the communication range of every sensor node during the journey along its path within A. Let us assume that the maximum distance between any sensor node from the path of the sink is D. So, for successful communication between the sink and any sensor node, the communication range, R_C, of sensor nodes can be calculated by using:

$$R_C \geq \sqrt{D^2 + (v \times t_{data}/2)^2}. \tag{11.19}$$

The value of R_C that satisfies equation (11.19) ensures that the sink will remain within the communication range of every node for at least t_{data} time. The relationship between D and R_C is shown in Figure 11.11.

In [17], the data collection by a sink is formulated as a queuing problem. As the sink forwards along its path, new sensors come within the range of sink while some sensors, previously able to communicate with sink, disappear from the communication range of the sink. In Δt time, the sink moves $v\Delta t$ distance. Sensor nodes that are within $2Dv\Delta t$ area appear within the range of the sink. While a new node appears within the communication range of the sink, the node has to wait if the sink is busy to communicate with others. There may be multiple nodes within the $2Dv\Delta t$ area. While the sink communicates with one, others have to wait for their turn. Each sensor, i, has a maximum waiting time and this waiting time depends on its distance, d_{path}, from the path traveled by the sink. The maximum waiting time of node i can be calculated as follows:

$$T_{wait} = \frac{2 \cdot \left(\sqrt{R_C^2 - d_{path}} \right) - vt_{data}}{v}. \tag{11.20}$$

The relationship between maximum waiting time of a node and its distance from sink's path is shown in Figure 11.12. Maximum waiting time of node 1 is more than the maximum waiting time of node 2, as node 1 is closer to the sink's path than node 2.

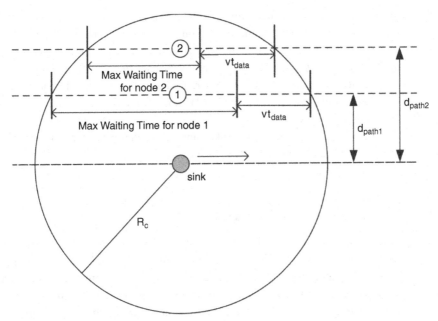

Figure 11.12   Relation between maximum waiting time of a node and its distance from the sink's path (adopted from [17], with minor modifications).

The outage value can be diminished by either increasing the value of R_C or reducing the value of t_{data}. But either of those tasks is achievable by increasing the transmission power. So, to get a desired outage value, R_C and t_{data} are selected in such a manner that the energy consumption is a minimum.

If the distance d between any two nodes is a minimum, then no outage is guaranteed while d satisfies equation (11.21):

$$d \geq \sqrt{(2D)^2 + (vt_{data})^2}. \qquad (11.21)$$

In the sink-initiated communication among the sensor nodes and the sink, the main tasks of a sensor node are sensing data and transmitting the data when requested. Other important matters such as dealing with failures and medium access control are managed by the sink itself. The sensor network lifetime can be partitioned into three phases: "start-up," "steady," and "failure-detection."

Start-up phase: In the start-up phase, the sink and the sensor nodes exchange information to get to know one another. The sink, while traveling along its path, repeatedly broadcasts a beacon signal. After receiving the beacon signal, a sensor node estimates the movement pattern of the sink and the duration of the staying period of the sink within its communication range. The sensors forward the estimated data to the sink after completion of those estimations. Using the received data, the sink calculates the priority of nodes when there is more than one sensor node waiting to transfer data to the sink. The start-up phase is repeated several times to get accurate data.

Steady phase: The sink acquires location information of sensor nodes during the start-up phase. On the basis of gathered information and its own location information, the sink initiates communication by sending wake-up calls to the sensor nodes that it estimates to be within its communication range. If there are several candidate nodes, the sink prioritizes them. The sink assigns higher priority to the sensor node that has the minimum waiting time. The sensor nodes predict the arrival of the sink within their communication range from the gathered data at the time of the start-up phase. For efficient power management, the sensor nodes monitor the channel only when they expect that the sink is nearby.

Failure-detection phase: The sink is able to detect node failures if some nodes do not respond to multiple wake-up calls. In such a scenario, the sink properly reschedules the rest of the nodes.

11.2.3   Unpredictable and uncontrollable sink

Wang *et al*. [18] proposed a mechanism where stationary sensor nodes are able to communicate with a randomly moveable sink. If the sink changes its location frequently and unpredictably, either sensor nodes may communicate with the sink through flooding, or the sink may periodically advertise its location information throughout the whole network. These two trivial solutions have several drawbacks. Flooding is used for both cases. During flooding, the sensor nodes lose valuable battery power because of (mostly unnecessary) message forwarding. Network performance is poor due to slow data rate and collisions. Scalability is also an issue here. The sink requires the flooding of the whole network frequently by its location update messages if the rate of location change of the sink is high. Two protocols, local update-based routing protocol (LURP) and adaptive local update-based routing protocol (ALURP), were proposed by Wang *et al*. [18].

In the initial phase of node deployment of the LURP and ALURP protocols, the sink broadcasts its location information within the whole network. Wang *et al*. [18] assumed that the nodes, both the sink and the sensor nodes, are able to estimate their own location and the locations of their one-hop neighbors by some means. Whenever a distant sensor node from the sink communicates with the latter, communication is divided into two phases. In the first phase, the communicating node forwards the data towards a small area known as the destination area, encircled around the sink, and when a sensor node within the destination area receives the data, it forwards them to the sink.

An example of two-phase packet forwarding mechanism of LURP is shown in Figure 11.13. The initial position of the sink is denoted by the virtual center (VC) and the sink broadcasts the location of VC throughout the whole network. A sensor node S forwards its data towards the VC by using some geocasting protocol such as GFG [19]. When a dissemination node DN_A within destination area A receives the data, the sink may already have changed its location. All the nodes of destination area A have the updated location information of the sink. DN_A forwards the data to the sink by using the updated location information. If the sink changes its location within the current destination area A, it floods only the destination area by the updated location information. All the dissemination nodes of the destination area modify their routing information to the sink

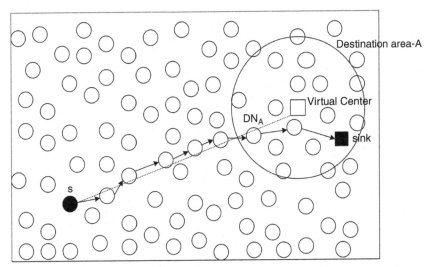

Figure 11.13   An example of message passing from a sensor node to sink in LURP (adopted from [18], with minor modifications).

after receiving the updated location information. But if the sink moves outside the current destination area, it recreates a new destination area by flooding the whole network with new location information.

Wang *et al.* [18] proposed ALURP, which is an improvement over the LURP approach. In ALURP, when a sink changes its location within the destination area, it does not broadcast its location information to the whole destination area. It restricts its update area and creates an adaptive area. The radius of the adaptive area is the distance between the virtual center (VC) and the sink's current location. An example of communication between a sensor node and sink in ALURP is shown in Figure 11.14. When a sensor node B, within destination area A, receives a packet to forward to the sink, it forwards the packet to a node DN_{AA}, which is inside the adaptive area AA. The node DN_{AA} then forwards the packet to the sink. The sink's initial location is its VC and the radius of adaptive area is zero. The size of the adaptive area increases as the sink moves away from the VC. The sink broadcasts its updated location information among all the nodes of the adaptive area only and the nodes of the adaptive area update their routing topology to the sink. Whenever a distant sensor node tries to send packets to the sink, the packets first reach any node within the adaptive area. The node acts as a dissemination node and forwards the packets to the sink. A problem occurs when the sink moves towards the VC. The size of the adaptive area shrinks. An example is shown in Figure 11.15.

The nodes that were in the previous adaptive area P, but not in the current adaptive area C, have outdated location information about the sink. As shown in Figure 11.15, the node DN_{P1} is not in C, but it was in P. So, when DN_{P1} receives a packet, it will erroneously forward that packet to DN_{P2}, instead of DN_C, as it has obsolete location information about the sink. To eliminate this problem, the sink informs the nodes that are not in C, but were in P, to remove the obsolete routing information about the sink.

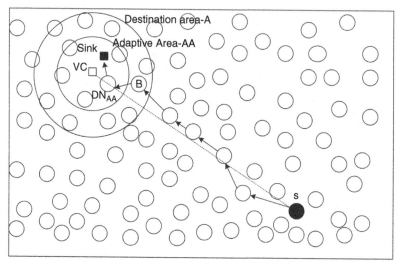

Figure 11.14   An example of message passing from a sensor node to sink in ALURP (adopted from [18], with minor modifications).

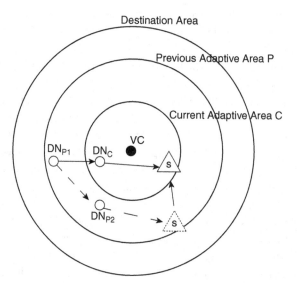

Figure 11.15   The size of adaptive area is reduced due to the sink's movement towards VC (adopted from [18], with minor modifications).

Estimating the value of the radius of the destination area is a critical issue for both LURP and ALURP. Wang *et al.* considered the monitoring area as a square field of length L, and with N sensor nodes distributed randomly all over the area. It is assumed that the velocity of the sink is v, and it changes its location p times within time period T_1. If the radius of destination area is R and the sink takes time T_2 to leave the destination area, then the maximum cost for location update in terms of energy can be expressed as:

$$E = pnh + \left(\frac{T_1}{T_2}\right) Nh. \tag{11.22}$$

The number of nodes in the destination area is represented by n, and nh is the maximum local update cost within the destination area. Inserting the value of T_2 and p, expressed by equations (11.23) and (11.24), the modified version of E is arrived at, as shown in equation (11.25):

$$T_2 = \alpha\left(\frac{R}{v}\right), \tag{11.23}$$

$$p = \beta T_1 v, \tag{11.24}$$

$$E = NhT_1 v\left(\frac{\beta \pi R^2}{L^2} + \frac{1}{\alpha R}\right), \tag{11.25}$$

where α, β are constants. It can be shown that the minimum value of E is obtained when the value of R is:

$$R = \sqrt[3]{\frac{L^2}{2\pi\alpha\beta}}. \tag{11.26}$$

An autonomous sink movement scheme is proposed by Bi *et al.* [20] to overcome the hot-spot problem. The network consists of one mobile sink and several static wireless sensor nodes. In this proposed scheme, the sink selects its location such that the sensor nodes with high residual energy are forced to forward data to the sink. The sensor nodes periodically transmit packets with sensed data and residual energy information, and after each data collection phase the sink adjusts its location on the basis of sensor nodes' remaining energy levels. The mobile sink is described as an energy mower by Bi *et al.* [20]. Owing to the movement of the sink, the sensor nodes with high residual energy consume more energy due to data transmission, whereas the nodes with low energy are able to preserve their energy. So, the energy levels of all the nodes of the network are balanced, and hence, the network lifetime is prolonged.

The sink collects data periodically from each sensor node. Each round of periodic data collection has three phases. In the first phase, the sink broadcasts its location information among sensor nodes. As the sink has restricted mobility, it is totally unnecessary to flood the location information to the whole network. The sensor nodes that are closer to the sink need to know the updated location information of the sink where the outlying sensor nodes may forward the data to the sink by using the outdated location information in the sink. By adjusting the time-to-live field of the location update message, the sink is able to restrict the spreading area of that message. In the second phase, all the sensor nodes send their sensed data using multi-hop communication. In the third phase, the sink estimates its next location from the received residual energy information of the network and reaches the estimated location before the next phase of data gathering begins.

In [20], the mobile sink moves according to the half-quadrant-based movement strategy (HUMS). Each data packet contains three types of data; the first is the sensed data, the second is the residual energy and location of the sensor node with the highest residual energy along the path from source sensor node to sink, and the third is the residual energy and location of the sensor node with lowest residual energy along the above mentioned path. After receiving data packets from all the sensor nodes, the sink or energy mower finds the node with the highest residual energy and selects the location of that node as a moving destination. The selected sensor node is named MoveDest [20]. The nodes with lowest residual energy are known as quasi-hotspots. In each data-gathering period, the sink selects its moving strategy on the basis of MoveDest and quasi-hotspots. Owing to the speed constraints of the mobile sink, the sink approaches MoveDest in steps to force it to forward data to the sink, while avoiding the quasi-hotspots, such that they are able to save their residual energy.

In HUMS, the sink generates a coordinate system, taking its current position as the origin of that system. The coordinate system is divided into eight half-quadrants, as shown in Figure 11.16. The half-quadrants, which are out of the network region, are marked by the sink as invalid. If there is at least one quasi-hotspot within a valid half-quadrant, it is regarded as a miry sector; otherwise, it is regarded as clean. The half-

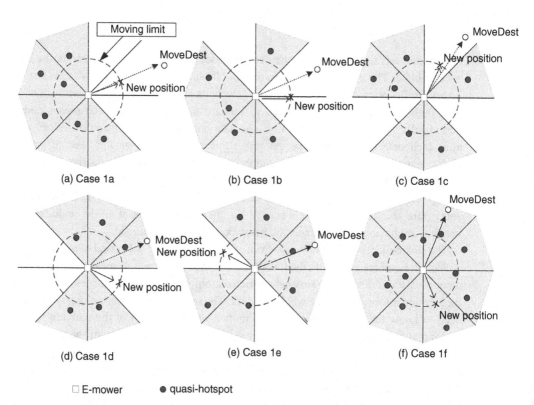

Figure 11.16 Different scenarios of half-quadrant-based moving strategy (adopted from [20], with minor modifications).

quadrant containing MoveDest is named the DestSector and the left and right half-quadrants of DestSector are known as the forward sectors.

The moving strategies of the mobile sink depend on how far it is from MoveDest. There are two different situations – in one, the sink is far from MoveDest, and in the other, the sink is within the communication range of MoveDest.

Case 1: distant MoveDest: The mobile sink moves, in steps, towards MoveDest when the distance between the mobile sink and the MoveDest is relatively high. Before arriving nearer to the target MoveDest, the sink selects another sensor node as the intermediate MoveDest and alters its movement plan accordingly. The sink selects, by using HUMS, one of the eight half-quadrants as the desired sector, and selects a suitable location by the minimum-influence position selection (MIPS) algorithm within the selected half-quadrant. The different movement scenarios, based on HUMS, of a mobile sink, can be classified into six categories depending on the locations of the MoveDest and quasi-hotspots.

Case 1a: If there are no quasi-hotspots present in the DestSector or in both forward sectors, the sink moves directly towards the MoveDest. Owing to velocity constraints of the sink, it may be assumed that the sink is able to cover maximum d distance in one step. The range of the mobile sink is denoted by a dotted circle in Figure 11.16. The sink selects the intersection point of the line towards MoveDest and the dotted circle, representing the range of sink, as its next destination point. The scenario is illustrated in Figure 11.16a.

Case 1b: If the DestSector and at least one of the forward sectors are clean, then the mobile sink selects the intersection point, as illustrated in Figure 11.16b, between the boundary of the dotted circle and the boundary line between the DestSector and clean forward sector as its destination point.

Case 1c: If the DestSector is clean, but quasi-hotspots are present in both the forward sectors, the sink selects its destination point within the DestSector, as specified in Figure 11.16c.

Case 1d: If DestSector is miry and at least one of the forward sectors is clean, then the sink moves to the clean forward sector. If both the forward sectors are clean, then the sink selects the forward sectors with highest residual energy. This scenario is illustrated in Figure 11.16d.

Case 1e: As illustrated in Figure 11.16e, if quasi-hotspots are present in the DestSector and both the forward sectors, but there is at least one clean half-quadrant or sector, the sink moves to the clean sector.

Case 1f: If all the eight sectors have quasi-hotspots, as shown in Figure 11.16f, the sink chooses the sector that has highest residual energy as its destination sector.

MIPS (minimum-influence position selection algorithm):   The mobile sink selects its destination point by using MIPS within the sector selected by HUMS. Initially, the sink chooses several points (generally four) as candidate positions, as illustrated in Figure 11.17. Each candidate position experiences a type of influence force by the nearby quasi-hotspots. The strength of the influence force of a quasi-hotspot depends on the location of that quasi-hotspot and its residual energy. The sink calculates the composite influence force by all the quasi-hotspots on each candidate position and selects the candidate

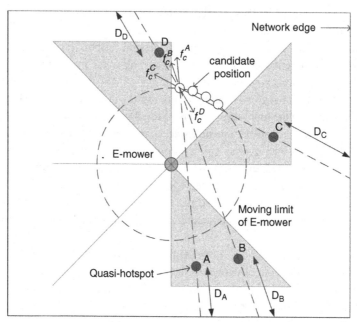

Figure 11.17 Influence of quasi-hotspots on a candidate position (adopted from [20], with minor modifications).

position with the lowest composite influence force as its destination position. The influence force by the quasi-hotspot q on candidate position c is shown in equation (11.27):

$$|\overrightarrow{f_c^q}| = k \cdot \frac{D_q^2}{E_q},$$
(11.27)

where k is a constant, D_q is the distance between q and network boundary, and E_q is the residual energy of q. The direction of the force aligns with the line from q to c. If it is assumed that the positions of q and c are (x_q, y_q) and (x_c, y_c), respectively, then the strengths of f_c^q along the x-axis and y-axis are shown as:

$$(\overrightarrow{f_c^q})_X = (\overrightarrow{f_c^q}) \cdot \frac{x_q - x_c}{\sqrt{(y_q - y_c)^2 + (x_q - x_c)^2}},$$

$$(\overrightarrow{f_c^q})_Y = (\overrightarrow{f_c^q}) \cdot \frac{y_q - y_c}{\sqrt{(y_q - y_c)^2 + (x_q - x_c)^2}}.$$
(11.28)

The composite influence force on candidate c can be calculated by using equation (11.29), where Q is the set of all quasi-hotspots:

$$|\overrightarrow{F_c}| = \sqrt{\left[\sum_{q \in Q} (\overrightarrow{f_c^q})_X\right]^2 + \left[\sum_{q \in Q} (\overrightarrow{f_c^q})_Y\right]^2}.$$
(11.29)

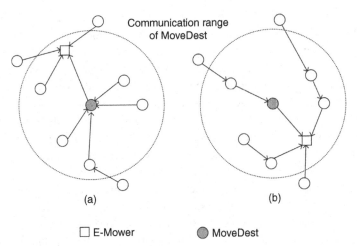

(a)   (b)

☐ E-Mower   ⬤ MoveDest

Figure 11.18 Example of different workloads of MoveDest depending on location of the sink (adopted from [20], with minor modifications).

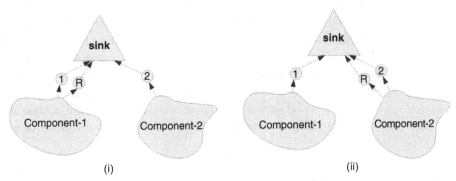

(i)   (ii)

Figure 11.19 One mobile relay node inherits the responsibilities of multiple bottleneck sensor nodes in different time periods (adopted from [20], with minor modifications).

Case 2: adjacent MoveDest: If the mobile sink is within the communication range of MoveDest, the sink selects a suitable position near the MoveDest to force the MoveDest to forward other's data and to consume more energy. An example of this scenario is shown in Figure 11.18. The workload of MoveDest in Figure 11.18a is more than that of MoveDest in Figure 11.18b.

11.2.4   Mobile relays and data mules

Mobile nodes may be used as relay nodes to increase the lifetime of a WSN. Mobile relay nodes are used in the scenarios where the sink and the sensor nodes are stationary. The mobile relay nodes give relief to a bottleneck sensor node by moving to the location of the node and inheriting its responsibilities. The responsibilities include sensing the surrounding environment, processing the data, and transmitting sensed or received data to the sink (see Figure 11.19).

Let us assume that a WSN is partitioned into two components, component 1 and component 2. These two components are connected with the sink through sensor node 1 and sensor node 2, respectively. As node 1 and node 2 exchange all the packets between the sink and component 1 and component 2, respectively, they will drain their energy more rapidly than the rest of the WSN. As a consequence, the WSN will be partitioned when either of nodes 1 or 2 dies. A mobile relay node may reduce the burdens of those bottleneck nodes. Alternatively, in one time period, the mobile relay, R, inherits the responsibilities of node 1 and in the second time period, it will reach the location of node 2 and perform the responsibilities of node 2. The presence and performance of mobile relay improves the network lifetime, and the lifetime of the WSN is doubled due to proper activities of the mobile relay R.

A framework for improving network lifetime using mobile relay node is described in [21]. N sensor nodes are distributed by a Poisson point process in an area of radius R. The sink, S, is located at the center of the monitored area. The transmission range of all the sensor nodes is assumed to be equal to unity, and we also assume that the nodes transmit their data to the sink at a fixed rate. The initial energy of a battery powered sensor node is denoted by E. It is assumed that the sink and mobile relay have unlimited energy. But the transmission range and sensing range of the mobile relay are the same as the sensor nodes'. The static nodes are partitioned on the basis of their distance from the sink. A node belongs to set P_i, if it is able to reach the sink in i hops. In other words, a sensor node s is in set P_i iff $i - 1 < d(s, S) \leq I$, where $d(s, S)$ is the Euclidean distance between s and S. All the nodes which can reach the sink within j hops are represented by the set $Q_j = \cup_{i \leq j} P_i$. Q'_j is the set of all nodes which are outside Q_j. An example of the described WSN is shown in Figure 11.20.

The authors of [21] show that the upper bound of the lifetime of a dense static network is $E/R^2 e$. They also prove that the lifetime of a dense static network can be improved by a

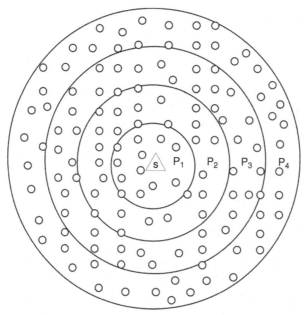

Figure 11.20 Partitions of sensor nodes in the circular network (adopted from [21], with minor modifications).

mobile relay. With one mobile relay, the upper bound of the lifetime of such a network is $4E/R^2e$. To maximize the network lifetime, the mobile relay needs to stay within two hops' distance to the sink.

In [21], a joint mobility and routing algorithm is proposed to maximize the network lifetime. It is assumed that the network is static and densely populated. The mobility pattern of the mobile relay is described as follows. The starting location of the mobile relay is same as the sink's location. The mobile relay traverses around the sink until it reaches the periphery of Q_2. The path of the mobile relay encircles the sink with concentric rings with increasing radii. After reaching the periphery of Q_2, it stays at each point of the path and relays messages to the sink. As already mentioned, Q_2 is the set of all nodes that can reach the sink within two hops.

In [21], the messages are forwarded by using the aggregation routing algorithm (ARA). An outline of ARA is described next. The sensor nodes are partitioned into two groups: nodes of Q_3 and nodes of Q'_3. The nodes of Q_3 and Q'_3 forward their messages differently to the sink. The nodes of P_1 directly forward their messages to the sink. The nodes of P_2 and P_3 forward their messages to the current aggregation node through the nodes that do not perform as aggregation nodes in P_3. Let us assume that the current position of the mobile relay is M, and the center of the network is C. A node, d distance apart from the sink, in P_k, $k > 3$, forwards its sensed data to a point P on the line CM. The distance between P and C is also d. After reaching P, the data packets are relayed along the points of CM to the current aggregation node. The aggregation node then forwards the packets to the sink with the help of the mobile relay and another static relay node from Q_2. Figure 11.21 describes the message forwarding mechanism for a node of Q'_3.

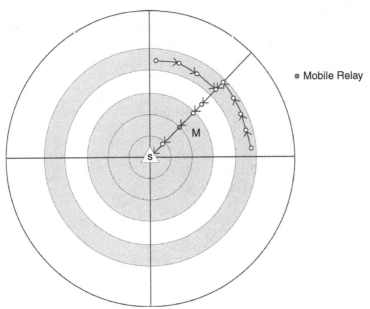

Figure 11.21 Message forwarding with ARA (adopted from [21], with minor modifications).

The mobile nodes are used as data collectors in [22, 23]. These nodes, also called data mules, collect data from the stationary sensor nodes of sparse sensor networks, buffer the data, and transfer the collected data to the sinks or access points at appropriate time. For many environmental monitoring applications such as air pollution monitoring in urban areas, and weather and terrain conditions monitoring for precision agriculture, sparse sensor networks are preferable. In such networks, either a node is out of the communication range of its neighbors, or it can communicate with at least one neighbor at the expense of high battery power. Using more number of nodes as relay nodes increases the cost of the network. In urban scenarios, pedestrians or vehicles such as buses and cabs may carry the transceivers to gather data from the sensor nodes. In other cases, robots and animals may be used as carriers. A three-tier MULE architecture was proposed by Shah *et al.* [23]. The mobility model adopted for the mules in [23] is the random walk model. The mules communicate with the sensors in the short range. The three-tier architecture saves the power of sensor nodes, as all the sensor communications are short range (see Figure 11.22). But there are two limitations of this approach – one is increased latency and the second is power consumption for mule discovery. The sensors have to wait for a mule before the actual transfer of sensed data can occur. This wait increases the latency of data delivery to the sink. The sensors also have to be continuously active to discover the presence of a mule within their communication range.

An energy-efficient data collection scheme is proposed in [22]. In the proposed scheme, the stationary sensor nodes have three states: sleep, discovery, and data transfer (Figure 11.23). As the appearance of a mule is unpredictable, a sensor node tries to discover the mule through idle listening in the discovery state. A mule periodically broadcasts beacon messages to alert a sensor node about its arrivals. After discovering a mule, a sensor node switches its state to the data transfer state from the discovery state. In the data transfer state, a sensor node uses a full duty cycle to transfer data and to reduce the time for that transfer. After the completion of data transfer, the node switches back to the discovery state. As the time gap between two consecutive visits of a mule to a sensor

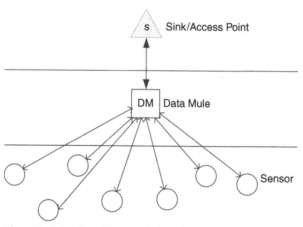

Figure 11.22   Three-tier MULE architecture (adopted from [23], with minor modifications).

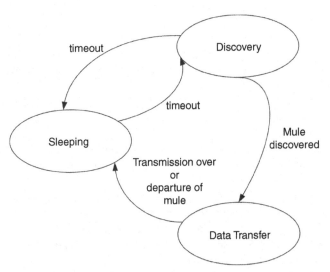

Figure 11.23   Three states of a stationary sensor node (adopted from [22], with minor modifications).

node is large, in the discovery phase, the sensor nodes follow a periodic wake up schedule. A node maintains low duty cycle in the discovery state. T_{active} and T_{sleep} represent a node's active and sleep times, respectively. The duration of a beacon broadcast by a mule is T_{B}, and the beacon period, the time between two subsequent beacon transmissions, is T_{BP}. The minimum duration of the active time of a sensor node is:

$$T_{\text{active}} = T_{\text{B}} + T_{\text{BP}}. \tag{11.30}$$

The value of the T_{sleep} depends on the desired level of the duty cycle.

11.3   Summary

This chapter illustrated the significance of wireless mobile sensor networks based on the resource-constrained considerations of static or stationary WSNs. We discussed the coverage issues of a mobile WSN and presented typical algorithms for ensuring coverage connectivity within such network. The major Voronoi diagram-based and virtual force-based approaches were discussed in this context. A grid-based approach was thoroughly studied, in which load balancing is ensured within a cell of the grid. Also, we discussed event coverage, where the mobile sensors sporadically arrive at the region of event detection. Finally, we presented the different approaches such as predictable and controllable mobile sink, predictable but uncontrollable mobile sink, unpredictable and uncontrollable sink, and mobile relays and data mules to improve the overall network lifetime [24].

References

[1] J. H. Jun, B. Xie and D. P. Agrawal, "Wireless mobile sensor networks: protocols and mobility strategies," in *Guide to Wireless Sensor Networks*, S. Misra, I. Woungang and S. C. Misra, Ed. Springer, 2009, pp. 607–634.

[2] N. Vlajic and D. Stevanovic, "Sink mobility in wireless sensor networks: when theory meets reality," *Proceedings of the 32nd International Conference on Sarnoff Symposium (SARNOFF'09)*, NJ, USA, 2009.

[3] B. Wang, H. B. Lim and D. Ma, "A survey of movement strategies for improving network coverage in wireless sensor networks," *Computer Communications*, Vol. **32**, Nos. 13–14, pp. 1427–1436, 2009.

[4] H. Zhang and J. Hou, "On deriving the upper bound of α-lifetime for large sensor networks," in *Proceedings of the 5th ACM International Symposium on Mobile ad hoc Networking and Computing (MobiHoc '04)*, New York, NY, USA, 2004, pp. 121–132.

[5] M. Berg, O. Cheong, M. Kreveld and M. Overmars, *Computational Geometry: Algorithms and Applications*, 3rd ed. Berlin: Springer-Verlag, 2008.

[6] A. Ghosh, "Estimating coverage holes and enhancing coverage in mixed sensor networks," in *Proceedings of the 29th Annual IEEE International Conference on Local Computer Networks (LCN '04)*, Washington, DC, USA, 2004, pp. 68–76.

[7] G. Wang, G. Cao and T. L. Porta, "Movement-assisted sensor deployment," *IEEE Transactions on Mobile Computing*, Vol. **5**, No. 6, pp. 640–652, 2006.

[8] H. J. Lee *et al.*, "Centroid-based movement assisted sensor deployment schemes in wireless sensor networks," in *Proceedings of the 70th IEEE Vehicular Technology Conference Fall (VTC 2009-Fall)*, Anchorage, Alaska, USA, Sep. 2009.

[9] G. Wang, G. Cao, P. Berman and T. F. La Porta, "Bidding protocols for deploying mobile sensors," *IEEE Transactions on Mobile Computing*, Vol. **6**, No. 5, pp. 563–576, 2007.

[10] A. Howard, M. J. Mataric and G. S. Sukhatme, "Mobile sensor network deployment using potential fields: a distributed, scalable solution to the area coverage problem," in *Proceedings of the 6th International Symposium on Distributed Autonomous Robotics Systems (DARS02)*, Fukuoka, Japan, June 2002.

[11] N. Ahmed, S. S. Kanhere and S. Jha, "Ensuring area coverage in hybrid wireless sensor networks," in *Proceedings of the 3rd international Conference on Mobile ad-hoc and Sensor Networks (MSN'07)*, H. Zhang, S. Olariu, J. Cao and D. B. Johnson, Ed. Berlin, Heidelberg: Springer-Verlag, pp. 548–560, 2007.

[12] P. Guo, G. Zhu and L. Fang, "An adaptive coverage algorithm for large-scale mobile sensor networks," *Ubiquitous Intelligence and Computing, Lecture Notes in Computer Science*, Vol. **4159**/2006, pp. 468–477, 2006.

[13] J. Wu and S. Yang, "SMART: a scan-based movement-assisted sensor deployment method in wireless sensor networks," in *Proceedings of 24th Annual Joint Conference of the IEEE Computer and Communications Societies (INFOCOM 2005)*, Miami, USA, Vol. **4**, 2005, pp. 2313–2324.

[14] T. P. Lambrou and C. G. Panayiotou, "Collaborative area monitoring using wireless sensor networks with stationary and mobile nodes," *EURASIP Journal on Advances in Signal Processing*, Vol. **2009**, 2009.

[15] Y. C. Wang, W. C. Peng, M. H. Chang and Y. C. Tseng, "Exploring load-balance to dispatch mobile sensors in wireless sensor networks," in *Proceedings of 16th International Conference on Computer Communications and Networks, (ICCCN 2007)*, Honolulu, HI, 2007, pp. 669–674.

[16] J. Luo and J. P. Hubaux, "Joint mobility and routing for lifetime elongation in wireless sensor networks," in *Proceedings of the 24th Annual Joint Conference of the IEEE Computer and Communications Societies (INFOCOM, 2005)*, Miami, FL, USA, Vol. **3**, 2005, pp. 1735–1746.

[17] A. Chakrabarti, A. Sabharwal and B. Aazhang, "Using predictable observer mobility for power efficient design of sensor networks," in *Proceedings of the 2nd International Conference on Information Processing in Sensor Networks (IPSN'03)*, F. Zhao and L. Guibas, Ed. Berlin, Heidelberg: Springer-Verlag, pp. 129–145, 2003.

[18] G. Wang, T. Wang, W. Jia, M. Guo, and J. Li, "Adaptive location updates for mobile sinks in wireless sensor networks," *The Journal of Supercomputing*, Vol. **47**, No. 2, pp. 127–145, 2009.

[19] K. Seada and A. Helmy, "Efficient and robust geocasting protocols for sensor networks," *Computer Communications*, Vol. **29**, No. 2, pp. 151–161, 2006.

[20] Y. Bi, L. Sun, J. Ma, *et al.*, "HUMS: an autonomous moving strategy for mobile sinks in data-gathering sensor networks," in *EURASIP Journal on Wireless Communications and Networking*, Vol. **2007**, 2007.

[21] W. Wang, V. Srinivasan and K. Chua, "Using mobile relays to prolong the lifetime of wireless sensor networks," in *Proceedings of the 11th Annual International Conference on Mobile Computing and Networking (MobiCom '05)*, ACM, New York, NY, USA, 2005, pp. 270–283.

[22] G. Anastasi, M. Conti and M. Di Francesco, "Data collection in sensor networks with data mules: an integrated simulation analysis," in *Proceedings of IEEE Symposium on Computers and Communications, 2008 (ISCC 2008)*, Marrakech, Morocco, 2008, pp. 1096–1102.

[23] R. C. Shah, S. Roy, S. Jain and W. Brunette, "Data MULEs: modeling a three-tier architecture for sparse sensor networks," in *Proceedings of the First IEEE International Workshop on Sensor Network Protocols and Applications*, Anchorage, Alaska, USA, 2003, pp. 30–41.

[24] Y. Yang, M. I. Fonoage and M. Cardei, "Improving network lifetime with mobile wireless sensor networks," *Computer Communications*, Vol. **33**, No. 4, pp. 409–419, 2010.

12 Wireless multimedia sensor networks

The wireless technology multimedia sensor network (WMSN) is a multidisciplinary technology. Researchers from diverse fields such as wireless and ad-hoc sensor networks, multimedia, distributed signal processing, control theory, and embedded systems have contributed new ideas to innovate real-life applications. Advances in embedded electronics and MEMS have resulted in greater processing power per unit size of the sensor nodes in such a way that today it is possible to bring them together in a more powerful, complex, and compact manner. With the increased processing power, it is now able to capture and process complex multimedia data such as video, audio, and images. For example, with the help of new sensing technologies, it is possible to provide proper images and audio-visual feeds in different surveillance applications, in addition to the scalar parameters of environment. The easy availability of inexpensive miniaturized hardware such as cameras and microphones, the advancements in distributed signal processing, and multimedia source coding allow much richer sampling of the environment than traditional WSNs. This trend opened up new opportunities for protocol designers to propose effective and energy-efficient solutions of the existing problems [1–62].

12.1 Network applications

Applications of WMSNs introduce new multimedia features in addition to the general capabilities of WSNs, to enrich them in terms of monitoring. Along with enhancing the existing sensor network applications, WMSNs use various multimedia-specific ideas to pioneer new opportunities in this field. We classify these applications of WMSNs into several categories and explain them briefly in this chapter.

12.1.1 Multimedia surveillance

The main limitation of normal surveillance applications based on scalar WSNs is that they are not able to get an accurate and detailed account of the environment they monitor. This highlights the importance of multimedia surveillance through WMSNs that are equipped with interconnected, battery-powered miniature video cameras. These cameras are integrated with low-power transceivers that are responsible for both local processing and wireless communication of monitored data. The ability to view images and video feeds from different angles overcomes the effects of occlusion and allows us to analyze the environment in a better way. Better surveillance systems for military operations, monitoring borders and disputed areas can be designed with the help of a wide-ranged network of camera sensors with varying capabilities and resolutions.

12.1.2   Traffic management

There has been steep rise in on-road traffic today, and thus, traffic management has become challenging. Applications of WMSN can satisfy the need for traffic management by streamlining traffic flow. Proper placement of multimedia sensors along the road-side offers traffic monitoring, traffic congestion avoidance suggestions, and routing advice. Mobility and traffic flow can be increased through proper use of WMSNs. For example, sensors can track vacant spaces in a parking lot and provide on-demand parking advice to the drivers. Even traffic law enforcement can be monitored by using WMSNs. Camera sensors can retrieve information such as the speed of a vehicle at a particular point and thereby detect violations of traffic rules along with the identity of the violator. Buffered images and video streams captured by the camera sensors turn out to be helpful to analyze the causes of road accidents.

12.1.3   Advanced health care

Technical advancements also lead to better and groundbreaking opportunities in the health care domain. People worldwide suffer from health-related problems due to the lack of proper infrastructure. WMSNs play an important role by providing innovative ideas for remote and ubiquitous health care of patients. With the help of telemedicine devices and WMSN technology, remote health monitoring is experiencing advances in techniques via motion and activity sensors, audio, and video sensors. Ubiquitous and periodic monitoring of patients' health can be achieved by deploying sensors in patients' rooms or houses. Images of patients' facial expressions, body movements, and other health information such as ECG pulse, blood pressure, respiratory conditions, and body temperature, can be gathered together through telemedicine devices. Finally, this information is forwarded to doctors periodically in the form of multimedia data, through the corresponding WMSN. In case of an emergency, the sensors can act as alarms and are able to automatically inform the hospitals.

12.1.4   Environmental monitoring

Low-power and energy-efficient sensor nodes are useful for environment monitoring applications, as these applications need reasonable power backup to provide long-time services. The sensor nodes are deployed over an area to be monitored. Then, they start sensing and gathering data autonomously. Reporting of the sensed video feeds or other multimedia data to the sink can be done periodically, or in an on-demand service. For example, to measure wind and ocean currents in a particular area, the sensor nodes can be deployed on the surface of an ocean. Before the occurrence of a storm, the sensors can detect the atmospheric change of that area and can report to the sink with the relevant information. This information in the form of multimedia data helps us to get a better idea of the environment.

12.1.5   Industrial process control

Mass production of products in factories generally undergoes several important and automated steps such as fabrication of several parts, inspection of the intermediate product, and measurement of accuracy of the final product. Machines are involved in most of the steps. Evidently, machines need to be accurate, robust, and reliable in nature. Multimedia data such as imaging and video feeds captured by a network of heterogeneous sensor nodes can be very useful to track the efficiency of these machines. To reduce maintenance costs, WMSNs can be deployed to guide machines properly to achieve automation and accuracy. Various types of cameras with new computer vision techniques can capture rich images of different resolutions. With the help of image-processing algorithms, WMSNs allow efficient sensing of the machine characteristics to enrich the process of mass production.

12.1.6   Virtual reality

By using advanced computer technology, virtual reality provides a simulated three-dimensional world that a user can explore, with the feelings that he/she is a part of that virtual world. In general, it involves placing several tiny sensor nodes and light-weight feedback devices on an individual body to form a small network of sensor nodes that communicate wirelessly with one another. To enrich this modern approach of virtual environment, WMSN applications provide real-time communication support in the form of complex multimedia data such as audio and video streams.

12.2   Challenges in WMSN

Integrating advanced multimedia technologies involves a lot of challenges in WMSNs. The applications should be designed in such a manner that wireless communication and multimedia data transfer satisfy a certain level of QoS (quality-of-service) requirements. In WSNs, one must consider resource constraints, in addition to an adverse communication medium. But in WMSNs, in addition to this, an acceptable QoS to the multimedia application is required. Some of the distinct challenges that present WMSNs are experiencing are described below.

12.2.1   Resource constraints

WMSNs are mainly different from scalar WSNs in terms of resource constraints. Multimedia sensors are mainly concerned with different types of content such as audio and video streams that require more battery power, storage, channel bandwidth and processing power. It is very difficult to recharge sensor nodes deployed in a dense network. Sometimes, when attempting to recharge the sensors when their battery gets discharged, they incur more cost than disposing them and deploying new sensors to the area. Therefore, the main challenge, as well as the primary concern of ongoing research on WSNs and WMSNs, is to design energy-efficient applications. In addition to this, WMSNs are more resource-constrained than WSNs, as they deal with multimedia data. Different image, video, and audio sensors capture multimedia data, which are several times larger in size than those generated by scalar sensors. Hence, storing locally or buffering these data is challenging, as the sensors generally do not have much storage space. Owing to the large data size, they also require large channel bandwidth for data transfer. For example, the data rate for a MicaZ mote is 250 kbps. For large multimedia data, the system may require higher data rates for successful transmission to the sink. Again, high data rate and limited memory space lead to the requirement of high processing power of sensor nodes, which is another challenge to implement.

12.2.2 Variable channel capacity

Capacity of the link channel is an important measure to design network architecture and protocols. In wired networks, the capacity of each link is assumed to be constant. But in multi-hop wireless networks, data pass over several links and finally reach the sink. Successful delivery of data packets depends on the link capacity and link quality at each hop. Several factors such as the transmission power of each hop, routing, rate policies, modulation, and error detection code chosen by the node control the attainable capacity of each link. These factors again depend on the local conditions of sensor nodes. Hence, capacity and delay at each link vary continuously and it is extremely difficult to predict the link quality at each hop. In the case of WMSNs, a robust and reliable application should have bounds on these parameters.

12.2.3 Multimedia coding technique

General video coding techniques for wired and wireless communications are based on predictive encoding. But this approach is not suitable for multimedia sensors, as it involves complex processing algorithms and high power consumption. Several compression algorithms are applied to the raw data, such as JPEG for images and MPEG in case of videos. These algorithms take advantage of statistical and perceptual characteristics to compress data. To handle multimedia content over wireless networks and to support real-time features, WMSN applications undergo different techniques for multimedia processing and source coding. Among them, the well-known one is distributed source coding, which, along with coding criteria, also meets the resource constraints for WMSNs.

12.2.4 Redundancy removal

Utilizing the available bandwidth efficiently with energy-efficient techniques is an important concern for WMSNs. In order to achieve this, the size of raw multimedia data must be reduced by removing redundancies. The raw data may consist of the following two types of redundancies.

- Statistical and perceptual redundancies: As already mentioned, this can be reduced by performing multimedia source coding and other types of in-node processing on the data.
- Spatial–temporal redundancies: As WMSNs contain a dense arrangement of sensor nodes, it is obvious that the sensors have overlapping fields of view. This overlap causes spatial and temporal redundancies in data. Data alignment by conciliating the different readings from sensors can remove this type of redundancy.

However, both multimedia source coding and data alignment techniques involve large processing power, which may not be available on all the sensor nodes. Therefore, these techniques are employed especially on special multimedia-enabled nodes that have the required processing power.

12.2.5 QoS requirements

Compared to WMSNs, scalar WSNs are more flexible in terms of QoS requirements. Developing energy-efficient algorithms and network protocols are the primary aims of ongoing research on WSNs. In most of the cases, the research efforts compromise throughput and latency to achieve an energy-efficient framework. However, it is very stringent in the case of WMSNs. In addition to energy efficiency, WMSN applications need to have a certain level of satisfaction in throughput and latency. As applications of WMSNs deal with video streaming, some varying delay (jitter) must be provided to the data packets transmitted from the source, in order to play the video consistently at the sink. To satisfy the necessary QoS requirements, the application level demands must be mapped to the latencies, permissible error rate, and other performance metrics of lower layers. In order to achieve these requirements, one has to apply the traditional scalar WSN-specific algorithms and protocols in a different way to WMSNs.

Another challenge regarding QoS is that different applications have different QoS requirements. For example, an application may be tolerant to loss of several consecutive frames in a video stream, but in some other applications dealing with images, the loss of any part of an image may lead to failure of that application. Therefore, application level requirements are also important and should always be conveyed to the lower layers, so that the lower layers can act accordingly, to meet the necessary requirements.

12.3 Different architecture of WMSNs

While designing different architectures for scalar WSNs, scalability and energy-efficiency should be the primary interests as the traditional scalar WSNs are composed of homogeneous, multifunctional, simple nodes. However, in the case of WMSNs, we must also take care of the service requirements of the resource-constrained multimedia

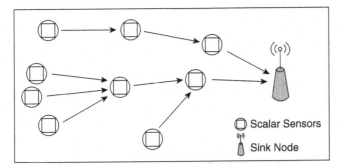

Figure 12.1   Traditional WSN architecture.

applications. Implementing the necessary abilities on each sensor node disfavors the deployment of sensor nodes on a large scale, as it is not at all effective in terms of size and cost of the nodes. Therefore, it is a challenge to design proper architecture that will efficiently utilize the resources as well as take care of the applications' service requirements. Different types of architectures and their advantages and limitations, including the traditional WSN architecture, are discussed below.

12.3.1   Traditional WSN architecture

Scalar WSNs are composed of several multi-operational homogeneous sensor nodes. Therefore, the scalar WSNs adopt a flat, non-hierarchical traditional framework, where all sensor nodes are equal in terms of hardware and software specifications. Sensor nodes sense the environment, gather data against several environmental parameters, perform multimedia processing, and report to the sink or base station over a multi-hop path, as illustrated in Figure 12.1.

This coordination among sensor nodes can be of two types. Sometimes it is distributed among the nodes throughout the network. It is also possible to have a centralized architecture, in which the sink acts as a master and volunteers the coordination among all other sensor nodes. Aggregation of data, which involves only simple computations, can take place on any node in the network.

12.3.2   Heterogeneous, single-tier, clustered architecture

In the presence of powerful multimedia-enabled nodes, the cost and resource consumption incurred in a WMSN increases drastically. To avoid this situation, it is possible to have a clustered architecture by deploying a few comparatively powerful, multimedia-enabled sensor nodes in a dense network of simple nodes. In this architecture, the network is organized into several clusters, in which each cluster has one multimedia-enabled node, and some simple nodes. All the simple nodes report to the multimedia-enabled node, which is responsible for storing and processing data on behalf of the simple nodes. On the other hand, the simple nodes capture multimedia data, perform some simple processing on them and report the semi-processed data to the multimedia-enabled node in that cluster for

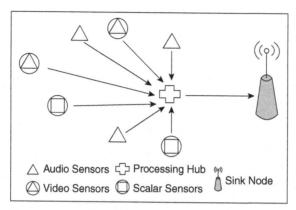

Figure 12.2   Heterogeneous, single-tier, clustered architecture.

further complex computation and processing. Together they coordinate to achieve the objectives of the application. This architecture is depicted in Figure 12.2.

12.3.3   Heterogeneous, multiple-tier architecture

It is not an efficient design approach to keep several nodes with different capabilities in a single tier. To achieve energy efficiency, the nodes need to be organized into multiple tiers, as shown in Figure 12.3. The comparatively simpler nodes comprise the lower tier and with the increment in complexity, the nodes should move on to the higher tiers. Generally, the complex nodes in the higher tiers consume more resources than the ones in the lower tier. Therefore, in order to reduce overall energy consumption of the network, the complex nodes in the higher tier are switched on, only when they are being requested for complex multimedia processing.

Various sensor hardware available on the market employs multi-tier WMSN architecture. For example, SensEye is a surveillance platform, which follows a three-tier camera sensor network architecture [11]. The complexity, ability, and resource consumption of the nodes increase from the lower tiers to the higher ones. In SensEye, the lower tier is composed of simple motes such as Mica2, which are equipped with low-resolution and low-power-consuming cameras such as Cyclops and CMUCam3. Tier 2 consists of powerful Stargate motes. These motes are equipped with medium-resolution cameras and 400 MHz XScale processors. Communication between tiers 1 and 2 is undertaken at the data rate of 900 MHz. These Stargate motes communicate with one another through 802.11 radios. Tier 3 is composed of high-resolution cameras such as Sony SNC-RZ30N.

The primary aim of any architecture for a WMSN, including this one, is minimizing resource consumption as well as providing the required service to the applications. Mapping the applications' requirements to the lowest tier is the main challenge. On the other hand, the network must satisfy resource constraints and performance issues. In order to furnish this, the network switches off the motes in the higher tier and only turns them on when required. The motes in tier 1 wake up periodically, as in a traditional WSN.

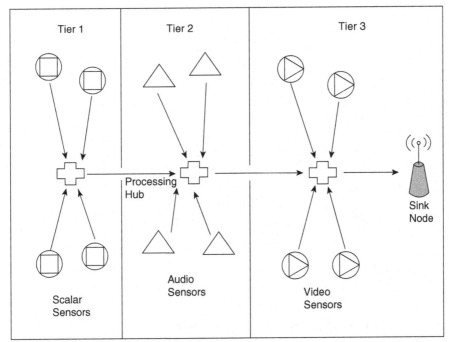

Figure 12.3   Heterogeneous, multiple-tier architecture (adopted from [43], with minor modifications).

When the motes detect an event, they attempt to focus on the object by analyzing the overlap of the cameras' field of view (FoV). The appropriate motes in tier 2 are turned on, depending on the result of the localization of the object in tier 1, and then perform the necessary processing on multimedia data to recognize the object. Motes in tier 3 are woken up by those in tier 2, to track the object.

12.3.4   Integrated architecture

Several different large-scale WMSNs can be integrated together using network gateways (Figure 12.4). Abstraction can be introduced further through hardware independent interfaces. User queries can be processed with more efficiency through this abstraction. For example, the WISEBED (WIreless SEnsor network testBED) project consists of several WSNs combined together that the user can access through the Internet as a federated test bed [45]. Each sensor network is connected to the WISEBED platform through its own portal server, which acts as a gateway between them. Users' queries to a particular network can be serviced through the collaboration of all sensor networks present in the test bed. IrisNet is another example of such integrated architecture [10].

The main advantage of integrated architecture is the users' abstraction of the architecture. Deploying new sensor nodes and integrating them into the system is simplified due to this abstraction, and the interface is independent of hardware used in different networks.

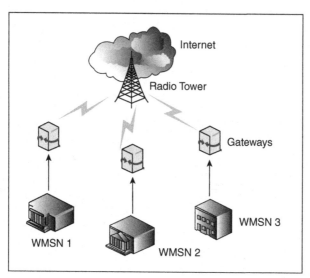

Figure 12.4   Integrated WMSN architecture.

12.4   Comparison of different architectures

Each of the architectures discussed above has its own advantages and limitations. A discussion regarding the usefulness of these architectures is presented in Table 12.1.

12.5   Multimedia sensor node architecture

Figure 12.5 shows a typical block diagram of a multimedia sensor node and its components [45]. Primarily, a single node is composed of some basic components such as a sensing unit, a processing unit, a storage unit, a communication unit, and a coordination subsystem. The tasks of each unit are briefly discussed below.

- The sensing units consist of different sensors (camera sensor, audio sensor, scalar sensor) and analog-to-digital converters (ADC). Sensor nodes observe and record phenomena and ADC convert these analog signals to digital form and store them in the memory unit.
- The processing unit coordinates sensing and communication operations. It also processes multimedia data stored in the memory unit.
- The communication unit consists of a transceiver and system software. The transceiver manages data communication and the system software provides the necessary support for the communication stack.
- The coordination subsystem coordinates operations such as network synchronization and location management. A power unit provides power backup to each unit and harvests energy from environment.

Table 12.1 Advantages and limitations of different architectures

Architecture	Advantages	Limitations
Traditional WSN architecture	• It is possible to extend the network by deploying additional sensor nodes. • The nodes are homogeneous in nature. • Implementation costs and maintenance costs are much less. • Balance resource consumption by distributing the workload.	• Dealing with multimedia content is difficult, as it involves huge resource consumption. • Handling multimedia data may result in reduction of network lifetime. • Transporting large amount of multimedia data may lead to rapid depletion of network resources.
Heterogeneous, single-tier, clustered architecture	• Dedicated storage and processing hubs are suitable for multimedia sensor networks. • Clustering of nodes balances the workload and utilizes the resources efficiently.	• Not scalable as the scalar WSNs. • Network management is difficult, as cluster formation depends on proper deployment of multimedia-enabled sensor nodes. • Not cost effective for certain applications.
Heterogeneous, multi-tier architecture	• Improved functionality, higher reliability and wide-range coverage of different applications. • Lower energy consumption in comparison to the other two architectures. • Storage and processing can be performed separately at each tier.	• Not scalable as the scalar WSNs. • Involves complex design.
Integrated architecture	• High scalability due to its hardware independent nature and easiness to integrate sensor networks of diverse types.	• Involves complex design.

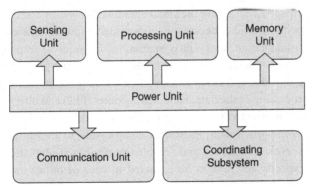

Figure 12.5   Block diagram of a wireless multimedia sensor node (adopted from [45], with minor modifications).

12.6   Existing sensor node platforms

There are several existing WMSN sensor node platforms available on the market. Panoptes, Cyclops, and SensEye are the most famous. Their architectures and designs are discussed below.

12.6.1   Panoptes

Panoptes is a scalable, low-power and single-tier architecture of sensor networks. It deals only with video-based sensors and can deliver high-quality video with very low power consumption. The following design requirements are considered to develop a video sensor node for Panoptes architecture, as presented in [38].

- Low power: Power is an important, but limited resource that needs to be consumed efficiently. Economical use of power can significantly increase the lifetime of the network, leaving the liberty to deploy new sensors in the network.
- Flexible buffering: As sensor nodes are associated with very low amount of intermediate buffer storage, the buffering technique should be changed dynamically taking account of the application requirements. There are applications that do not allow data older than some predefined time in the buffer. Again, some applications transmit too much data, without considering buffer limitations. Therefore, dynamic decisions must be taken to manage buffered data.
- Power management: Some applications need to sense as much data as possible within the lifetime of the sensor nodes. On the other hand, there are applications that require the sensor nodes to sustain long durations of time by harvesting energy from any possible source.

Panoptes sensor hardware is composed of the following components [38].

- Intel StrongARM 206 MHz embedded platform with 802.11 card.
- Logitech 3000 video camera (USB-based).
- 64 Mb of memory.
- Linux 2.4.19 kernel operating system.

The Panoptes sensor nodes capture and transmit a 320×240 resolution video at 18–20 frames per second (fps) by consuming the small amount of power of nearly 5.5 W.

Panoptes deals with several processes such as video capture, data compression, filtering, buffering, adaptation, and video streaming. In order to implement these features, Panoptes uses single-threaded asynchronous architecture [38]. The processes start with capturing data through video camera, then decompression takes place at the kernel before the processing of the data. Panoptes supports JPEG and differential JPEG for spatial compression of video frames. According to the rules specified by the user, Panoptes filters less-important data, instead of compressing or transmitting them as they are to the next level. An efficient and flexible priority-based adaptation algorithm is used in [38] to specify the data to be removed in case of buffer overflow. This algorithm maintains two mark levels to manage the buffer; one is the low-water mark level, and another is the high-water mark level. When buffer data run beyond the high-water mark level, then data need to be removed until the low-water mark level is reached, to make free space for new data. If priorities are similar, then this algorithm chooses data to be removed according to the timestamp of the data. Older data are removed prior to the newer data in this case.

12.6.2   Cyclops

Cyclops [39] is a camera-based sensor hardware device that differentiates itself from other sensors in terms of energy efficiency, simple interfacing techniques, and its versatility with a variety of sensor network platforms. It is mainly designed to support advanced video-based sensor applications.

The following set of rules is followed by Cyclops to achieve the specific characteristics of this hardware device.

- Cyclops efficiently utilizes memory and other resources to reduce power consumption. It uses external SRAM for memory access on demand.
- Cyclops introduces parallelism into sensing computations.
- To reduce energy consumption, Cyclops keeps the sub-parts of the hardware device in sleep mode when their services are not needed.

Cyclops' sensor hardware mainly consists of the following components [39].

- An imager.
- A microcontroller unit (MCU).
- A complex programmable logic device (CPLD).
- An external SRAM.
- An external flash.

The MCU instructs the imager to capture frames after setting the imager parameters. CPLD helps the imager by providing a high-speed clock, clock synchronization, and memory control. As CPLD activities need more power, it is the responsibility of the MCU to terminate the CPLD services when they are not needed. In addition to the internal MCU memory, external SRAM also helps to store captured images for further manipulation. Local storing and template matching are handled by the flash memory. Each of the components shares a common data bus for ease of data communication among one another. Cyclops uses three MOSFET switches to activate and shut down the components.

It is evident from the experimental results that this approach of putting Cyclops in the sleep mode with the help of these MOSFET switches, when no resources are in use, is very energy efficient.

The most distinctive characteristic of Cyclops is that it implements some principles in designing necessary algorithms. They are the following.

- Data-reduction mechanisms for keeping captured images at a minimum size by maintaining proper aspect ratio and output resolution.
- Rapid reduction in data size as well as in its execution flow, by proper selection of computation procedures on image operations.
- Reusing stored computation results whenever they are required again.

Like hardware, the Cyclops software design is also focused on issues such as transparent usage of resources, support for long computations involved in image processing, and, quite importantly, the synchronized access of MCU and CPLD to the shared resources. Cyclops' software runs on the TinyOS operating system, and is written in the NesC [40] language. Drivers, libraries, and sensing application are the main components of Cyclops' software design. Drivers are responsible for establishing communication between the MCU and peripherals (such as the host mote). Processing of raw images is handled by the libraries. The sensor application receives commands from the host and performs the required sensing operation.

12.6.3   SensEye

SensEye [11] is a sensor hardware-based network with multi-tier heterogeneous architecture intended mainly for video-based applications. This type of architecture provides considerable accuracy in surveillance applications, by mitigating energy consumption by sensor nodes. The heterogeneous nature of networks and better performance of cameras than in the homogeneous sensor networks result in increased effectiveness of this kind of architecture.

Several camera sensor tasks are handled by SensEye. Examples include the detection of objects, recognition of objects, and object tracking [11]. Object detection indicates the identification of new species into the sensing range of a camera sensor node. SensEye employs proper object-detection algorithms to identify objects. The detected object needs to be classified to determine its type. The system understands the distinctions of objects belonging to the same type (e.g. distinctly identifying a lion and a goat), through proper and efficient object recognition. Finally, if the object is of interest to the application, then the movement of the object can be tracked. It involves different tasks such as computing current location, trajectory of current object, and sending video or images of tracked objects to the nearby station. As the object moves through a particular path, the tracking responsibility also moves from one sensor node to another.

Each tier of SensEye is made up of different components. They are as follows [11].

- A camera sensor.
- A microcontroller unit.
- A radio.
- On-board RAM.
- Flash memory.

SensEye [11] follows a three-tier architecture, in which the tiers are formed based on the capabilities of sensor nodes. The lowest tier consists of several mote nodes [40] with 900 MHz radio and CMUCam camera sensors. The next tier is composed of Stargate [41] nodes equipped with web cameras and 400 MHz XScale processor that runs on the Linux operating system. High-resolution pan-tilt-cameras with embedded PCs are present in the uppermost tier. The processing power, networking, and imaging capabilities are improved in this tier at the cost of increased power consumption.

In order to provide energy-efficient results and to design network protocols for intra-node and inter-node communications, SensEye maintains some important principles [11]. They are as follows [11].

- Involvement of the higher layers incurs more energy than in the lower layers to execute a specific task. SensEye schedules tasks starting from the lowest tier that is still capable of meeting the application requirements of the task to mitigate overall energy consumption.
- Improving network lifetime and reducing power consumption by putting high-resolution radios in the sleep mode until any new interesting object is detected.
- Avoiding redundancy due to overlapping regions of camera sensors by taking intelligently waking-up neighbor nodes based on the object locations and trajectory.

12.7   Communication layers

The communication procedure used by WMSNs and the corresponding challenges behind its implementation are described in this section. WMSNs are composed of five communication layers, which are physical, link, network, transport, and application. One of the key concerns is to ensure proper resource utilization and mitigate energy consumption while designing the layers.

12.7.1   Physical layer

Providing the basic framework for data communication and error controlling of transmitted data are the main goals of the physical layer. While designing this layer, the standard techniques must take care of the characteristics of the physical signal such as frequency of the carrier, choice of modulation technique, and channel encoding. Currently available notable protocols are IEEE 802.11 (Wi-Fi), IEEE 802.15.1 (Bluetooth), IEEE 802.15.4 (ZigBee) and IEEE 802.15.3a (UWB) [44]. Among them, ZigBee is most frequently used in commercial multimedia sensors due to its low power consumption. Apart from ZigBee, ultra-wide band (UWB) technology is also very useful for short-range radio communications. We discuss these two protocols briefly in this section.

12.7.1.1   ZigBee

ZigBee is a suitable choice to satisfy the general requirements of WSNs and WMSNs such as low power consumption, low cost, infrastructure-less ubiquitous communication among sensor nodes. It uses IEEE 802.15.4 as a standard for MAC and physical layers. This standard operates on different bandwidth range with different data rates, which is described in Table 12.2.

Simple and energy-efficient dealings of IEEE 802.15.4 with different traffic types such as periodic data, intermittent data, and low-frequency data have made this standard a favored choice in commercial sensor motes.

12.7.1.2   Ultra-Wide Band (UWB)

As the name suggests, UWB technology operates in a wide bandwidth range of the order of 100 MHz to several GHz for network communication. It offers high data transfer rates as well as low power consumption within the short range in the order of tens of meters.

Table 12.2 Bandwidth range and data rates standards [43]

Area	Bandwidth range (MHz)	Data rate (kbps)
Europe	868.0–868.6	20
America	902–928	40
Worldwide	2400–2483.5	250

Though initially this technology was used in radars and environments having low coordination among radios, now-a-days it is used in short-range communication. There is also current research to make use of it in "wireless USB" technology, which will enable a data rate of several Gbps in short-range transmission. According to the rules specified by the Federal Communications Commission (FCC) of the USA, UWB operates in the spectrum ranged between 3.1 GHz and 10.6 GHz [43]. Twenty percent of the center frequency has been designated to the UWB by this commission. If the center frequency is above 6 GHz, then UWB gets a bandwidth of 500 MHz. With this large range of bandwidth, it is itself a challenge for this technology to avoid interference in other narrowband systems. Thus, the FCC has mandated strict limits on the maximum power spectral density. Certainly, the power spectral density of some of the techniques is lower than that of some common electric devices, whose spurious radiation, if produced at an undesired frequency, can cause interference. It is also true that the UWB devices are allowed to interfere with the existing systems through overlay, in such a way that they do not experience any degradation in their performance.

There are two different UWB techniques available. They are direct sequence UWB (DS-UWB) and multiband OFDM UWB (MB-OFDM). The first one is proposed by the UWB Forum, while another leading industry group WiMedia Alliance proposed the second one. Brief description of these two techniques is provided below.

Direct sequence UWB
According to the developer of DS-UWB (the UWB Forum), this technology has certain useful capabilities such as low cost, low power consumption, and, most importantly, optimal interference with other traditional carrier-based techniques [47, 49]. DS-UWB is different from conventional wireless systems in respect of its data transmission policy. It transmits data by sending low-power Gaussian-shaped pulses of energy generated at very high rates. Each of these pulses has very short duration, typically 10–100 ps (pico-seconds). Short pulse duration leads to a large range of bandwidth for UWB transmission. A frame consists of several hundreds of these pulses. The number of pulses per frame is also restricted to some extent. Consequently, the power spectral density acquired by the entire bandwidth is very small. This results in a much reduced chance of interference with other transmissions using narrowband systems.

There are several ways to modulate DS-UWB transmissions, among which pulse position modulation (PPM) and binary phase shift keying (BPSK) are the two that provide the best performance in terms of modulation efficiency and spectral performance.

- PPM modifies the position of a pulse to represent a binary 1 or a 0.
- BPSK reverses the phase of a pulse by 180 degrees to represent binary 1 or 0.

Owing to better scalability, this new technology is ideal for power-constrained applications that need high-rate data transfer as well as a simple implementation technique. Thus, DS-UWB is quite suitable to be part of the physical layer for personal area networks (PANs).

Multiband OFDM UWB

Unlike DS-UWB, the multiband technique uses orthogonal frequency division multiplexing (OFDM) for data transmission [48]. This approach divides the entire bandwidth into several bands, each with a bandwidth of greater than 500 MHz. It transmits data simultaneously over these bands, which are precisely placed at different frequencies all over the range and bypass interference with the other systems. The transmitter transmits a single bit on one band, and it follows a round-robin fashion and hops to the other bands. It employs the fast Fourier transform (FFT) algorithm to provide better efficiency in capturing energy in a multipath environment.

Comparison of DS-UWB and MB-OFDM

Both of the techniques have their unique features involving some advantages and limitations; we show them in Table 12.3.

In a nutshell, the UWB technology leads to several advantages such as its low cost, low power consumption, higher data-transmission rate, and better capability to avoid interference. However, this technology is still under research to find areas to which it can be incorporated. For example, prolific research on integrating UWB with an advanced cognitive radio technique to increase spectrum utilization is on the verge of success.

Table 12.3 Advantages and limitations of DS-UWB and MB-OFDM

UWB technique	Advantages	Limitations
DS-UWB	• It can be used to locate nodes, due to its short pulse and low duty cycle. • Provides high data rates that scale to 1 Gbps or more. • Low cost and long battery life make it suitable for PANs.	• Sending and receiving activities should be synchronized, and achieving precise synchronization is a bit complex in this technique.
MB-OFDM	• Location of each band can be chosen precisely with respect to other narrowband systems, to mitigate interference with them. • Involves high spectral efficiency and efficient technique to collect multipath energy using a single RF chain.	• Overall implementation is complex and difficult. • Use of FFT may not be cost effective for some applications.

12.7.2   Link layer

Satisfying the QoS requirements is a crucial issue for WMSN applications. The most important challenge is enabling the link layer is to ensure energy-efficient, reliable, and error-free data transmission to the higher layers without compromising the specific QoS requirements of the applications. High energy consumption by sensor nodes is a result of several conventional activities such as energy spent on operating the radio, transmission and reception of data packets, energy lost due to packet collisions and retransmissions. Overhearing and idle listening of sensor nodes are also responsible for unnecessary energy consumption, and, thus, they make the link layer design more challenging for WMSNs. Another important consideration in WMSNs is to satisfy end-to-end service requirements of the multimedia applications. Providing the same service to all data packets in real-time applications is not an efficient approach to satisfy QoS requirements. Therefore, packet-specific categorization and offering different policies for channel access, scheduling, transmissions, and error control is more justified in order to palliate overall energy expenditure as well as to achieve the requirement goals [43].

The major functions of the link layer are channel access, scheduling, and error control. We discuss these three operations briefly in this section.

12.7.2.1   Channel access policies

Channel access policies of the link layer mostly depend on contention over the channel. When two nearby sensor nodes try to access a channel at the same time, then contention occurs. This leads to packet collisions, followed by retransmission of data packets, which results in unnecessary wastage of energy. Therefore, based on contention, the channel access mechanisms of link layer protocols can be classified into three different techniques. They are contention-free channel access, contention-based channel access, and hybrid channel access. Each of these techniques is now briefly discussed.

Contention-free channel access
This mechanism is based on the allocation of a channel to the sensor nodes through a time division technique. When a node wants to use a channel, it reserves the channel for a particular time duration. This node is not allowed to reserve any other time interval, and similarly, the reserved interval is not available for further reservation by other nodes. The contention-free protocol transmits data packets without any collision. TDMA, FDMA and CDMA are the common examples of contention-free channel access techniques. We explain this technique through the discussion of TDMA mechanism.

Time division multiple access (TDMA)
This protocol divides the entire communication channel into several time slots. The time slots are again grouped into frames. The sensor nodes consider the QoS requirements of their data streams and reserve time slots for data transmission. At a particular time, only one node can transmit data through the channel using its own reserved time slot. The most common reservation policy is the centralized approach,

in which a node volunteers the whole process by querying other nodes for their channel requirements, and accordingly allocates the time slots to them. However, it is also possible to implement the reservation procedure in a decentralized manner. The TDMA protocol is essentially energy efficient, as a sensor node can turn off its transceiver until its reserved time slot appears. But the synchronization of its turning on and off with the corresponding time slot is very necessary.

Caccamo *et al.* [2] proposed a TDMA-like scheme that divides the whole network into several cells. Each neighboring cell operates on a different frequency. Scheduling within a cell follows the earliest delay first (EDF) procedure, which ensures real-time data having strict bounds on data delivery time. Another work by Liu *et al.* [3] provides significant modification to this scheme. It employs CDMA as the intra-node link layer protocol and performs better than traditional TDMA.

Contention-based channel access

Unlike TDMA, contention-based techniques do not employ time or frequency division and reservation policies. In contention-based mechanisms, the shared channel is served in an on-demand fashion. The sensor nodes compete with one another to grab the channel when it is free. The node that conquers this contention uses the channel for its transmission. The carrier sense multiple access (CSMA) scheme is mostly used for contention-based channel access.

Carrier sense multiple access (CSMA)

This technique involves continuous monitoring of a channel by the sensors to track its availability for data transmission. If the channel is busy, i.e. it is already in use by some other node, then all other nodes must refrain from transmitting data, to avoid collision in the channel. When the active node finishes its transmission and makes the channel free, then only another node can start its transmission. There are several types of CSMA techniques such as non-persistent, 1-persistent, and p-persistent [63]. But in a multi-hop environment, all these mechanisms are not at all energy efficient for the reasons described below [55–62].

- Idle listening: Continuous sensing of the availability of the channel includes unnecessary power consumption by the sensor nodes. The transceiver circuits of the sensor nodes must be kept "on" till the channel becomes available to the transceiver.
- Collisions: In the case of WMSNs there may be several sensor nodes waiting to access the channel for transmission. So, when it becomes available, all waiting nodes try to transmit their data simultaneously, which leads to collision followed by retransmission of data packets.
- Overhearing: Sometimes sensor nodes receive packets that are meant for other nodes, i.e. overhearing takes place among the sensor nodes. It costs a significant amount of energy wastage in the case of a dense network of multimedia sensors.

- Overheads by protocol: Protocol headers that are used for signaling are sometimes considered as overheads, as they do not represent data directly. Overheads and energy wastages due to inefficient sending, receiving, and listening associated with protocol headers and control packets can be significant sometimes.

The carrier sense multiple access with collision avoidance (CSMA/CA) protocol [63] was introduced to increase energy efficiency in the general CSMA technique. In CSMA/CA, after detection of a busy channel, the sensors stop sensing, wait for a random amount of time, and then sense the channel again. Therefore, it reduces the chances of idle listening. Again, its three-way handshaking procedure, by using short request-to-send (RTS) and clear-to-send (CTS) frames, ensures the avoidance of collisions to a large extent. But the random waiting time of nodes in this protocol leads to an unpredictable latency and jitter, which is a barrier to achieving the necessary service requirements.

Sensor medium access (S-MAC) protocol

The sensor MAC protocol increases the energy efficiency of an application. The introduction of new techniques such as virtual clustering and sleep–awake cycles makes it suitable for a single application, which can tolerate some latency. In this scheme, a sensor node periodically turns off its radio and enters the "sleep" state, and data transmission via this node occurs only when it is in the "awake" state. The sensor nodes choose their own sleep–awake schedules and share them with their neighbors to form a cluster and coordinate their sleep–awake schedules to achieve well-synchronized communication with each other. This protocol also deals with the problem of overhearing. In this protocol, each packet has a specific field, which indicates the duration of that packet. Therefore, if a node overhears any packet, it can check the duration of this packet and, accordingly, it will be inactive for that amount of duration, to mitigate energy consumption.

The problem with the S-MAC protocol is that it achieves energy efficiency at the cost of increased latency. Each time a node waits for its next "awake" cycle to transmit data, it introduces unnecessary latency. Therefore, to provide application-specific service requirements, it is necessary to modify the existing protocols. For example, we present a scheme described in [4] that modifies the protocols in order to achieve energy efficiency as well as QoS requirements. The proposed scheme controls unpredictable latency by putting restrictions on the random wait time of CSMA/CA. It dynamically adjusts the contention window size, duty cycle and backoff interval time, according to the traffic intensity and channel conditions, without compromising the QoS requirements.

Similarly, apart from the general S-MAC protocol, there exist several variants of S-MAC. Among them, dynamic S-MAC (DSMAC) [56], timeout MAC (T-MAC) [57], AC-MAC [58], and MS-MACs [59] are notable contributions in contention-based access protocols.

Each of these techniques has its own advantages and limitations. We summarize these properties and compare the contention-free and contention-based protocols in Table 12.4.

Hybrid channel access

From the above comparison, we conclude that both pure contention-free and pure contention-based protocols are not sufficient to provide both QoS requirements and

Table 12.4 Comparison of contention-free and contention-based protocols

Channel access protocols	Advantages	Limitations
Contention-free protocols	• Reservation of time slots guarantees collision-free transmission. • Overhearing is reduced, as every node uses its own dedicated time slot.	• When a node completes its task within the allotted time slots, then the remaining time period is left idle. Bandwidth is not utilized efficiently. • Synchronization of nodes becomes difficult for shorter time slots. It also spends considerable amount of energy to synchronize nodes. • Not scalable. More number of nodes involves more difficulties.
Contention-based protocols	• Energy consumption is very low due to its simple implementation techniques. • Scalable in nature; thus, suitable for large and dense networks.	• Latency is an unavoidable obstacle for these protocols. • It is difficult to provide proper bounds to latency due to the randomness of waiting times and backoff times of nodes. • Improper and inefficient scheduling of sleep–awake cycles can lead to traffic bursts and congestions.

methods to reduce power consumption. Therefore, adopting a hybrid access policy is beneficial for WMSN applications. For example, Akkaya and Younis [5] proposed a scalable hybrid protocol in which the channel is divided into several frames and each frame consists of two parts, i.e. a reservation period (contention-free) and a transmission period (contention-based). The whole network behaves like a grid where all nodes in a grid can perform communication and local synchronization among themselves.

Normal sensor nodes that are commercially available on the market provide very low data rates. The MICAz node provides a data rate of 250 kbps, which is not suitable for multimedia streaming. A solution to this problem is transmission through multiple channels. For example, Mica2 provides two channels, Rockwell's WINS provides 40 channels, and IEEE 802.15.4 or ZigBee provide 16 channels. Therefore, in WMSN, the MAC protocols can operate in parallel in multiple low-rate channels, providing better throughput than operating on a single high-rate channel.

12.7.2.2   Scheduling procedures

It is not possible to achieve the QoS requirements of multimedia applications only by improving network parameters. Along with channel access techniques, proper scheduling of different tasks is also equally important. Scheduling in WMSNs is different from normal networks in many respects – latency bounds, access control, and high channel-error conditions. A particular multimedia application deals with data having different priorities. Generally the data with higher priority get a higher level of services from the network. For example, in the case of image coding of a JPEG image, an important region, known as the region of interest (ROI), is chosen. Similarly, the

I-frames in MPEG video coding have more priority than the other frames. Thus, the ROI in an image and I-frames in a video get greater preference than the other parts of the multimedia data to fulfill the application requirements. But because of improper scheduling, this approach may sometimes lead to starvation of data with low priority, which is undesirable. Therefore, packet-specific service differentiation and ensuring proper scheduling is necessary in WMSNs.

The proposal mentioned in [6] provides a solution to scheduling in WMSNs by using priority queuing and weighted round-robin algorithms. According to the proposal, multimedia applications mainly deal with two types of data – expedited forward (EF) for real-time traffic, and assured forward (AF) for other traffic. AF can be further subdivided into three classes – high priority (AF1), medium priority (AF2), and low priority (AF3). Priority queuing is used in this study for EF traffic, and AF traffic is handled by the weighted round-robin scheduling algorithm. A variant of weighted fair queuing (WFQ) is described in [60], which considers residual energy and latency trade-off in the network. Another interesting study involving network calculus is proposed in [61, 62]. This study deals with various network parameter bounds such as delay and queue length. The work mostly addresses the QoS flow for wired networks, but it can be extended to serve the benefits of this work also in WMSNs.

Imposing strict rules for access control is an important consideration, in addition to scheduling. For example, if the link condition is poor for allowing transmission within certain latency bounds, then an application that requires stricter latency bounds should not be considered for scheduling. If delay of some data packets exceeds their real-time delay bounds, then it is better to drop such packets. In a nutshell, the applications whose service requirements cannot be satisfied in any manner should not be included in the system, as their presence will not serve any purpose. Instead, they may affect the network flow by creating congestion.

12.7.2.3   Error control techniques

Serving QoS requirements along with proper and flawless multimedia streaming is a challenging task of the link layer. This layer always tries to provide an error-free bit stream to the higher layers. However, as mentioned earlier, energy is an important resource constraint that needs to be taken into consideration in WMSNs. Errors in the link layer may be caused by several issues such as multipath fading, shadowing, attenuation at physical layer, collisions of data packets, and co-channel interference at the link layer. There are several efficient techniques for error control in this layer. The two most important are automatic repeat request (ARQ) and forward error correction (FEC). These two techniques are briefly described below.

Automatic repeat request (ARQ)
In this technique, the sender sends packets to the receiver and a sequence number associated with each packet is used to identify them. After receiving a packet, the receiver checks whether the packet is corrupted or not. If the packet does not lose any data during transmission, then the receiver receives it perfectly in its original condition. Then the receiver sends an acknowledgement message (ACK) to the sender. This ACK implies

that the receiver has correctly received the previous packet and is ready to receive the next one. Therefore, the sender sends the next packet. However, if the packet reaches the receiver in corrupted condition, then it sends a negative acknowledgement message (NAK) to the sender. On receiving the NAK message, the sender retransmits the erroneous packet.

Whenever the sender sends a data packet for transmission or retransmission, it sets a timer to the timeout interval. If the sender does not get an ACK or NAK from the receiver within this timeout interval time, then it retransmits the corresponding packet and resets the timer. Similarly, if the sender receives a NAK within the timeout interval, then it retransmits the previous packet and resets the timer.

Forward error correction (FEC)

Forward error correction (FEC) involves different techniques that require adding redundant data to the packet, and finally, getting the correct data in the receiving end, even in the presence of a certain amount of error in the data. Reed Solomon (RS), erasure codes (EC), and cyclic redundancy check (CRC) are different types of existing FEC techniques. Because of the high bit error rate of wireless channels and low energy consumption of these techniques, FEC is preferred in WMSN applications. On the other hand, ARQ is more suitable for general networks (see Table 12.5).

Error concealment (EC), a technique for error resilience, is presented in [7]. It encodes image data using discrete wavelet transform (DWT). This scheme needs interactions between the application and the link layers, so that information about the channel condition can be passed to the source coder, who can handle issues accordingly. The main advantage of this technique is that it provides an efficient collaboration of source coding functions of application layer and error control techniques of the link layer, which, in turn, increases coding efficiency and reduces retransmission rate. On the other hand, the limitation of this scheme is its greater power consumption to perform DWT.

Table 12.5 Advantages and limitations of ARQ and FEC

Error control technique	Advantages	Limitations
ARQ	• Overhead in transmitting control information is lower than FEC.	• Dependence on the timeout interval to detect fault increases latency and the amount of delay is also unpredictable. • Retransmission of packets causes significant energy consumption. • ARQ is not free from overhearing.
FEC	• It provides lower latency and greater delay predictability than ARQ.	• Owing to addition of redundant data, coding efficiency is much less. • It cannot handle node failure.

12.7.3   Network layer

Designing this layer for any kind of network always involves several challenges. This layer deals with efficient delivery of data between the source and the destination, and, simultaneously, it must consider network parameters such as link quality, fault tolerance, scalability, and, above all, energy efficiency, especially in case of WMSNs. There exist some network routing protocols for traditional WSNs. However, due to the multimedia nature of sensed data – video streams, audio, and images – the existing protocols for WSNs are not directly applicable for WMSNs. Thus, it opens up a new research area to design a network layer that is balanced in terms of providing efficient routing algorithms, as well as the QoS requirements of WMSNs. We now discuss several existing routing protocols for WSNs and the limitations to applying them in WMSNs. We further explain some of the WMSN-specific routing protocols and their functionalities.

12.7.3.1   Routing protocols for WSNs

As seen in the previous chapters, there exist different routing protocols for WSNs. Examples include low-energy adaptive clustering hierarchy (LEACH), sensor protocol for information via negotiation (SPIN), and directed diffusion. In this section, we briefly discuss these protocols and their inapplicability to WMSNs.

LEACH

LEACH is a cluster-based wireless routing protocol, as explained in [12]. Densely deployed wireless micro-sensor nodes form several local clusters, depending on their positions and communication ranges. One node from each cluster is selected as the cluster head or local base station, which collects local information from other member nodes belonging to that cluster, aggregates information, and finally transmits it to the main base station. One of the important features of LEACH is that it dynamically selects the cluster heads in order to maximize the network lifetime by not draining the energy of a single node. However, this approach is not suitable for WMSNs, as they are very sensitive to their application requirements, which may not be achieved as requested due to the overheads of dynamic formation of clusters and cluster heads.

SPIN

Avoiding redundant data transmission by advertising meta-data is the primary feature of this protocol, as described in [13, 14]. SPIN employs three types of messages. They are ADV, REQ, and DATA. When a node gathers data, it broadcasts an ADV message containing the meta-data. The nodes that are interested in receiving the data respond with a REQ message. Finally, the advertising node transmits the data to the interested nodes upon receiving this message. The meta-data in the ADV message is a high level data descriptor that is used to identify the data. This approach reduces flooding and gossiping of data across the network. However, this protocol broadcasts the ADV message to a range of single-hop nodes from the sender. Thus, it is not suitable for WMSNs due to the overheads of message broadcasts and the significant amount of power consumption by the sensor nodes.

Directed diffusion

This is a data-centric protocol, in which a node that is interested in receiving data broadcasts an interest message. Data that match the interest message are directed back to the node that generates the interest message. Data propagation and aggregation are determined through message exchanges among the neighbors, as described in [22]. Though it has several useful features, this approach cannot be directly used in WMSNs, where application-specific QoS requirements must be considered.

12.7.3.2   WMSN-specific routing protocols

Both new WMSN-specific proposals and modified versions of previously discussed WSN routing protocols are available in this domain. Ant-based service aware routing (ASAR), landmark routing protocol (LANMAR), and multimedia-enabled improved adaptive routing protocol (M-IAR) are notable among the new methodologies. The following subsections describe some of these routing protocols that are designed in order to confront WMSN-specific challenges.

ASAR

In real life, ants follow a certain path by sensing a chemical substance called a pheromone that is left by other ants along the path. This specific behavior of ants is employed conceptually in the design of ant-based routing algorithm [16], by Dorigo *et al*. This ant-based routing algorithm is the basis of ASAR model discussed in [15]. To achieve diverse QoS requirements, this model mainly addresses two kinds of basic service modes. They are: (i) the event-driven service mode and (ii) the query-driven service mode. The first mode contains only one type of service, i.e. event-driven service (R service), where the multimedia sensor nodes filter out and compress actual data from a pool of raw environmental data before transmission. The second mode consists of two types of services: data query service (D service) and stream query service (S service). D services tolerate query-specific delay, whereas the S services deal with delay-intolerant, but query-specific error-tolerant applications [15].

ASAR is a cluster-based routing algorithm, and it executes on all the cluster heads in a WMSN. There are two types of tables: optimal path tables (OPtables) and pheromone tables (PRtables), which are being monitored by the cluster heads. The cluster heads maintain three OPtables and three PRtables for the above-mentioned three types of services (R, D, and S). The computed probability values [15] of moving from one cluster head to another for a specific service help to find the next node on the way to the sink. The sink node changes the table values for both the tables and suggests a modification to the corresponding cluster heads by sending back the ants. Experimental results provided in [15] indicate that ASAR is a better approach than Dijkstra's and DD algorithms, in the case of S services. For R services, ASAR delivers better results than the DD algorithm, and for D services. The effectiveness of these three algorithms is the same. In a nutshell, the service-aware optimal routing algorithm followed by this model leads to improved network performance and better QoS for WMSNs.

LANMAR

The landmark ad hoc routing (LANMAR) protocol is a combination of fisheye state routing (FSR) and landmark routing, as described in [18]. The study described in [17] extended this work further for multimedia streaming in large-scale sensor networks, by introducing the concept of mobile "swarm" nodes into the sensor network [17]. These nodes can perform better than the normal sensor nodes, as they are equipped with high-quality hardware in order to transmit more detailed and accurate information about the intrusion prone areas. These intrusion prone areas are known as "hotspots" according to the works in [17, 18]. Each swarm contains a swarm leader, which establishes a mobile backbone network (MBN) with the help of powerful backbone radios with longer ranges to communicate with other swarm leaders. As soon as intrusion is detected over the "hotspots," the LANMARK routing algorithm starts deploying swarm nodes, which are supposed to transmit the detailed report of the "hotspots" to the base station. Experimental results available in [17] show that this algorithm performs better than the traditional AODV algorithm.

M-IAR

The M-IAR protocol [19] is an extension of the improved adaptive routing protocol (IAR). As described in [20], IAR borrows the biological concept of helping ants to find appropriate paths. M-IAR extends this work by adding two additional multimedia QoS parameters – end-to-end delay, and jitter. In this protocol, the next hop is decided by analyzing the information retrieved from the neighborhood. Thus, this protocol is effective for WMSNs in terms of scalability. M-IAR decides the shortest path between the source and the sink nodes, based on the geographical location of each sensor node. This approach helps to achieve low end-to-end delay and low jitter. Piggybacking of end-to-end acknowledgements guarantees the delivery of packets, and retransmission takes place in the case of packet losses. Shortest paths can be found by visiting a smaller number of nodes in this approach, and power consumption to achieve QoS requirements is also very low.

Modified directed diffusion

Directed diffusion is a network layer protocol used in traditional WSNs, as discussed in [22]. A study described in [21] employs a modified version of this protocol to achieve the benefits of directed diffusion in multipath transmission over WMSNs. This work incorporates a new metric, termed the path metric or Costp, which is defined as a product of the expected transmission count (ETX) and delay. The algorithm proposed in this work finds multiple disjoint paths with minimum Costp. It also introduces a concept of timeline and deadline to discard data packets that cross this transmission deadline. Initially, the sink node broadcasts interest packets with a timestamp value T0. When the source receives this interest packet, it transmits exploratory data packets. Every intermediate node keeps track of the Costp value of packets that are being forwarded. The forwarding of packets is done based on the lowest Costp value. The sink node records the timestamp T1 when it receives the first exploratory data packet. Packets get dropped if the difference between two timestamp values is more than the pre-assigned deadline. These WMSN-specific modifications on directed diffusion maximize throughput and minimize delay. Simulations in [21] compare the throughput and delay with the throughput produced by the EDGE protocol described in [42], and the experimental results prove that it exhibits twice as much throughput as EDGE. The best-case end-to-end delay is also one-fourth of the delay incurred by the EDGE protocol. Thus, it proves that the modifications on directed diffusion technique lead to better service to achieve throughput requirements of multimedia data.

12.7.4   Transport layer

In any network, providing end-to-end reliable data transfer and congestion control are the main functionalities of the transport layer. However, in order to achieve the QoS requirements of multimedia applications with very low power consumption, the transport layer of a WMSN must provide highly reliable data delivery and must consider the presence of multipaths in the network. The two traditionally used transport layer protocols are transmission control protocol (TCP) and user datagram protocol (UDP). Neither of these protocols is suitable for any WMSN. For real-life multimedia applications such as streaming media, UDP is preferred over TCP due to its low latency and jitter. However, if we compare them with respect to reliability, then UDP is not much reliable for WMSNs, as WMSNs are prone to link failures and transmission errors. Therefore, modification of UDP and TCP protocols is required for efficient design of the transport layer for WMSNs. The disadvantages of the traditional UDP and TCP protocols and the issues that should be taken care of while designing the transport layer are discussed below in three parts – modification of UDP, modification of TCP, and developing new protocols.

12.7.4.1   Modification of UDP
Packet dropping
The packet dropping technique followed by UDP is not suitable for multimedia applications. UDP simply drops an erroneous packet in the case of any error detection, which may lead to quality degradation of the final output of those applications. As we have discussed earlier, WMSNs treat different packets with different importance. This technique provides a special focus on some packets, which are more important to reproduce/retrieve than the other packets on the actual multimedia stream, in case of link failures. For example, I-frames in an MPEG video stream and region of interest (ROI) in a JPEG image are the important parts of their respective data streams. Loss of these frames causes significant damage to the corresponding video stream or image. Therefore, the packet dropping technique used by UDP is not efficient as it does not consider the importance of data packet while dropping it. Hence, to achieve a more accurate outcome, the transport layer should consider an efficient packet dropping technique.

Heterogeneous traffic
Communication in WMSNs deals with diverse data sets. Several network mechanisms such as congestion and reliability control can be improved by taking dynamic decisions

depending on specific data types. Like the option field in the TCP header, the UDP header does not contain any bit or set of bits that can represent data type. Therefore, managing heterogeneous data types involves a lot of difficulties in the traditional UDP protocol.

12.7.4.2 Modification of TCP

Handshaking

The TCP protocol establishes end-to-end connection through a three-way handshaking mechanism, which is very efficient in terms of reliability. However, this mechanism can cause intense latency while establishing connections. Especially in the case of event-driven WMSNs, an event that uses this kind of connection model may lead to a sudden spurt of messages associated with the execution of that event. This type of sudden burst of message entirely disrupts the data flow through communication channel and gradually leads to severe latency and congestion in the network.

Jitter

The TCP protocol employs different congestion-avoidance strategies that involve unpredictable sudden changes in node-to-node latency. For example, the additive increase multiplicative decrease algorithm constantly adjusts the data transmission rate to achieve a congestion-free channel. But these mechanisms cause difficulties for predicting a fixed bound on end-to-end jitter.

Congestion control mechanism

TCP congestion control mechanisms sometimes create confusion, and, thereafter, they take wrong decisions that can lead to unnecessary latency in data transmission. For example, TCP congestion detection algorithms treat persistent packet loss as congestion. However, in WMSNs, one of the most frequent reasons for packet loss is link failure. But the traditional TCP protocol invokes congestion control strategies such as the slow start algorithm in situations of link failure. Therefore, due to the effect of this algorithm, when the link comes up, it cannot get a high transmission rate to clear the backlog of packets while the link was down. So, applying these traditional TCP mechanisms to WMSN-specific issues is not an efficient approach to mitigate transmission latency and congestion.

Reliability mechanism

The reliability mechanism adopted by TCP is not energy efficient, and it causes latency due to the frequent use of timeout timers and retransmissions of data packets. In particular, repeated retransmission of data packets in a high-rate data path leads to a large amount of energy deficiency in the last-hop forwarders that are closest to the sink. Therefore, proper balancing of the data load is also important, while ensuring reliability in the whole network. Distributed TCP (DTCP) [50] is a solution to overcome this problem. DTCP distributes the burden of the sink over several intermediate nodes in the network. The source establishes a connection with the intermediate nodes in place of the sinks. They also take the responsibility of acknowledging the status of data delivery to the source. DTCP effectively distributes the load of sink nodes to the intermediate nodes by breaking down one big high-data-rate link into multiple smaller ones.

Dealing with heterogeneous data is a specific property of multimedia networks. Each connection may transfer different types of data. These diverse data types and different packets of the same data type may have different requirements to achieve a certain level of reliability. Therefore, it is necessary to design protocols that consider the type of data and, accordingly, decide on the transmission strategy to ensure maximum reliability in data delivery. The sensor transmission control protocol (STCP) [51] is an example of a protocol that achieves the above-stated requirement by modifying the existing TCP header. To represent the particular data type, it includes some relevant fields in the TCP header, which satisfies the need for packet-based reliability in WMSNs.

12.7.4.3 Design consideration
Congestion control
Designing proper congestion-control mechanisms is one of the primary objectives of the transport layer. Congestion in communication channels leads to a series of problems throughout the network. Frequent congestion in the transport layer has several implications on the other layers such as a very busy physical channel in the physical layer, and collision of packets in the link layer. One of the ways to mitigate the effects of collisions and packet droppings is retransmission. In a busy channel, the retransmission of packets may collide with one another. Owing to this repetitive retransmission of the packets, the power consumption by the whole network rises abruptly.

Congestion in the transport layer can be categorized into two classes: node-level congestion and link-level congestion. Each node in the network has a certain amount of buffer capacity. Exceeding this limit causes a bottleneck. In node-level congestion, data arrive in a node with an arrival rate greater than the service rate of that node. The frequent occurrence of such a situation gradually fills the buffer and, after the limiting capacity, it causes bottleneck and congestion. In link-level congestion, persistent collisions at the link layer disrupt the data flow through the channel and certainly exceed the channel capacity at a point of time. Therefore, proper congestion-control mechanisms should be designed, considering these two types of congestion.

Multipath and multistream
To achieve QoS requirements, the transport layer handles multimedia streams with large amounts of data by splitting them into several sub-streams, and then transmitting these sub-streams. The benefit of such an approach is that we always get an acceptable output even if some of the sub-streams are delayed or lost during transmission. There are two techniques by which this can be achieved. The first technique is to divide the paths between the sender and receiver nodes into multiple end-to-end paths, so that one stream with a demand of a high-data-rate channel can be transmitted over several low-data-rate paths. This approach increases the robustness of data transmission against problems such as congestion and link failure. As an example, load repartitioning-based congestion control (LRCC) uses this strategy [52]. In cases of congestion, LRCC generates notification in order to inform other nodes about congestion. After receiving

such a notification, the upstream nodes reduce the data rate for that path and increase the rate for another path to maintain the data rate over all the paths.

The second strategy is a collaboration of the application and transport layers. It divides the data into several sub-data and then transmits them separately. The benefit of this approach is that it is possible to vary the priorities of the sub-data according to the requirement prior to transmission. The application layer takes on this responsibility to generate sub-data of varying importance, and then the transport layer processes them, depending on the types of data stream. The reliable synchronous transport protocol (RSTP) [53] is an example of this mechanism. RSTP uses progressive JPEG to encode data streams. This encoding scheme performs multiple scans, and finally generates discrete cosine transform (DCT) coefficients. These coefficients are responsible for coarse-grained data that get transmitted with maximum priority.

12.7.5   Application layer

Real-life streaming applications are much more challenging than other normal applications due to the presence of multimedia content such as audio and video streams. The work described in [33, 34] conveys that traditional WSNs can provide support for image coding. Compression techniques such as JPEG and JPEG2000 [32] are also supported by WSNs. Traditional WSNs help in audio transmission through pulse code modulation, but in the case of video streaming they are not efficient. MPEG or H.26x [33] are currently the most common formats used for video coding.

The main functionalities of the application layer are support for source coding techniques, collaboration with other multimedia processing applications, traffic management, and admission control. We discuss these functionalities briefly in this section.

12.7.5.1   Support for source coding techniques

Different source coding techniques are required to deal with diverse types of multimedia data in WMSNs. In order to deal with several challenges in WMSNs, the application layer must use algorithms that consider resource constraints. A coder must be aware of the following design considerations.

- An efficient compression mechanism is required. Uncompressed data consume resources such as bandwidth and power at a great rate.
- Complexity of multimedia encoders and decoders should be low in order to reduce power consumption and cost of sensor devices.
- The coder must have to ensure robustness of the coding technique.

We now discuss some of the existing source coding techniques.

Layered coding (LC)

According to this technique, total multimedia data are distributed into several layers depending on their importance. There are two types of layers – base layer and enhancement layer. The actual important data are encoded into the base layer and all other information gets stored into the enhancement layers. Receivers use different decoders for the base and enhancement layers to retrieve data. If the base layer is corrupted or lost, then the actual data cannot be retrieved. This layered coding technique is used currently in some studies [25, 26, 27, 28].

Predictive video coding (PVC)

PVC is a well-known video coding technique used by the MPEG-x and H.26x coding standards. It requires complex encoders, efficient algorithms, and comparatively simpler decoders. In MPEG-x and H.26x, the encoder is five to ten times more complex than the decoder. According to the study demonstrated in [29], video is a temporal sequence of frames. Each frame is composed of several blocks that are encoded in the following two basic modes.

(i) *Intra-coding mode (I)*: This mode utilizes spatial correlation to improve the frames having poor compression, and is used to reduce redundancy within one frame. It does not exploit temporal correlation in video streaming.

(ii) *Inter-coding mode (P)*: This mode uses both spatial and temporal correlation to reduce data redundancy and achieve high compression ratio. This mode is also known as the motion-compensated predictive mode.

Multiple description coding (MDC)

MDC efficiently utilizes resource and control packet losses, especially for video applications to avoid retransmission of data packets. In this technique, coders generate two or more independent and equally important streams or descriptions. The benefit of this technique is that each of the descriptors provides some low-quality, but generally acceptable, streams. Therefore, it is possible to obtain a slightly distorted, but overall acceptable, video streams in spite of losing one or more descriptors during transmission. Hence, using multiple descriptions, one can avoid retransmission and reduce energy consumption.

The work by Wang *et al.* [30] describes the usefulness of MDC for video streaming. Apart from error resilience, MDC with a multiple path transport (MPT) mechanism provides traffic dispersion and load balancing in the network [30]. MDC also employs motion-compensated prediction when the system finds that the signal used by the encoder for prediction is not available for the decoder, i.e. when a "mismatch" occurs.

Distributed video coding (DVC)

In this technique, the complexity of the encoder gets transferred to the decoder, thereby making the task of decoding more efficient and resource-rich in terms of power consumption and hardware capabilities. The DVC technique is robust and it efficiently distributes computational complexities between the encoder and the decoder. There are two different techniques to implement the DVC codec for multimedia content. They are discrete cosine transform (DCT) and the pixel-based scheme. DCT results in high compression spending more power in complex computations. But due to highly compressed data, it needs low transmission power. The other technique, i.e. the pixel-based scheme, involves less computation power, but more transmission power. A practical

DVC encoding technique, PRISM, its architecture, and the benefits over MPEG-x and H.26x technique, are described in [29].

12.7.5.2   Collaborative multimedia processing

In WMSNs, it is possible to get redundant information gathered by neighboring nodes in several ways, such as overlapping of fields of view (FoV) of camera sensor nodes. Efficient data fusion and aggregation techniques are required to explore the redundant information to produce more accurate and robust output. This aggregation of data can be designed hierarchically in several levels. The aggregated data of one level undergo further aggregation and fusion in the upper level of the hierarchy.

Distributed filtering architecture supports scalable data collection and it reduces the bandwidth demands of sensor nodes. The specific study in [35] shows how IrisNet (Internet-scale resource-intensive sensor network), a WMSN test bed, follows such architecture to reduce repeatedly processing overheads and bandwidth requirements by mitigating data redundancy. IrisNet addresses the following challenges in order to achieve accurate and robust output.

- It uses rich and shared data sources.
- Scalability of resources up to Internet size.
- Application-specific filtering of sensor feeds.
- Effective use of bandwidth.

IrisNet avoids sending the gathered raw data. It only sends a small amount of processed data, which is sufficient to serve the demand. For example, in the Parking Space Finder application service, the video sensors send a small amount of information regarding the availability of parking space, instead of sending the whole video feed.

One of the most important features of IrisNet is the sharing of results between sensing services that run on the same node. This approach reduces the computational overhead generated in a single node. However, it is difficult to provide security in such a scenario where application services share the results with each other. Malicious nodes can feed incorrect information, and detecting malicious nodes is another troublesome challenge. To confront this challenge, IrisNet distributes the privileges by categorizing the nodes into two classes – trusted and untrusted. But this approach also involves additional overhead to the system.

12.7.5.3   Traffic management and admission control

The functionalities of network traffic management and admission control deal directly with application-specific requirements. Delay tolerance and loss tolerance of data vary with application requirements. WMSNs need to provide differentiated services to satisfy such a diverse range of application requirements. This section explains a few traffic classes that are supported by application layer in WMSNs. Brief descriptions of the traffic classes are as follows.

- Real-time, loss-tolerant multimedia streams: This class deals with audio and video streams and other scalar data such as temperature, humidity, and light intensity. The real-time streams in this class involve high bandwidth consumption and they have strict bounds on latency.

- Delay-tolerant, loss-tolerant, multimedia streams: The strictness of delay for the data belonging to this class is comparatively less than that in real-time streams. Large amounts of data and limited buffer in sensors may sometimes cause buffer overflow. Effective transmission policy is required to overcome this problem.

- Real-time, loss-tolerant data: This class can tolerate some loss of data depending on their nature. For example, the data loss due to high special correlation between video sensors, and consequently, the overlapping FoV of video images, can sometimes be allowed. Certain bound of delay is essential for data transmission.

- Real-time, loss-intolerant data: This class of data is mainly gathered from distributed sources. Their bandwidth demand generally varies in between low and moderate.

- Delay-tolerant, loss-intolerant data: Mainly the data that are meant for post-processing or storage belong to this class. Delayed transmission is allowed for this class of data.

- Delay-tolerant, loss-tolerant data: This class allows delayed data as well as lossy data with low or moderate bandwidth demand.

In [36, 37], some admission control algorithms are discussed. Their main focus is to maximize network lifetime and admissions based on added energy loads, respectively. However, there are some limitations of these algorithms. They do not consider tight balancing between complex computations in sensor nodes and communication techniques. Some important QoS parameters such as delay, reliability, and energy consumption are also not considered in these works. Therefore, providing efficient admission control schemes is still a challenge for the application layer of WMSNs.

12.7.6   Cross-layer issues

Several protocols associated with each layer in the network architecture attempt to optimize the layer-specific performance metrics. However, the traditional approach to break down into distinct layers does not consider the system as a whole, and thus it suffers from the lack of optimal network performance. The interdependence between several layers, and its impact on the whole network, are described briefly below.

Each layer has some relevant interdependencies with the other layers. For example, interference at the receiver, which is decided by the transmission power selected by the physical layer, affects the link layer performance. Similarly, the frequency at the link layer decides the sensitivity at the physical layer. The network and link layers have their respective effects on the operations managed by the transport layer. The routing decision taken by the network layer affects the network parameters associated with the chosen path or link, which has an impact on the congestion of that link. Similarly, frequent collisions at the link layer may lead to congestion problems at the transport layer. Therefore, strict categorization of network functionalities into different layers, especially in a WMSN, is not sufficient to achieve desirable performance. We also need to treat the cross-layer issues with equal importance.

As it is important to meet the QoS requirements in a WMSN, it is suggested to employ a cross-layer approach to overcome the inefficiencies in the individual layers. It is important to map the service requirements with lower layer performance metrics such as latency and data rate. The main advantage of this approach is that it optimizes the parameters at a lower layer to meet the service requirements of the higher layer. Also, the dynamic nature of mapping should be ensured so that each layer optimizes the network parameters dynamically, according to the network or channel conditions. For this the lower layers should also notify the higher layers about the channel condition. For example, depending on the severity of the link layer errors, the application layer can choose the source coding technique by balancing the efficiency and robustness of that technique.

A study discussed in [17] views cross-layer consideration as an optimization problem. The authors have shown that a cross-layer solution can be achieved by first finding out the most efficient protocols in each layer.

12.8   Summary

In this chapter, we have discussed the diverse application areas and challenges associated with WMSNs. We elaborated on the different architectures for WMSNs, and discussed the existing WMSN platforms, such as Panoptes, Cyclops, and SensEye. In the latter half of the chapter, we presented a layer-wise analysis of the wireless protocol stack that supports communications among the WMSNs, and a comparative study among these techniques. This involved a comparative study of ZigBee and UWB in the physical layer; the channel access policies, scheduling procedures, and error control techniques in the data link layer. This was followed by elaboration of different routing protocols for WSNs and WMSNs in the network layer, and WMSN-specific modification of TCP and UDP in the transport layer. Our discussion came to a closure as we discussed the different coding techniques along with collaborative multimedia processing (aggregation and fusion of data) and traffic management policies in the application layer.

References

[1] C. Li, P. Wang, H.-H. Chen and M. Guizani, "A cluster based on-demand multi-channel MAC protocol for wireless multimedia sensor networks," in *Proceedings of IEEE International Conference on Communications*, Beijing, China, pp. 2371–2376, May 2008.

[2] M. Caccamo, L. Zhang, S. Lui and G. Buttazzo, "An implicit prioritized access protocol for wireless sensor networks," in *Proceedings of 23rd IEEE Real-Time Systems Symposium*, pp. 39–48, Dec. 2002.

[3] X. Liu, Q. Wang, L. Sha and W. He, "Optimal QoS sampling frequency assignment for real-time wireless sensor networks," in *Proceedings of 24th IEEE Real-Time Systems Symposium*, pp. 308–319, Dec. 2003.

[4] N. Saxena, A. Roy and J. Shin, "Dynamic duty cycle and adaptive contention window based Qos-MAC protocol for wireless multimedia sensor networks," *Computer Networks*, Vol. **52**, pp. 2532–2542, 2008.

[5] K. Akkaya and M. Younis, "An energy-aware QoS routing protocol for wireless sensor network," in *Proceedings of the 23rd International Conference on Distributed Computing Systems*, pp. 710–715, May 2003.

[6] M. Yaghmaee and D. Adjeroh, "A model for differentiated service support in wireless multimedia sensor networks," in *Proceedings of 17th International Conference on Computer Communications and Networks*, Virgin Islands, USA, pp. 1–6, Aug. 2008.

[7] P. Sarisaray, G. Gur, S. Baydere and E. Harmanc, "Performance comparison of error compensation techniques with multipath transmission in wireless multimedia sensor networks," in *15th International Symposium on Modeling, Analysis, and Simulation of Computer and Telecommunication Systems*, Istanbul, Turkey, pp. 73–86, Oct. 2007.

[8] I. F. Akyildiz, D. Pompili and T. Melodia, "Underwater acoustic sensor networks: research challenges," *Ad Hoc Networks*, Vol. **3**, No. 3, pp. 257–279, May 2005.

[9] S. Hengstler and H. Aghajan, "WiSNAP: a wireless image sensor network application platform," in *Proceedings of 2nd International Conference on Testbeds and Research Infrastructures for the Development of Networks and Communities*, Barcelona, Spain, pp. 6–12, Mar. 2006.

[10] J. Campbell, P. B. Gibbons, S. Nath, *et al.*, "IrisNet: an internet-scale architecture for multimedia sensors," in *Proceedings of the 13th Annual ACM International Conference on Multimedia*, New York, USA, pp. 81–88, 2005.

[11] P. Kulkarni, D. Ganesan, P. Shenoy and Q. Lu, "SensEye: a multi-tier camera sensor network," in *Proceedings of the 13th Annual ACM International Conference on Multimedia*, New York, USA, pp. 229–238, 2005.

[12] W. R. Heinzelman, A. Chandrakasan and H. Balakrishnan, "Energy-efficient communication protocol for wireless microsensor networks," in *Proceedings of the Hawaii International Conference on System Sciences*, Vol. **2**, p. 10, Jan. 2000.

[13] J. Kulik, W. Heinzelman and H. Balakrishnan, "Negotiation-based protocols for disseminating information in wireless sensor networks," *Wireless Networks*, Vol. **8**, pp. 169–185, 2002.

[14] W. R. Heinzelman, J. Kulik and H. Balakrishnan, "Adaptive protocols for information dissemination in wireless sensor networks," in *Proceedings of the ACM/IEEE International Conference on Mobile Computing and Networking*, pp. 174–185, 1999.

[15] Y. Sun, H. Ma, L. Liu and Y. Zheng, "ASAR: an ant-based service-aware routing algorithm for multimedia sensor networks," *Frontiers of Electrical and Electronic Engineering, China*, Vol. **3**, pp. 25–33, 2008.

[16] M. Dorigo, D. G. Caro and L. M. Gambardella, "Ant algorithms for discrete optimization," *Artificial Life*, Vol. **5**(2), pp. 137–172, 1999.

[17] M. Gerla and K. Xu, "Multimedia streaming in large-scale sensor networks with mobile swarms," *Special Interest Group on Management of Data*, Vol. **32**, pp. 72–76, 2003.

[18] P. Guangyu, M. Gerla and X. Hong, "LANMAR: landmark routing for large scale wireless ad hoc networks with group mobility," in *Mobile and Ad Hoc Networking and Computing (MobiHOC)*, Boston, USA, pp. 11–18, Aug. 2000.

[19] M. Rahman, R. G. Aghaei, A. E. Saddik and W. Gueaieb, "M-IAR: biologically inspired routing protocol for wireless multimedia sensor networks," in *Proceedings Instrumentation*

and *Measurement Technology Conference (IMTC)*, Victoria, British Columbia, Canada, pp. 1823–1827, October 2008.

[20] R. G. Aghaei, M. A. Rahman, W. Gueaieb and A. E. Saddik, "Ant colony-based reinforcement learning algorithm for routing in wireless sensor networks," in *Instrumentation and Measurement Technology Conference (IMTC)*, Warsaw, Poland, 2007.

[21] S. Li, R. Neelisetti, C. Liu and A. Lim, "Delay-constrained high throughput protocol for multipath transmission over wireless multimedia sensor networks," in *International Symposium on a World of Wireless, Mobile and Multimedia Networks (WoWMoM)*, Newport Beach, USA, pp. 1–8, June 2008.

[22] C. Intanagonwiwat, R. Govindan and D. Estrin, "Directed diffusion: a scalable and robust communication paradigm for sensor networks," in *Proceedings of the ACM MobiCom'00*, Boston, MA, pp. 56–67, 2000.

[23] A. Seema and M. Reisslein, "Towards efficient wireless video sensor networks: a survey of existing node architectures and proposal for a flexi-WVSNP design," *Communications Surveys & Tutorials, IEEE*, Vol. **13**, Issue 3, pp. 462–486, 2011.

[24] I. T. Almalkawi, M. G. Zapata, J. N. Al-Karaki and J. M. Pozo, "Wireless multimedia sensor networks: current trends and future directions," *Sensor*, Vol. **10**, No. 7, pp. 6662–6717, Jul. 2010

[25] H. Radha, M. Schaar and Y. Chen, "The MPEG-4 fine-grained scalable video coding method for multimedia streaming over IP," *IEEE Transactions Multimedia*, Vol. **3**, pp. 53–68, 2001.

[26] A. Boukerche, J. Feng, R. Werner, Y. Du and Y. Huang, "Reconstructing the plenoptic function from wireless multimedia sensor networks," in *Proceedings of 33rd IEEE Conference on Local Computer Networks (LCN)*, Montreal, Quebec, Canada, pp. 74–81, Oct. 2008.

[27] V. Lecuire, C. D. Faundez and N. Krommenacker, "Energy-efficient image transmission in sensor networks," *International Journal of Sensor Networks*, Vol. **4**, pp. 37–47, 2008.

[28] W. Wang, D. Peng, H. Wang, H. Sharif and H. H. Chen, "Energy-constrained distortion reduction optimization for wavelet-based coded image transmission in wireless sensor networks," *Multimedia*, Vol. **10**, pp. 1169–1180, 2008.

[29] R. Puri, A. Majumdar, P. Ishwar and K. Ramchandran, "Distributed video coding in wireless sensor networks," *IEEE Signal Processing Magazine*, Vol. **23**, pp. 94–106, 2006.

[30] Y. Wang, A. Reibman and S. Lin, "Multiple description coding for video delivery," *Proceedings of the IEEE*, Vol. **93**, pp. 57–70, 2005.

[31] B. Girod, A. Aaron, S. Rane and D. Monedero, "Distributed video coding," *Proceedings of the IEEE*, Vol. **93**, pp. 71–83, 2005.

[32] A. Skordas, C. Chirstopoulos and T. Ebrahimi, "The JPEG 2000 still image compression standard," *IEEE Signal Processing Magazine*, Vol. **18**, No. 5, pp. 36–58, Sept. 2001.

[33] C. Chiasserini and E. Magli, "Energy consumption and image quality in wireless video-surveillance networks," in *Proceedings of 13th IEEE International Symposium on Personal, Indoor and Mobile Radio Communications (PIMRC)*, pp. 2357–2361, Sept. 2002.

[34] W. Feng, B. Code, E. Kaiser, M. Shea and W. Feng, "Panoptes: scalable low-power video sensor networking technologies," in *Proceedings of the Eleventh ACM International Conference on Multimedia*, pp. 90–91, Nov. 2003.

[35] S. Nath, Y. Ke, P. B. Gibbons, B. Karp and S. Seshan, "A distributed filtering architecture for multimedia sensors technical report IRP-TR-04–16," in *First Workshop on Broadband Advanced Sensor Networks (BaseNets)*, Aug. 2004.

[36] M. Perillo and W. Heinzelman, "Sensor management policies to provide application QoS," *Ad Hoc Networks (Elsevier)*, Vol. **1**, No. 2–3, pp. 235–246, 2003.

[37] A. Boulis and M. Srivastava, "Node-level energy management for sensor networks in the presence of multiple applications," in *Proceedings of IEEE International Conference on Pervasive Computing and Communications (PerCom)*, Dallas, USA, pp. 41–49, 2003.

[38] W. C. Feng, E. Kaiser and M. LeBaillif, "Panoptes: scalable low-power video sensor networking technologies," *ACM Transactions on Multimedia Computing, Communications, and Applications*, Vol. **1**, No. 2, pp. 151–167, May 2005.

[39] M. Rahimi, R. Baer, J. Warrior, D. Estrin and M. Srivastava, "Cyclops: in situ image sensing and interpretation in wireless sensor networks," in *Proceedings of ACM Conference on Embedded Networked Sensor Systems (SENSYS)*, 2005.

[40] Crossbow, Wireless Sensor Platform, website: http://www.xbow.com/Products/Wireless SensorNetworks.htm

[41] Stargate Platform, Crossbox, website: http://www.xbow.com/Products/XScale.htm

[42] S. Li, A. Lim, S. Kulkarni and C. Liu, "EDGE: a routing algorithm for maximizing throughput and minimizing delay in wireless sensor networks," in *Proceedings of the 26th Military Communications Conference (MILCOM'07)*, Oct. 2007.

[43] I. F. Akyildiz, T. Melodia and K. R. Chowdhury, "A survey on wireless multimedia sensor networks," *Computer Networks*, Vol. **51**, No. 4, p. 921, 2007.

[44] A. Sharif, V. Potdar and E. Chang, "Wireless multimedia sensor network technology: a survey," in *7th IEEE International Conference on Industrial Informatics*, pp. 606–613, June 2009.

[45] I. F. Akyildiz, T. Melodia and K. R. Chowdhury, "Wireless multimedia sensor networks: applications and testbeds," *Proceedings of the IEEE*, Vol. **96**, No. 10, pp. 1588–1605, Oct. 2008.

[46] WISEBED, website: http://wisebed.eu/site/

[47] DS-UWB, website: http://www.radio-electronics.com/info/wireless/uwb/ds-uwb.php

[48] MBOFDM-UWB, website: http://www.radio-electronics.com/info/wireless/uwb/mb-ofdm-uwb.php

[49] DS-UWB, website: http://www.networkworld.com/news/tech/2004/0614techupdate.html

[50] A. Dunkels, T. Voigt, J. Alonso and H. Ritter, "Distributed TCP caching for wireless sensor networks," in *Proceedings of the Mediterranean Ad Hoc Networking Workshop (MedHoc-Net)*, June 2004.

[51] Y. G. Iyer, S. Gandham and S. Venkatesan, "STCP: a generic transport layer protocol for wireless sensor networks," in *Proceedings of IEEE International Conference on Computer Communications and Networks (ICCCN)*, USA, pp. 449–454, 2005.

[52] M. Maimour, C. Pham and J. Amelot, "Load repartition for congestion control in multimedia wireless sensor networks with multipath routing," in *Proceedings of 3rd International Symposium on Wireless Pervasive Computing, ISWPC*, pp. 11–15, Oct. 2008.

[53] A. Boukerche, Y. Du, J. Feng and R. Pazzi, "A reliable synchronous transport protocol for wireless image sensor networks," in *Proceedings of IEEE Symposium on Computers and Communications, ISCC 2008*, Marrakech, Morocco, pp. 1083–1089, July 2008.

[54] Wireless LAN Medium Access Control (MAC) and Physical Layer (PHY) Specification, IEEE Std. 802.11.

[55] I. F. Akyildiz, W. Su, Y. Sankarasubramaniam and E. Cayirci, "Wireless sensor networks: a survey," *Computer Networks*, Vol. **38**, pp. 393–422, March 2002.

[56] P. Lin, C. Qiao and X. Wang, "Medium access control with a dynamic duty cycle for sensor networks," in *Proceedings of the IEEE Wireless Communications and Networking Conference (WCNC)*, Vol. **3**, pp. 1534–1539, March 2004.

[57] T. V. Dam and K. Langendoen, "An adaptive energy-efficient MAC protocol for wireless sensor networks," in *Proceedings of International Conference on Embedded Networked Sensor Systems (SenSys)*, pp. 171–180, Nov. 2003.

[58] J. Ai, J. Kong and D. Turgut, "An adaptive coordinated medium access control for wireless sensor networks," in *Proceedings of the International Symposium on Computers and Communications*, Vol. **1**, pp. 214–219, July 2004.

[59] H. Pham and S. Jha, "An adaptive mobility-aware MAC protocol for sensor networks (MS-MAC)," in *Proceedings of IEEE International Conference on Mobile Ad-hoc and Sensor Systems (MASS)*, pp. 214–226, Oct. 2004.

[60] S. Lu, V. Bharghavan and R. Srikant, "Fair scheduling in wireless packet networks," *IEEE/ACM Trans. Network*, Vol. **7**, pp. 473–489, Aug. 1999.

[61] J.-Y. L. Boudec and P. Thiran, "Network calculus," in *LNCS*, Vol. **2050**, New York: Springer-Verlag, 2001.

[62] R. Cruz, "A calculus for network delay. I. Network elementsin isolation," *IEEE Transactions on Information Theory*, Vol. **37**, pp. 114–131, Jan. 1991.

[63] G. I. Papadimitiou, A. S. Pomportisis, P. Nicopolitidis and M. S. Obaidat, *Wireless Network*, Wiley, 2002.

13   Underwater sensor networks

Underwater sensor networks (UWSNs) are wireless networks of autonomous sensor-aided devices, called *motes* or *sensor nodes*, deployed over a region of water for the collaborative execution of a given task. Nearly 70% of the earth's surface is covered by water, mainly oceans. The vast majority of this area remains unexplored. The advent of UWSNs provides a new direction in the field of oceanic exploration and information collection. Major applications of UWSNs exist in both the military and civilian fields. Oceanographic data collection, environmental monitoring, pollution monitoring and control, intrusion detection, mapping of underwater area, detection of explosives, mines, oil and minerals, and guided navigation of rescue teams by collaboration with autonomous underwater vehicles (AUVs) and remotely operated vehicles (ROVs) are a few such potential applications [1].

Recently, a lot of real-world short-term deployments have been performed and long-term projects have been undertaken using UWSNs for various applications [2]. One such initial experiment was done in Seaweb [3]. Seaweb was targeted for military applications with a goal of designing specific protocols for detection of submarines and communication between them. In this case, UWSN was deployed in a coastal area, and experiments were carried out for several days. Various institutes have taken such initiatives for designing autonomous and robotic vehicles to be used in underwater exploration. In an underwater data-collection experiment undertaken by Massachusetts Institute of Technology (MIT) and Australia's Commonwealth Scientific and Industrial Research Organisation, both fixed nodes and autonomous vehicles were used [4]. Another recent initiative was undertaken by IBM and Beacon Institute jointly [5]. The project concerns on environmental monitoring application to study and collect the biological, chemical, and physical information of the Hudson River in New York.

The use of UWSNs for various applications is inspired by the ability of wireless sensor networks (WSNs) to explore and monitor terrestrial environments. WSNs have been in use in terrestrial environments, and considerable research had been put into the development of WSNs [6–9]. On account of several similarities of the WSNs and UWSNs, such as multi-hop network and energy constraints, solutions for UWSNs draw inspiration from WSNs. On account of the similarities with WSNs, UWSNs may be visualized as special cases of WSNs. A general model for the UWSNs is depicted in Figure 13.1.

UWSNs pose challenges that are different from those of the terrestrial wireless sensor networks [10–12]. These challenges mainly emerge from the differences in the deployment environment and the communication medium. Radio frequency (RF) waves are not

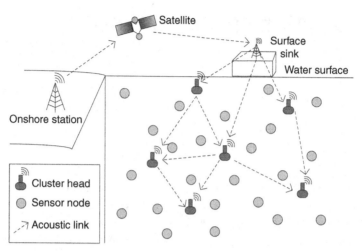

Figure 13.1   A typical underwater sensor network.

suitable for underwater environments, and only low-frequency waves (~30–300 Hz) can propagate over long distances in such environments. However, for transmission of such low-frequency waves, high transmission power and large sized antennas are required. Owing to the saline nature of the oceanic communication medium, RF signals are affected by severe attenuation. Optical waves do not suffer from the problem of attenuation; however, they require high-precision pointing beams, and are affected by scattering. Therefore, acoustic signals are used in underwater communication. The acoustic communication in underwater environments is affected by multipath and fading losses, Doppler effects, high propagation delay, and low available bandwidth.

There are a few advantages of using UWSNs instead of the traditional approaches of ocean-bottom and ocean-column monitoring. Besides reduction in complexity, cost, and time, the use of UWSNs leads to the following benefits.

(a)   *Real-time monitoring*: In the traditional methods of ocean exploration and monitoring, the data collected were stored in the on-board memory of the equipment. The collected data were recovered from the memory only after the equipment reached the on-shore control station. This time lag between when the data were collected and when they were read could be as large as weeks. In case of UWSNs, however, the data are relayed to the on-shore control station through a multi-hop network as soon as the data are measured. Hence, real-time monitoring of the oceanic environment is possible.

(b)   *Online system reconfiguration*: In the traditional methods, the control station cannot monitor the state of the equipment. Generally, the control station cannot send commands to the equipment either. Hence, it is not possible to reconfigure the equipment during the experiment. In a UWSN, however, the control station can monitor the data sent by the sensor nodes, and can send commands to the individual nodes. This allows online reconfiguration of the UWSN during the experiment.

(c) *Failure detection and resolution*: As the control station is unable to monitor and reconfigure the equipment used in the traditional methods, failure of the equipment cannot be detected until it is recovered. In a UWSN, fortunately, the failure of a sensor node can be detected by the neighboring sensor nodes. As a UWSN generally has some redundancy, it can operate even if a few nodes fail.

(d) *Ability to measure greater amount of data*: As the on-board memory on equipment is limited, the amount of data that can be measured in an experiment through traditional methods is limited. UWSNs, however, relay the data to the control station as quickly as possible. Hence, the amount of data measured in an experiment is not limited by the memory on the sensor nodes.

13.1   Characteristics, properties, and applications of UWSNs

In this section, the characteristics of UWSNs are discussed in detail. UWSNs share many common properties with terrestrial sensor networks. The properties similar to the WSNs are as follows.

- *Autonomous*: Once deployed, the sensor nodes form a network among themselves autonomously. The sensor nodes communicate with one another to set up the routing tables, data paths, and other communication parameters. Once the network is formed, the sensor nodes begin to perform their given task. Information obtained by the sensor nodes is forwarded to the control station through a multi-hop path. Hence, the entire process, from formation of the network to collection of data, is completely autonomous, and does not require any human intervention.

- *Low maintenance*: A UWSN is designed to have little or no maintenance. Similar to sensor networks, UWSNs are deployed in harsh areas where maintenance may not be possible. The nodes are used until they fail or their batteries get discharged.

- *Ability of continuous operation*: UWSNs are envisioned to work continuously for several months in applications such as environmental and pollution monitoring. Once deployed, the nodes can gather data periodically, continuously, or on-demand. The choice of a particular data model is made based on the application and the energy constraints. For example, in the case of a surveillance application, the nodes need to send data to the sink only on the occurrence of an event (such as detection of intrusion). In this case, an event-driven data model is used. In a reservoir-monitoring application, on the other hand, the nodes need to fetch data continuously for as long as the acoustic source is kept on.

The UWSNs are different in many ways from the conventional sensor networks like WSNs. The major differences with WSNs are as follows.

- *Three-dimensional space*: UWSNs are generally deployed in three-dimensional space, unlike the terrestrial sensor networks, which are mostly deployed in two-dimensional space. However, applications such as sea-bed monitoring and shallow-water pollution monitoring are possible with two-dimensional UWSN deployments.

- *Sparse deployment*: UWSN leads to sparse deployment of sensor nodes compared to the node density in terrestrial sensor networks because of the high cost of sensor nodes, and deployment and maintenance costs.
- *Acoustic communication signal*: Radio waves with high frequency cannot travel long distances in the sea due to the salty nature of water; and low-frequency radio waves (~30–300 Hz) require high transmission power and large-sized antennas for transmission [13]. On the other hand, optical waves suffer from the scattering effect, and thus require high precision of the pointing beam as well. Thus, these are not appropriate tools for long-distance communication under water. Therefore, the use of acoustic waves (sound waves) is the best choice for underwater communication.
- *High power consumption*: Power consumption in UWSNs is higher than terrestrial networks' due to larger distance coverage and complex signal processing. The use of acoustic waves steps up the power requirement ten times in transmission than in the reception of signal. Furthermore, a node's battery cannot be recharged easily and the use of solar power is also not possible in UWSNs [14].
- *Node mobility*: The unique feature of UWSNs is that the deployed nodes do not remain fixed in their initial deployment position, if they are not anchored on the sea bed, due to underwater current. They can even flow 2–3 knots or 3–5 km/h in some typical underwater circumstances [10, 12].
- *Failure proneness*: On account of the highly corrosive environment, nodes may malfunction due to fouling and corrosion. From the protocol point of view, this evidences the need for a robust UWSN capable of handling node failures.

Based on different characteristics, the potential applications of UWSNs can be classified into the following few types [1].

- *Environmental monitoring*: UWSNs can perform pollution monitoring (chemical, biological, nuclear, and oil-leakage) in bays, lakes, rivers, or oceans. They can also be used to monitor ocean currents and winds, and the effect of human activities on marine ecosystem, temperature change (e.g. the global warming effect).
- *Habitat monitoring*: UWSNs can be used for biological monitoring of underwater animals, fish, or micro-organisms.
- *Surveillance applications*: Autonomous underwater vehicles (AUVs) and under-water sensors can collaboratively monitor areas for surveillance, reconnaissance, targeting, and intrusion detection systems.
- *Disaster prevention*: Sensors measure seismic activity from remote locations to provide warnings of ocean-related disasters such as tsunamis and sea-quakes (effect of submarine earthquakes) to coastal areas in real-time as they happen.

- *Undersea exploration*: UWSNs can help in the exploration of minerals and oil-fields or reservoirs, determine routes for laying undersea cables, and assist in the exploration for valuable minerals.
- *Ocean sampling*: Networks of sensors and AUVs can perform synoptic, and cooperative adaptive sampling of the three-dimensional (3D) coastal ocean environment.
- *Assisted navigation*: Sensors can be used to identify hazards on the sea bed, locate dangerous rocks or shoals in shallow waters, mooring positions, submerged wrecks, and to perform bathymetry profiling.
- *Mine detection*: A UWSN can detect a mine efficiently by using acoustic sensors and optical sensors together.

13.2   Underwater physics and dynamics

Temperature, salinity, and pressure are the three most important properties of sea water, as they determine the other physical properties associated with sea water, for instance, the density of sea water is a function of temperature, pressure, and salinity [15]. This differs from pure water, in which only pressure and temperature determine the physical properties. The presence of dissolved and suspended particles in sea water increases the scattering of radiation, thereby absorbing more radiation than a similar layer of pure water. Wave motion causes a change in the processes of chemical diffusion, heat conduction, and transfer of momentum from one layer to another. These variables of waves, dissolved and solid particles in sea water, although important, cannot be measured. Additionally, velocity of sound, viscosity, propagation and absorption losses, electrical conductivity, specific heat, and the freezing point of water are also important. These features of sea water can be calculated only by using complex mathematical formulation of one or more common physical properties.

(a)   *Temperature*: The ocean, like the atmosphere, is heated by the Sun's incoming radiation. Some of this heat is given back to the atmosphere and some part of it is retained. Considering the average conditions, the sea-surface temperature is more than the air temperature, as the sea retains some portions of this heat. However, whether the sea temperature is warmer or cooler than the air passing above it depends upon a few factors, e.g. geographic location, the season of the year, the nature of the ocean currents, and the atmospheric conditions.

Usually, the ocean temperature ranges from –2 °C to 30 °C, The sea temperature near the coast is higher than in the open sea, where the water is free to move around. The open sea hardly heats above 30 °C, because of the fact that the ocean currents distribute the heat all over the sea by flowing from one zone to another, thereby balancing the temperature.

The ocean column can be assumed as a 3D model, where the temperature gradually decreases with the increase in the ocean depth. In the mixed and deep isothermal layers, the water temperature falls gradually and a sharp change is noticed in the thermocline layer. This type of temperature change can be seen in all latitudes around the world.

(b)   *Pressure*: Pressure is computed as force per unit area acting on a surface. The Standard International (SI) unit is the Pascal, based on the MKS (meter–kilogram–second) method:

1 Pascal = 1 Newton/m^2, where 1 Newton = 1 kg-m / s^2.

Other units commonly used are based on the Bar, where

1 Bar = 10^5 Pascal,
1 dBar = 10^{-1} Bar.

The pressure in the sea is created by the weight of the sea water above. The pressure of the sea water increases regularly with the increase of height of the water column or the depth of the ocean. Again, the weight per unit volume of sea water varies with temperature and salinity. In a column of water of constant depth, the pressure increases as the temperature of the sea decreases, or the salinity increases [15]. Pressure exerted by the water layer of height 1 m is nearly equal to 1 dbar. Therefore, pressure ranges from zero at the surface to about 10^5 dbar at the bottom of the sea.

(c)   *Salinity*: In oceanography, salinity is defined as "the total amount of dissolved solids in sea water." Salinity is measured in parts per thousand (ppt) by weight, and is symbolized as ‰. Using this information, we can find the grams of dissolved material per kilogram of sea water [15]. In an open ocean, salinity is changed by the flow of water, decreased by precipitation, and increased by evaporation. The surface salinity value varies in the same fashion in all the oceans through the latitudes. In coastal areas, salinity values get reduced by the flow of the fresh water poured by the rivers. In high latitudes, salinity generally increases during periods of ice formation and decreases during periods of ice melt.

Typical salinity values of ocean water typically range between 33‰ and 37‰, with an average of about 35‰. Salinity values are found to be maximum in latitudes 20°–23° North and South, whereas minimum salinity values occur at near the equator and high latitudes. In general, salinity decreases slowly with the increase in depth in all the oceans.

(d)   *Density*: Density can be defined as mass per unit volume. Sea-water density depends on its temperature, pressure, and salinity values. At constant temperature and pressure, density varies with salinity. The vertical column structure of the ocean is controlled by density of water, as less-dense water has a tendency to float above the more-dense water [15].

The density at the surface of the water varies significantly due to precipitation, melting of ice, or heating. The density versus depth profile of sea water is the same as the salinity profile and has a sharp change in the thermocline layer.

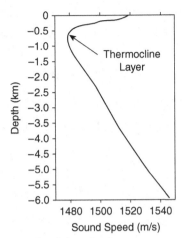

Figure 13.2   Sound speed vs. depth.

(e)   *Viscosity*: Viscosity or shear viscosity is the property of a liquid that resists flow through it. The level of resistance is determined by its temperature, pressure, and salinity. Viscosity increases when salinity increases or the water temperature decreases. However, the change in temperature is more deterministic for viscosity. The shear viscosity of sea water is 1.4 mPa s at 10 °C, at salinity values of 35‰ [16].

(f)   *Variable sound speed*: Sound speed in sea water is dependent mainly on temperature, pressure, and salinity or density. An increase in these parameters increases the velocity of sound. Again, pressure is a function of depth in underwater scenarios. So, the sound velocity increases from surface to bottom. However, the sound wave velocity decreases to a minimum value at the thermocline layer, and again it increases with the increase in the depth, which is shown in Figure 13.2 [17]. The variable speed of sound has effects on node localization, as reported in [17].

The average value of sound velocity found in the sea is 1445 m/s; and sound velocity increases by 1.3 m/s for every 1 ‰ increase in salinity, by 4.5 m/s for every 1 °C temperature and by 1.7 m/s for every 100 m increase in depth [15]. Owing to the decrease of sea-water temperature with depth, while both the pressure and salinity values increase, the sound speed value shows a minimum value at a depth of several hundred meters, called the thermocline region, then increases again with increasing depth [18]. Mackenzie proposed a formula to calculate the speed of sound waves in sea water, as given in equation (13.1) [19]:

$$c(T, S, z) = a_1 + a_2T + a_3T^2 + a_4T^3 + a_5(S - 35)$$
$$+ a_6z + a_7z^2 + a_8T(S - 35) + a_9Tz^3, \qquad (13.1)$$

where, T, S, and z are temperature in °C, salinity in ‰ and depth in meters. The other values are: $a_1 = 1448.96$, $a_2 = 4.591$, $a_3 = -5.304 \times 10^{-2}$, $a_4 = 2.374 \times 10^{-4}$, $a_5 = 1.340$, $a_6 = 1.630 \times 10^{-2}$, $a_7 = 1.675 \times 10^{-7}$, $a_8 = -1.025 \times 10^{-2}$, $a_9 = -7.139 \times 10^{-13}$.

There are several other proposals for underwater sound speed calculation. Chen and Millero [20] proposed the following formula for estimating underwater sound speed:

$$(U^P - U^P_{H_2O}) - (U - U_{H_2O}) = AS(\%) + BS(\%)^{\frac{3}{2}} + CS(\%)^2, \tag{13.2}$$

where U and U_{H_2O} denote the speeds of sound in sea water and pure water, respectively, P is the applied pressure, and S (‰) is the salinity in parts per thousand. The values of A, B and C are:

$$\begin{aligned} A &= (9.4742 \times 10^{-5} - 1.2580 \times 10^{-5}t - 6.4885 \times 10^{-8}t^2 + 1.0507 \times 10^{-8}t^3 - 2.0222 \times 10^{-10}t^4) \\ &+ (-3.9064 \times 10^{-7} + 9.1041 \times 10^{-9}t - 1.6002 \times 10^{-10}t^2 + 7.988 \times 10^{-12}t^3)P^2 \\ &+ (1.100 \times 10^{-10} + 6.649 \times 10^{-12}t - 3.389 \times 10^{-13}t^2)P^3, \\ B &= (7.3637 \times 10^{-5} + 1.7945 \times 10^{-7}t)P, \\ C &= -7.9836 \times 10^{-6}P. \end{aligned}$$

Interestingly, sound waves under water have a tendency to bend towards the region of less velocity. Sound waves are refracted inside water due to this reason. If there is a certain boost in sound velocity with increase in the depth, then the sound waves will be refracted towards the surface. When the sound velocity shrinks with an increase in depth, the sound waves propagate on the way to the bottom of the sea. This wave bending and the different sound speed zones create low-intensity sound regions – "shadow zones" [21] – and high-intensity sound zones – "Caustics." Acoustic waves travel great distances with minimum velocity in the ocean, as they get trapped in a waveguide-like channel in the thermocline region. This channel is named the SOFAR channel (SOund Fixing And Ranging channel) or the deep sound channel [22]. The SOFAR channel and the creation of the shadow zone are shown in Figure 13.3 [23].

(g) *High and variable latency*: The sound speed in water is five orders of magnitude slower than the speed of RF waves. Hence, the time required for an acoustic wave

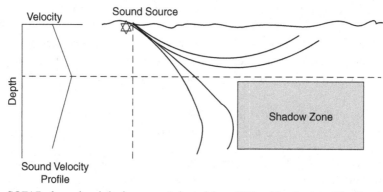

Figure 13.3   SOFAR channel and shadow zone (adopted from [23], with minor modifications).

to travel a particular distance is five orders of magnitude larger than the time taken by an RF signal to travel the same distance. This creates another problem called the bandwidth-delay product problem, as the propagation time of the signal is greater than the signal transmission time [10].

(h) *Propagation loss*: According to the Colossus model for propagation loss in shallow water proposed by Marsh and Schulkin [24], the attenuation factor in sea water at frequencies between 3 kHz and 0.5 MHz is given by

$$\beta = 8.68 \times 10^3 \left(\frac{SAf_T f^2}{f_T^2 + f^2} + \frac{Bf^2}{f_T} \right) (1 - 6.54 \times 10^{-4} P) \ [\text{dB/km}], \qquad (13.3)$$

where

$$A = 2.34 \times 10^{-6},$$
$$B = 3.38 \times 10^{-6},$$
$$f_T = 21.9 \times 10^{6 - \frac{1520}{T + 273}},$$

S is the salinity in percent (%),
P is the hydrostatic pressure in kg/cm^2,
T is the temperature in °C,
f is the frequency in kHz, and
f_T is a relaxation frequency in kHz.

At low frequencies (100 Hz–3 kHz), the attenuation factor is better represented by Thorp's formula [25], which is shown in equation (13.4):

$$\beta = \frac{0.11f^2}{1 + f^2} + \frac{44f^2}{4100 + f^2} \ [\text{dB/km}], \qquad (13.4)$$

where f is the sound frequency in kHz.

It can be seen that the attenuation for low-frequency sounds is very low, as compared to the high-frequency waves. It is clear that communication through acoustic waves is far better than that for any other kind of radiation. However, the use of acoustic waves for underwater communication leads to lower bandwidth for long-range communications. The available bandwidth at different ranges in the underwater communication channel is given in Table 13.1 [1].

Table 13.1 Available bandwidth for different ranges in UW-A channels

	Range [km]	Bandwidth [kHz]
Very long	1000	< 1
Long	10–100	2–5
Medium	1–10	~10
Short	0.1–1	20–50
Very short	< 0.1	>100

(i)   *Transmission loss*: Transmission loss (TL) varies according to the depth of water. Transmission loss in the sea emerges due to the spreading loss and absorption or attenuation loss [26]. According to the height of the water column, transmission loss may be calculated.

   (i)   *Shallow water*: In shallow water, it is assumed that the acoustic signal propagates within a cylinder bounded by the sea surface and the sea floor. As a consequence, cylindrical spreading occurs [27]. Mathematically, it can be expressed as:

$$TL = 10 \log r + \alpha \times r \times 10^{-3}$$

   where α is the absorption coefficient having unit db/km, and r is the range expressed in meters.

   (ii)   *Deep water*: Unlike shallow water, in this case, there is no feasible bounded source, having a bounded surface. In this case, TL can be expressed as

$$TL = 20 \log r + \alpha \times r \times 10^{-3}$$

   where, again, α is the absorption coefficient having unit db/km, and r is the range expressed in meters.

(j)   *Noise*: Underwater acoustic communication is mainly affected by two types of noise – natural and man-made. The sources of natural noise are various hydro-dynamic activities (such as waves, currents, tides, and wind), and thermal and seismic activities. Noises created by machinery and ships are the causes of man-made noise. Power spectral density values of these noises (dB or µPa per Hz) are calculated for any specific frequency (kHz) by these formulas expressed in equations (13.5–13.8) [28, 29]:

$$10 \log N_t(f) = 17 - 30 \log f, \tag{13.5}$$

$$10 \log N_s(f) = 40 + 20(s - 5) + 26 \log f - 60 \log(f + 0.03), \tag{13.6}$$

$$10 \log N_w(f) = 50 + 7.5 w^{0.5} + 20 \log f - 40 \log(f + 0.4), \tag{13.7}$$

$$10 \log N_{th}(f) = -15 + 20 \log f. \tag{13.8}$$

Here, N_t, N_s, N_w, and N_{th} denote the *turbulence*, *shipping*, *wind*, and *thermal* noises, respectively. The total noise power spectral density is represented by:

$$N(f) = N_t(f) + N_s(f) + N_w(f) + N_{th}(f). \tag{13.9}$$

(k)   *Doppler spread*: Underwater signals may also suffer from the Doppler effect. This effect arises during the situation when there is a relative motion between the source and the receiver. The platform motion may lead to the fluctuation in the propagation path length, which may lead to a Doppler shift of the received signal [30]. The Doppler shift of the received signal is given by [30]:

$$f_d = \frac{(f_c \times v)}{c},$$

where f_c = original frequency of the signal,
v = platform velocity, and
$c = 3 \times 10^8$m/s; speed of light in free space.

Owing to the Doppler effect, the frequency of the signal received might differ significantly from the frequency of the signal transmitted. It is observed that, for $v = 2$ m/s and f_o= 25 kHz, the Doppler shift is approximately 33 Hz. So, the Doppler effect distorts the initial frequency band [30].

13.3   UWSN design: communication model and networking protocols

A UWSN, in general, is an autonomous network of distributed intelligent devices. In this section, the design of the UWSN components is discussed. The communication model of such networks for performing various tasks, and the required specific protocols in each layer of the protocol stack, are also presented.

13.3.1   UWSN components

In a UWSN, different types of devices are seen taking part in the whole network. These devices include sensor nodes, AUVs, drifters and floaters. Among these, the components of sensor nodes and autonomous underwater vehicles are discussed.

13.3.1.1   Sensor nodes
The change in the underlying communication medium from homogeneous air to complex and heterogeneous oceanic environment introduces a set of modifications in the design of sensor nodes for the underwater environment. The general components of sensor nodes or motes, as shown in Figure 13.4, include a microcontroller which acts as the central processing unit (CPU), on-board memory, battery power supply, acoustic transceiver unit, and oceanographic sensors. The power supply unit is mainly battery based, and very much essential for most other components of the sensor node. The transceiver part of the sensor node is for communication with other nodes via acoustic modem. However, software modems have recently emerged as a potential alternative for the hardware modems. The advantage of a software modem [31, 32] is in the trade-off of cheap computation power for costly hardware [33]. Sensor nodes that are placed on the water surface are generally also powered with a radio transceiver facility to communicate with the on-shore nodes or the control station. These motes have multiple on-board sensors such as oceanographic sensors for temperature, pressure, and salinity measurement. To protect the instruments, PVC coating is allied on a sensor node.

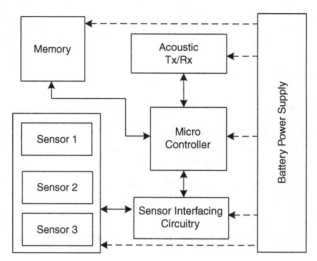

Figure 13.4   Architecture of an underwater sensor node (adopted from [1], with minor modifications).

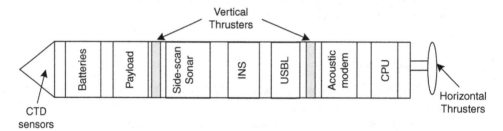

Figure 13.5   Components of an AUV (adopted from [36], with minor modifications).

13.3.1.2   Autonomous underwater vehicles (AUVs)

AUVs are intelligent devices that can travel through the water without any external control input. The presence of an autonomous system differentiates AUVs from remotely operated vehicles (ROVs). Consequently, AUVs are not required to be cabled with the controller, and are generally used for applications such as conducting underwater surveys for submerged rocks, explosives, checking any submerged oil pipeline, search missions, or guided navigation.

From a system point of view, an AUV includes side-scan sonar, various sensors such as conductivity–temperature–depth (CTD) sensors, magnetometers, thermistors, inertial navigation systems (INS), and an acoustic Doppler current profiler (ADCP). AUVs use a technique called dead-reckoning [34, 35]. The inertial navigation system continuously monitors the acceleration and position of the AUV. With the help of the sonar system, the speed of the AUV is measured by calculating the Doppler shift of the acoustic signals that are reflected from the ocean floor. The schematic diagram representing the components of an AUV is shown in Figure 13.5.

13.3.2   UWSN architecture

Underwater sensor network architectures may be classified into two major categories – two-dimensional (2D) and three-dimensional (3D). Based on the communication model of the architectures, researchers have studied different UWSN topology and deployment challenges [37, 38]. In this regard, two metrics are used to measure coverage of a UWSN – sensing coverage and communication coverage. Sensing coverage means the coverage of the target by the deployed sensors, and communication coverage indicates the connectivity of each sensor node to the base station via a multi-hop path.

13.3.2.1   Two-dimensional architecture

As the name suggests, in the two-dimensional (2D) architecture, the sensor nodes are deployed in a 2D plane such as the ocean bottom [39]. In an ocean-bottom monitoring application, the sensor nodes are deployed on the sea bed, and the sink nodes are deployed on the surface of the water. The nodes in the bottom of the sea and the surface based ones communicate via acoustic signals. The sea-bed embedded nodes communicate among themselves via a horizontal-transceiver, and a vertical-transceiver is used for communicating with the surface-located sinks. The problem with this type of scheme is that it requires long-range communication between the sea-bed embedded sensor nodes and the surface sinks. In Figure 13.6, one model of such 2D architecture is depicted for ocean-bottom monitoring applications.

Multipath virtual sink architecture [40], is a hierarchical network architecture proposed to reduce the loads of the sink nodes and to enhance robustness in the network. The authors proposed grouping of sensor nodes into few clusters, and each cluster having one local aggregator named a virtual sink. The virtual sinks are interconnected with the local sinks via a high-speed mesh network. This type of design distributes the load of the local sinks, but increases data redundancy due to the use of a multipath data delivery scheme.

In [41], the sensor nodes are grouped into clusters and each cluster has a cluster head (CH). The CHs are elected in the configuration phase, and these CHs form a tree-like

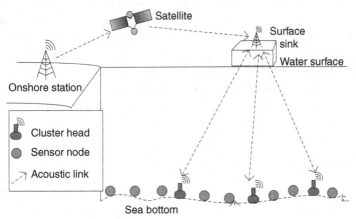

Figure 13.6   Two-dimensional UWSN architecture.

structure among themselves. The schedule of data transmission between the nodes and the corresponding CH, and the CH and its parent CH (in the tree structure) is also decided in this phase.

13.3.2.2   Three-dimensional architecture

In [37, 38], the problem of achieving maximum sensing and communication coverage with minimum number of deployed sensor nodes in a three-dimensional (3D) architecture was studied. The authors proposed three strategies for sensing and communication coverage, namely 3D-random, bottom-random, and bottom-grid [42]. In 3D-random strategy, the sensor nodes are randomly deployed on the sea bed, and they adjust their depth randomly. Similarly, in bottom-random, the nodes are deployed in a random fashion; however, the surface station finds the depths of each node such that the targeted coverage is achieved. In the bottom-grid technique, an AUV allots the depth of each node, and helps to attain target coverage.

One of the challenges in UWSNs is to guarantee that each underwater sensor node has connectivity towards the sink nodes located on the surface, in the presence of node mobility. Towards this goal, Ojha *et al.* [43] proposed an architecture named "tic-tac-toe-arch." In this scheme, the authors proposed an algorithm for calculating the "duration of connectivity" between underwater nodes, depending on the speed of the underwater current present in different layers of water. Each source node selects a "best" neighbor, based on the metric "duration of connectivity." The virtual topology is formed out of the 3D deployment of the sensor nodes, by repeating the process of "best" neighbor selection. The sensor nodes follow the virtual topology for the event reporting to the surface sink. The *virtual topology* is restructured such that the connectivity between the source and the sink nodes is always maintained. This automatic restructuring self-organizes the selection of active nodes to maintain the virtual topology.

UWSN architectures are proposed where sensor nodes are deployed with AUVs. This type of architecture is referred to as the hybrid architecture. Wang *et al.* [44] proposed one such hybrid architecture consisting of mobile actors and sensor nodes. The targeted application model is an underwater data collection scenario with high temporal resolution. The sensor nodes form virtual clusters among themselves, and these nodes send their sensed data to the local cluster head at regular time intervals. The mobile actors have their own navigation capability, and collect data from the local cluster heads with high temporal resolution. The authors proposed three algorithms to carry out the application: (a) area partitioning and actors scattering algorithm, (b) subregion optimizing algorithm, and (c) virtual cluster formation algorithm. One model of a 3D architecture is illustrated in Figure 13.7, where sensor nodes are grouped into different clusters with cluster heads, and AUVs are also shown to be taking part in the network.

13.3.3   Localization services

Localization is an important service that is often required by the higher-layer protocols. It is an essential requirement for UWSN applications such as target tracking, disaster prevention, environmental monitoring, and surveillance. The geographic routing schemes require

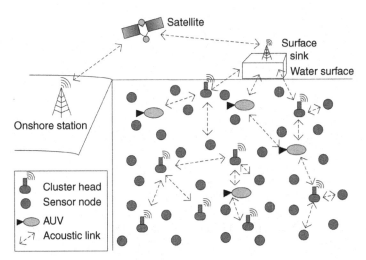

Figure 13.7   Three-dimensional UWSN architecture.

the location information of nodes. To localize nodes, sometimes, time-synchronization between the nodes is assumed; however, for accurate time-synchronization, the skew between two clocks is calculated. In the following subsections, a few localization schemes for UWSNs are discussed.

13.3.3.1   Dive'N'Rise localization (DNRL)

The Dive'N'Rise localization (DNRL) scheme [45] is based on special types of beacon nodes called the Dive'N'Rise beacons or DNR beacons. In this scheme, the mobile DNR beacons dive and rise through the ocean column, and announce their location. Initially, the DNR beacons receive their coordinates using GPS, while they float above the water surface. The underwater nodes "silently" localize themselves by listening to several beacons. The scheme is "silent," that is, the sensor nodes are only required to receive messages, and they are not required to send any messages. The sensor nodes calculate their x- and y-coordinates based on triangulation or bounding box, and the z-coordinate value is know by the on-board pressure sensor.

However, DNRL is criticized for the requirement of large number of DNR beacons for large-scale networks. This increases the cost of the network implementation. Another disadvantage of this scheme is that, node mobility greatly affects the localization performance due to the slow speed of the beacons. The authors also analyzed the case with mobile networks, where both DNR beacons and the sensors are drifted with currents. Figure 13.8 depicts the DNRL procedure.

13.3.3.2   Three-dimensional underwater localization (3DUL)

The 3DUL [46] localization scheme was proposed for achieving network-wide localization in UWSNs. Only three surface-based anchor nodes are required to initiate the localization procedure. The overall procedure is divided into two phases – "ranging" and "projection and dynamic trilateration." The procedure is initiated by the anchor nodes,

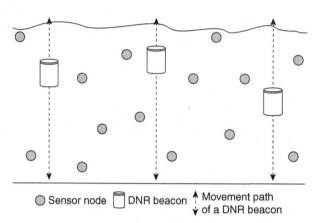

Sensor node □ DNR beacon ↑↓ Movement path of a DNR beacon

Figure 13.8   Illustration of the Dive'N'Rise (DNR) localization scheme [45].

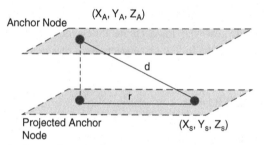

Anchor Node (X_A, Y_A, Z_A)

d

r

Projected Anchor Node (X_s, Y_s, Z_s)

Figure 13.9   Anchor node projection technique of 3DUL.

which broadcast an anchor ranging packet. Upon receiving the packet, any unlocalized node sends a ranging packet to initiate the "ranging" phase. In the ranging phase, a sensor node estimates its distance from the anchor nodes, and collects their depth information. The distance-estimation process is based on two-way message transfer. Let us assume that a node sends the ranging packet at time T_1, and at time T_2 the message is received by the anchor node. The anchor node replies with an ACK packet at time T_3, which is received by the node at T_4. Therefore, the propagation delay is calculated as $t_{prop} = [(T_2 - T_1) + (T_4 - T_3)]/2$. The distance between these two nodes is $d = t_{prop} \times v$, where v is the velocity of the sound in water = 1500 m/s.

Once the unlocalized node receives three ACK messages, it starts the second phase of the algorithm. The authors proposed a technique to select three anchors such that a robust virtual anchor plane is formed. Once the combination of anchors is found, a node uses the projection technique (shown in Figure 13.9), and the distance measure for the rest of the process.

Similarly, for three anchors, a node calculates the projected distance and estimates the location by the trilateration technique.

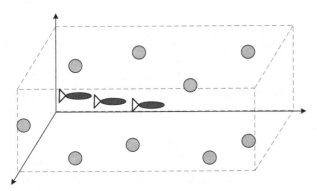

Figure 13.10   High-speed AUV-based silent localization scenario.

13.3.3.3 High-speed AUV-based silent localization (HASL)

The high-speed AUV-based silent localization scheme proposed in [47] utilizes the flexibility of a mobile beacon provider, an AUV. The main goal of this scheme is to reduce the effects of passive node mobility in increasing localization error of underwater sensor nodes (see Figure 13.10). Maintaining energy efficiency of the sensor nodes during localization process was another concern. Thus, the authors make use of the "silent" messaging technique. Three AUVs were assumed to cover the deployment region broadcasting their locations. The trajectory may be preplanned based on the deployment scenario. The AUVs move at a fixed depth, while maintaining their velocity. The sensor nodes employ trilateration technique to estimate their coordinates.

The three AUVs are time-synchronized, i.e. they send beacons at the same time in the first phase of the algorithm. The beacon messages contain the following fields: AUV id, x_A, y_A, z_A, and timestamp, where x_A, y_A, z_A are the locations of the AUV at that time. This synchronized beacon sending ensures reduced delay in between the reception of three beacons. Localization error due to node mobility is reduced in this fashion. A sensor node calculates its own coordinates after receiving beacons from three AUVs indicating the same timestamp. This set of beacons is called effective beacon messages. However, the sensor nodes are able to acquire their z-coordinate values with the assistance of a pressure sensor attached to them. Let us assume that the three AUV position locations are (x_{A1}, y_{A1}, z_{A1}), (x_{A2}, y_{A2}, z_{A2}), and (x_{A3}, y_{A3}, z_{A3}), respectively, at time t_1. The unlocalized node's location, then, can be known by using the following equations [47]:

$$(x - x_{A1})^2 + (y - y_{A1})^2 + (z - z_{A1})^2 = d_1^2,$$
$$(x - x_{A2})^2 + (y - y_{A2})^2 + (z - z_{A2})^2 = d_2^2,$$
$$(x - x_{A3})^2 + (y - y_{A3})^2 + (z - z_{A3})^2 = d_3^2,$$

where, d_1, d_2, and d_3 may be calculated by RSSI measurement of the received beacons.

13.3.3.4 Mobility-assisted localization (MobiL)

An inherent feature of UWSNs is that the nodes move according to the underwater currents. Owing to the effect of this inherent passive node mobility, the location estimation error in the sensor nodes increases. Greater amount of messaging between the unlocalized node and the reference node increases the protocol overhead, energy consumption, and delay, and, thus, increases the estimation error in the presence of node mobility.

Mobility-assisted localization or MobiL [48] is a localization method for the sensor nodes in UWSNs. In this scheme, the localization process starts from the surface-based anchor nodes. The overall procedure is divided into two phases: (a) mobility prediction and (b) ranging and localization. In the first phase, the authors used a velocity estimation technique proposed by [49, 50]. The sensor nodes update their velocity based on the velocities of their neighbors, as in equation (13.10). This is considered based on the fact that the movement of underwater nodes shows a group-like behavior:

$$
\begin{aligned}
v_j^x(d) &= \sum_{i=1}^{n} \xi_{ij} v_i^x(d), \\
v_j^x(d) &= \sum_{i=1}^{n} \xi_{ij} v_i^x(d).
\end{aligned}
\tag{13.10}
$$

The second phase utilizes trilateration-based localization to localize the nodes. "Silent" messaging was considered to keep the energy consumption of the sensor nodes low. As the underlying localization procedure considered was trilateration, a node requires three beacons from three reference nodes. Assume that the node S receives three beacons at times t_1, t_2, t_3. The displacement of node S is from (x, y, z) to (x', y', z') to (x'', y'', z'') in the meantime. The assumption is that the node can determine its position in the z-coordinate and have a negligible movement in the z-direction, i.e. $z = z' = z''$. Thus, using the velocity of node S, $x' = x + (t_2 - t_1) \times v_S^x$ and $y' = y_1 + (t_2 - t_1) \times v_S^y$. Similarly, x'' and y'' are calculated:

$$
\begin{aligned}
(x - x_1)^2 + (y - y_1)^2 + (z - z_1)^2 &= d_1^2, \\
(x' - x_2)^2 + (y' - y_2)^2 + (z - z_2)^2 &= d_2^2, \\
(x'' - x_3)^2 + (y'' - y_3)^2 + (z - z_3)^2 &= d_3^2.
\end{aligned}
$$

Replacing these values in the above equation, the location of node S is calculated.

13.3.4 UWSN protocol design

13.3.4.1 Physical layer

The physical layer is concerned with the transmission of signals over the communication medium. As we know, in UWSNs, acoustics are used as the signal, and the communication medium is the ocean water. The ocean water itself is a heterogeneous communication medium due to variation of the physical properties throughout the ocean. This makes the effects of the communication medium on the signal highly spatially and temporally dependent. Further, the ocean bottom, ocean surface, and various obstacles reflect, refract, and block acoustic waves, thereby impairing communication over the physical channel. Moreover, the use of an acoustic signal increases the propagation delay by five orders of the RF channels. The available bandwidth for communication is also low. Therefore, a robust modulation scheme is required, such that the receivers can exploit the limited bandwidth available in the underwater acoustic channel [5].

The initial modulation techniques used in UWSNs were mainly non-coherent frequency shift keying (FSK). The use of non-coherent FSK reduces the complexity of phase tracking, which is undesirable in underwater channels due to Doppler spread of acoustic signals [1]. The effect of multipath (due to reflection, and scattering) was eliminated by introducing guard time between two successive pulses. However, the introduction of guard time reduces the achieved data rate. Thus, selection of guard time is a trade-off between the data rate and the inter symbol interference (ISI). The disadvantage of non-coherent FSK is low bandwidth efficiency. The frequency channels are required to be separated by a coherence bandwidth $B_c = 1/T_m$ for simultaneous use, where T_m is the multipath delay spread [51, 52]. This type of modulation scheme was used in the design of the Teledyne-Benthos modem [53] and WHOI micro modem [54].

The advancements in phase-tracking algorithms empowered the coherent modulation schemes such as phase shift keying (PSK) and quadrature amplitude modulation (QAM) to achieve high data rates in different underwater channels [51]. Following this, the researchers designed coherent schemes capable of achieving the data rate of the non-coherent schemes [52]. Differential phase shift keying (DPSK) based schemes can detect the carrier frequency easily; however, the bit error rate (BER) also increased significantly over PSK at the same data rate [1].

13.3.4.2 Medium access control layer

Because of the inherent shared nature of the wireless medium, we need specific medium access techniques for efficient utilization of the medium. In designing the medium access control (MAC) techniques, the properties of the physical layer need to be kept in mind. In particular, the frequency-dependent attenuation of the acoustic signal, the limited available bandwidth, and the high propagation delay affect the MAC design mostly. The MAC protocols are generally classified into two broad categories – scheduling based and random access based. Frequency division multiple access (FDMA) and time division multiple access (TDMA) are scheduling based, whereas Aloha and Carrier Sense Multiple Access (CSMA) are random access based.

Scheduling-based MAC protocols
The limited bandwidth communication channel and its vulnerability to fading and multipath effects make the FDMA-based schemes inappropriate for UWSNs. Also, decisions are often taken in UWSNs in a distributed fashion, which makes FDMA unsuitable as a UWSN MAC.

TDMA-based MAC protocols provide a solution for avoiding collisions by sharing the total time slot, but reducing the channel utilization. Owing to increased delay in communication, long time guards are required in TDMA to maintain the collision-free nature in an underwater channel. Moreover, due to the distributed nature of the control, and the variable propagation delay, guard time determination and finding a common reference are also challenging [1]. STUMP [55] is a TDMA-based scheme that increases channel utilization by exploiting the propagation delay information.

Hybrid reservation-based MAC (HRMAC) protocol is proposed for UWSNs by Fan *et al.* [56]. For contention-based MAC protocols, a major reason for packet collisions is the increased packet transmission time for the large length of the packets. Also, the competition phase length increases if more nodes are shared in the same channel. After deployment, the sensor nodes interchange their coordinates. A node is selected as the network coordinator, which decides the order of the transmission of the nodes. The nodes that have data for transmission declare and reserve the channel. The nodes that already reserve the channel send data in the pre-decided order.

13.3.4.3 Random access-based MAC protocols

Aloha MAC protocols provide low utilization of the channel, despite being simple [5]. The CSMA-based schemes provide fewer collisions in the transmitter side by introducing a carrier sense facility. However, the carrier sense mechanism may not be able to detect an ongoing transmission due to the high propagation delay of the underwater channel. Thus, receiver side collisions may still occur.

To avoid collision, contention-based schemes such as MACA [57], and those available for IEEE 802.11 introduce the RTS/CTS mechanism. However, these schemes have some problems: (a) carrier sense may fail due to high propagation delay, and (b) RTS/CTS introduces additional delay in transmission, and so the throughput degrades.

Molins and Stojanovic proposed slotted floor acquisition multiple access (SFAMA) [58] by extending the original FAMA [59] protocol. Before sending data, the sender and the receiver exchange control packets. In SFAMA, the total time is divided into slots of fixed lengths. Each slot length is equal to the sum of the CTS transmission time and the maximum propagation delay. A transmission starts at the beginning of the slot. As shown in Figure 13.11, node B receives an RTS packet from node A at time slot 1. However, it starts sending CTS at the beginning of slot 2. Another node, C, a hidden terminal of node A, defers its transmission by knowing that an active transmission exists after receiving the CTS packet from B. Data transmission between nodes A and B also starts at the beginning of the next slot, which is similar for the acknowledgement packet. One of the advantages of the SFAMA protocol is that it is also suitable for mobile networks, where the propagation delay is variable in nature.

CDMA-based MAC protocols

Code division multiple access (CDMA) techniques have immunity against the selective fading and multipath effects of the underwater communication channel. Using CDMA, each transmitted signal is associated with a pseudo noise (PN) sequence, which is used to allocate the signal to spread over the whole bandwidth. The advantages of CDMA

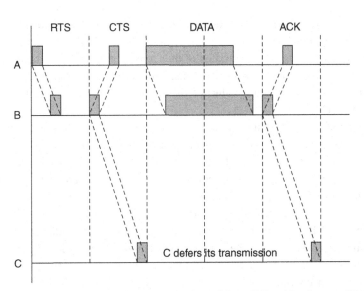

Figure 13.11   Illustration of the SFAMA protocol (adopted from [58], with minor modifications).

are: (i) the nodes are not bothered by what the other nodes are sending, (ii) the total number of nodes is minimized, and (iii) the effects of propagation delay are minimized. Although each user throughout decreases compared with the single user case, CDMA-based schemes provide a solution to many MAC-related problems in UWSNs [60].

13.3.4.4   Network layer

The following sections describe some of the schemes that are introduced for routing data from the source node to the target node (sink node) in UWSNs. Comparatively, routing in underwater networks is more challenging than in terrestrial ones. The following are some of the challenges in underwater routing: (i) low bandwidth, (ii) mobility of the nodes, (iii) latency, (iv) 3D space, and (v) high error rate. Therefore, routing protocols that are currently available for terrestrial sensor networks are not suitable for UWSNs.

Vector-based forwarding (VBF) routing

VBF [61] is a geographic routing protocol, i.e. the protocol requires the location information of sensor nodes. Only a limited number of nodes that fall in the path between the source node and the sink node participate in routing; the rest of the nodes simply discard the data packets to save energy. Each data packet has three fields for the coordinates of the sender, the sink, and the forwarder. The algorithm finds a routing vector $(\overrightarrow{S_1 S_0})$ formed between the sender (S_1) and the sink node (S_0), and further data are transferred along this path. Any node (A) forwards any packet received by it only after calculating the desirableness factor (α), as given below [61]:

$$\alpha = \frac{p}{W} + \frac{(R - d \times \cos\theta)}{R},$$

where p is the projection of A to the routing vector $(\overrightarrow{S_1S_0})$, d is the distance between node A and forwarder node F, θ is the angle between vector $(\overrightarrow{FS_0})$ and vector $\overrightarrow{FA}$, R is the transmission range, and W is the radius of the "routing pipe" [61]. If the desirableness factor for a node is 0, then this node is considered to be the optimal and best node for packet forwarding. For higher value of the desirableness factor, the node does not forward the packet.

In data transfer, either the sink or the source node shows interest first. If the sink node is interested in the data, then it asks for data from a specific location or, alternatively, it asks only for specific data, where the location is immaterial, by sending an "INTEREST" packet. In the first case, the intermediate nodes simply forward data packets if the location mentioned in the packet does not match their own location. On the other hand, for the second case, intermediate nodes send the data back to the sink node if the data of interest are available with them; otherwise, they simply forward data.

Data may be readily available for the source node, after which it broadcasts the "DATA_READY" packet. The intermediate nodes simply forward the packet by updating the forwarder coordinate field in the packet. Upon receiving the "DATA_READY" packet, the sink node sends an "INTEREST" packet directed to the source location for further communication. Any node, after receiving a packet, checks its own position, whether it is in the routing pipe or not. The node present in the routing pipe forwards the packet after holding it for a period of $T_{adaptation}$. This $T_{adaptation}$ is calculated based on the α value, as shown below [61]:

$$T_{adaptation} = \sqrt{\alpha} \times T_{delay} + \frac{R-d}{v_0},$$

where T_{delay} is the maximum propagation delay for the network, v_0 is the sound propagation speed in water (considered to be 1500 m/s), and d is the distance between the node and the forwarder.

Depth-based routing (DBR) protocol

Another simple and popular routing scheme for UWSNs is the depth-based routing (DBR) [62] protocol. Unlike VBF, DBR requires only the depth information of a sensor node for finding the packet-forwarding condition. Getting the depth information in a sensor node is cheap and may be attained with the help of a low-cost on-board pressure sensor. Thus, the scheme is free from the requirement of localizing the sensor nodes prior to routing.

As the name suggests, DBR requires only the depth information of a node. For this, the packet format of DBR has the following fields: sender id (s), packet sequence number (PSN) and depth (d_s). Any sensor node, upon receiving a packet, first compares the depth value in the packet header (d_s) with its own depth value (d_r). If its depth is smaller than that of the previous forwarder, i.e. $d_r < d_s$, only then does it forward the packet.

In the above case, the data redundancy problem arises when multiple packets arrive from different paths or from the same sender. In order to stop wastage of energy in handling unnecessary packets, DBR proposes a solution in the form of priority queues.

DBR introduces a holding time for deciding which node is eligible for forwarding packets. The holding time is a function of the depth difference between the node and the previous forwarder. Nodes with less depth get a chance to forward packets first, which makes routing easier and saves energy by stopping other nodes from sending packets (redundant packets).

To limit multiple transmissions of a packet by the same node, a node maintains two data structures: a priority queue Q_1 and a packet history buffer Q_2. Each packet to be sent is stored in Q_1 with its sending time. The sending time for a packet is calculated based on its arrival time added to its holding time. After sending a packet, DBR stores the unique id of the packet, i.e. the sender id and the PSN, in the packet history buffer Q_2. Q_2 is updated periodically. Prior to forwarding a packet, a node checks if the packet is in its history buffer. This process limits multiple transmission of a packet.

Void-aware pressure routing (VAPR)
Void-aware pressure routing (VARP) [63] is a geographic routing protocol where a node forwards a packet based on its depth information. A node may find its depth from the on-board pressure gauges. In the 3D underwater space, pressure routing should carefully handle the 3D void spaces. The countermeasures proposed for this problem require extensive flooding, which is a greedy approach. In VARP, the next hop direction is chosen, based on the information about sequence number, hop count, and depth. In this manner, a directional trail is found with the closest sonobuoy located on the surface. Earlier, the sonobuoys broadcast their reachability information in the enhanced beaconing stage. Each node updates additional information about its depth and hop count in its packet with the packet sequence number. Upon receiving multiple packets of this type, the nodes calculate the minimum hop to the surface, the data forwarding direction, and the next hop data forwarding direction.

13.3.4.5 Transport layer
The transport layer is required to provide end-to-end reliability, congestion control and flow control. It is required to ensure that the sensor data collected by a node are received by the control station. Transmission control protocol (TCP) is the most popular transport protocol in the conventional IP-based networks. However, in a UWSN, the propagation delay is high and the bandwidth is low. Also, the propagation delay increases by a far greater amount than the reduction in bandwidth. These reasons make TCP inappropriate for UWSNs. The greater propagation delay results in rapid depletion of the sender's window, which leads to poor utilization of the channel. Also, the high propagation delay leads to large round-trip-time (RTT), the main cause of difficulty in timely synchronization between the sender's and receiver's operations. Similarly, the existing rate-based schemes are also inapplicable in UWSNs due to the difficulty in maintaining timely synchronization of the packet and its acknowledgement on account of the high RTT.

In a UWSN, hop-to-hop reliability is considered over end-to-end reliability. Owing to the highly adverse communication channel, the probability of a correct packet traveling from one end to another over a long path is very low. Hence, it is preferable to have an intermediate node-check for reliable packet reception, and then forward the packet to the next hop. This reduces the path length and, hence, the probability of an erroneous packet. Also, shorter paths have lower RTT values. Hence, it is easier to maintain timely synchronization between the sender and the receiver over shorter links.

One general solution to ensure reliability and flow control is to have the receiver provide feedback to the sender. However, the use of feedback in UWSNs has its disadvantages. As seen from the example of TCP, excessive reliance on feedbacks results in lower channel utilization and greater difficulty in maintaining timely synchronization between the sender and receiver. Also, feedback has several overheads such as timeout counters and synchronization between the packet and its acknowledgement. In dense networks, this may be a significant overhead. Further, the use of feedback in environments with a high probability of error leads to more retransmissions, which increase energy consumption significantly.

On account of this, the segmented data reliable transport (SDRT) protocol [64, 65] attempts to find an appropriate trade-off between pure feedback-based and pure feedback-free schemes. SDRT is a hybrid approach of automatic repeat request (ARQ) and forward error correction (FEC). In SDRT, several packets are grouped into one block. Transmission between the sender and the receiver is done hop-by-hop and block-by-block. The receiver sends only one acknowledgement for an entire block. Within the block, the FEC codes are used to detect and rectify any errors within the block. The number of packets in a block can be controlled to adjust the trade-off between the feedback-free and feedback-based schemes.

Within each block, SDRT uses a pure feedback-free scheme. SDRT uses a special erasure code called the random forward error correction code, which is based on the Tornado code [66] (simple variant of Tornado code, or SVT). The protocol operation is shown in Table 13.2. Once the sender has determined the block size to be used, it collects the requisite number of packets for the block. It then encodes the packets using the random forward error correction code. The encoded packets (along with the redundant packets) are designated as a block. The sender transmits all the packets in a block in one burst, although they are transmitted in a random order. The sender keeps transmitting the packets in the block till it receives an acknowledgement for the entire block from the receiver. The receiver, on the other hand, keeps accepting the packets until it has

Table 13.2 SDRT protocol operation

Sender	Receiver
(1) Encode individual packets into a block using the random forward error correction erasure code.	(1) Keep accepting the packets until sufficient packets are received to recover the original packets.
(2) Keep sending the packets in an individual block in random order till an acknowledgement is received from the receiver.	(2) Send an acknowledgement for the entire block.
	(3) If the receiver is not the final destination of the packets, forward the packets to the next hop.

sufficient packets to recover (decode) the original packets. After one block is successfully decoded, the receiver sends a positive acknowledgement to the sender of the packet block. If the receiver is the final destination of the packets, it forwards the packets to its application layer; otherwise, it becomes the sender and repeats the procedure to send the packets to the next hop.

13.3.4.6   Cross-layer design (CLD)

The layered structure of the protocol stack is well established in wired networks. The layered design has many problems – it is non-adaptive and non-optimized [67]. In this regard, the network architecture and protocol design based on cooperative approaches between layers called cross-layer design (CLD), show a promising alternative to help solve some of the inefficiencies of the layered design. The issues are even more challenging in WSNs because of limited energy and computation power [68, 69]. Unlike the traditional layered structure, in a cross-layer technique, interaction between two non-adjacent layers is also possible. Cross-layer design is capable of increasing energy efficiency of nodes in various wireless protocols [70]. For example, the energy efficiency of a reference node selection algorithm of any iterative localization scheme may be increased by exploiting the residual battery status of the nodes. It is inferred that cross-layer design enables flexibility in protocol design. The routing and medium access decisions, taken with cross-layer information, can reduce the power consumption of UWSNs. Reference [67] surveys the existing literature on cross-layer design in WSNs.

13.4   Summary

In this chapter, we first discussed the characteristics, properties, and major application areas of UWSNs. The effects of different factors such as temperature, pressure, salinity, density, and viscosity on the underwater communication medium were discussed in the following section. The communication architecture for UWSNs was illustrated along with the depiction of different localization services such as DNRL, 3DUL, HASL, and MobiL. Finally, we designed the protocol stack for UWSNs and also focused on designing a cross-layer communication protocol. We discussed the MAC protocols and routing algorithms that are designed specifically for the underwater environment.

References

[1] I. F. Akyildiz, D. Pompili and T. Melodia, "Underwater acoustic sensor networks: research challenges," *Ad Hoc Networks*, Vol. **2**, No. 3, pp. 257–279, 2005.
[2] J. Heidemann, M. Stojanovic and M. Zorzi, "Underwater sensor networks: applications, advances and challenges," *Philosophical Transactions of the Royal Society A*, Vol. **370**, pp. 158–175, 2012.

[3] J. Proakis, E. Sozer, J. Rice and M. Stojanovic, "Shallow water acoustic networks," *Proceedings of IEEE Communication Magazine*, Vol. **39**, pp. 114–119, 2001.

[4] I. Vasilescu, K. Kotay, D. Rus, M. Dunbabin and P. Corke, "Data collection, storage, and retrieval with an underwater sensor network," in *Proceedings of ACM SenSys*, San Diego, CA, pp. 154–165, Nov. 2005.

[5] T. Melodia, H. Kulhandjian, L. C. Kuo and E. Demirors, "Advances in underwater acoustic networking," in *Mobile Ad Hoc Networking: Cutting Edge Directions*, Second Edition, S. Basangni *et al.*, Ed. John Wiley & Sons, 2013, pp. 804–852.

[6] I. F. Akyildiz, W. Su, Y. Sankarasubramaniam and E. Cayirci, "Wireless sensor networks: a survey," *Computer Networks*, Vol. **38**, No. 4, pp. 393–422, 2002.

[7] K. Akkaya and M. Younis, "A survey on routing protocols for wireless sensor networks," *Ad hoc Networks*, Vol. **3**, No. 3, pp. 325–349, 2005.

[8] I. F. Akyildiz, W. Su, Y. Sankarasubramaniam and E. Cayirci, "A survey on sensor networks," *IEEE Communications Magazine*, Vol. **40**, No. 8, pp. 102–114, 2002.

[9] G. Anastasi, M. Conti, M. Di Francesco and A. Passarella, "Energy conservation in wireless sensor networks: a survey," *Ad Hoc Networks*, Vol. **7**, No. 3, pp. 537–568, 2009.

[10] J.-H. Cui, J. Kong, M. Gerla and S. Zhou, "Challenges: building scalable mobile underwater wireless sensor networks for aquatic applications," *IEEE Network*, Vol. **20**, No. 3, pp. 12–18, 2006.

[11] J. Heidemann, W. Ye, J. Wills, A. Syed and Y. Li, "Research challenges and applications for underwater sensor networking," in *Proceedings of IEEE Wireless Communication and Networking Conference*, Las Vegas, NV, USA, pp. 228–235, 2006.

[12] J. Kong, J.-H. Cui, D. Wu and M. Gerla, "Building underwater ad-hoc networks and sensor networks for large scale real-time aquatic applications," in *Proceedings of IEEE Military Communication Conference*, Atlantic City, NJ, USA, pp. 1535–1541, 2005.

[13] M. Ayaz and A. Abdullah, "Underwater wireless sensor networks: routing issues and future challenges," in *Proceedings of the 7th ACM International Conference on Advances in Mobile Computing & Multimedia*, Kuala Lumpur, Malaysia, pp. 370–375, 2009.

[14] I. F. Akyildiz, D. Pompili and T. Melodia, "State of the art in protocol research for underwater acoustic sensor networks," *Mobile Computing and Communications Review*, Vol. **11**, No. 4, 2007.

[15] P. R. Pinet, *Invitation to Oceanography*, 5th Edition, Jones & Bartlett Learning, ISBN: 9780763759933, 2008.

[16] M. A. Ainslie, *Principles of Sonar Performance Modelling*, Springer, ISBN: 9783540876625, 2010.

[17] S. Misra and A. A. Ghosh, "The effects of variable sound speed on localization in underwater sensor networks," in *Proceedings of Australasian Telecommunication Networks and Applications Conference*, pp. 1–4, 2011.

[18] B. D. Dushaw, P. F. Worcester, B. D. Cornuelle and B. M. Howe, "On equations for the speed of sound in seawater," *Journal of the Acoustical Society of America*, Vol. **93**, No. 1, pp. 255–275, 1993.

[19] K. V. Mackenzie, "Discussion of sea water sound-speed determinations," *Journal of the Acoustical Society of America*, Vol. **70**, No. 3, pp. 801–806, 1981.

[20] C. T. Chen and F. J. Millero, "Speed of sound in seawater at high pressures," *Journal of the Acoustical Society of America*, Vol. **62**, No. 5, pp. 1129–1135, 1977.

[21] C. S. Clay and H. Medwin, *Acoustical Oceanography: Principles and Applications*. New York: John Wiley & Sons, ISBN: 978-0471160410, 1977, pp. 88 and 98–99.

[22] M. W. Denny, *How the Ocean Works: an Introduction to Oceanography*, Princeton University Press, ISBN: 9780691126470, 2008.

[23] Technical Report: International Association for Oil and Gas Producers, "Fundamentals of underwater sound," Report No. 406, May 2008.

[24] M. Schulkin and H. W. Marsh, "Sound absorption in sea water," *Journal of the Acoustical Society of America*, Vol. **34**, pp. 864–865, 1962.

[25] W. H. Thorp and D. G. Browning, "Attenuation of low frequency sound in the ocean," *Journal of Sound and Vibration*, Vol. **26**, pp. 576–578, 1973.

[26] W. L. Au Whitlow, *The Sonar of Dolphins*, Springer, ISBN: 9780387978352, 1993.

[27] L. M. Brekhovskikh, *Fundamentals of Ocean Acoustics*, 3rd ed. Springer, ISBN: 9780387954677, 2003.

[28] R. F. W. Coates, *Underwater Acoustic Systems*, John Wiley & Sons, ISBN: 9780470215449, 1989.

[29] M. Stojanovic, "On the relationship between capacity and distance in an underwater acoustic channel," in *Proceedings of ACM Workshop on Underwater Networks*, pp. 41–47, 2006.

[30] J. Preisig, "Acoustic propagation considerations for underwater acoustics communications network development," *ACM SIGMOBILE Mobile Computing and Communications Review*, Vol. **11**, No. 4, Oct. 2007.

[31] R. Jurdak, C. V. Lopes and P. Baldi, "Software acoustic modems for short range mote-based underwater sensor networks," in *Proceedings of IEEE Oceans*, Singapore, pp. 1–7, May 2006.

[32] R. Jurdak, P. M. Q. Aguiar, P. Baldi and C. V. Lopes, "Software modems for underwater sensor networks," in *Proceedings of Oceans*, pp. 1–6, June 2007.

[33] R. Jurdak, A. G. Ruzzelli, G. M. P. O'Hare and C. V. Lopes, "Mote-based underwater sensor networks: opportunities, challenges, and guidelines," *Telecommunication System*, Vol. **37**, No. 1–3, pp. 37–47, 2008.

[34] M. F. Fallon, G. Papadopoulos, J. J. Leonard and N. M. Patrikalakis, "Cooperative AUV navigation using a single maneuvering surface craft," *International Journal of Robotics Research*, Vol. **29**, pp. 1461–1474, 2010.

[35] H. Woithe, D. Boehm and U. Kremer, "Improving Slocum glider dead reckoning using a Doppler velocity log," in *Proceedings of MTS/IEEE OCEANS*, Waikoloa, HI, pp. 1–5, 2011.

[36] J. B. Paduan, D. W. Caress, D. A. Clague, C. K. Paull and H. Thomas, "High-resolution mapping of mass wasting, tectonic, and volcanic hazards using the MBARI mapping AUV," in *International Conference on Seafloor Mapping for Geohazard Assessment*, Forio d'Ischia, Italy, May 11–13, 2009.

[37] D. Pompili, T. Melodia and I. F. Akyildiz, "Deployment analysis in underwater acoustic wireless sensor networks," in *Proceedings of ACM Workshop on Underwater Networks*, Los Angeles, California, USA, pp. 48–55, September 2006.

[38] D. Pompili, T. Melodia and I. F. Akyildiz, "Three-dimensional and two-dimensional deployment analysis for underwater acoustic sensor networks," *Ad Hoc Networks*, Vol. **7**, No. 4, pp. 778–790, 2009.

[39] D. Pompili and T. Melodia, "An architecture for ocean bottom underwater acoustic sensor networks (UWASN)," in *Proceedings of Mediterranean Ad Hoc Networking Workshop*, Bodrum, Turkey, 2004.

[40] W. K. G. Seah and H.-X. Tan, "Multipath virtual sink architecture for underwater sensor networks," in *Proceedings of IEEE Oceans*, Singapore, pp. 1–6, 2006.

[41] S. Climent, J. V. Capella, N. Meratnia and J. J. Serrano, "Underwater sensor networks: a new energy efficient and robust architecture," *SENSORS*, Vol. **12**, pp. 704–731, 2012.

[42] W. Lin, D. Li, Y. Tan, J. Chen and T. Sun, "Architecture of underwater acoustic sensor networks: a survey," in *Proceedings of Intelligent Networks and Intelligent Systems*, pp. 155–159, 2008.

[43] T. Ojha, M. Khatua and S. Misra, "Tic-tac-toe-arch: a self-organizing virtual architecture for underwater sensor networks," *IET Wireless Sensor Systems*, Vol. **3**, No. 4, pp. 307–316, Dec. 2013.

[44] J. Wang, D. Li, M. Zhou and D. Ghosal, "Data collection with multiple mobile actors in underwater sensor networks," in *Proceedings of IEEE Workshop on Delay/Disruption-Tolerant Mobile Networks*, pp. 216–221, 2008.

[45] M. Erol, L. F. M. Vieira and M. Gerla, "Localization with DiveNRise (DNR) beacons for underwater acoustic sensor networks," in *Proceedings of ACM WUWNet*, pp. 97–100, 2007.

[46] M. T. Isik and O. B. Akan, "A three dimensional localization algorithm for underwater acoustic sensor networks," *IEEE Transactions on Wireless Communications*, Vol. **8**, No. 9, pp. 4457–4463, 2009.

[47] T. Ojha and S. Misra, "HASL: high-speed AUV-based silent localization for underwater sensor networks," in *Proceedings of the International Conference on Heterogeneous Networking for Quality, Reliability, Security and Robustness, LNICST 115*, Greater Noida, India, pp. 128–140, 2013.

[48] T. Ojha and S. Misra, "MobiL: a 3-dimensional localization scheme for mobile underwater sensor networks," in *Proceedings of National Conference on Communications*, New Delhi, India, pp. 1–5, 2013.

[49] A. Novikov and A. Bagtzoglou, "Hydrodynamic model of the lower Hudson River estuarine system and its application for water quality management," *Water Resources Management*, Vol. **20**, No. 2, pp. 257–276, 2006.

[50] A. Bagtzoglou and A. Novikov, "Chaotic behavior and pollution dispersion characteristics in engineered tidal embayments: a numerical investigation," *Journal of American Water Resources Association*, Vol. **43**, No. 1, pp. 207–219, 2007.

[51] M. Stojanovic, "Recent advances in high-speed underwater acoustic communications," *IEEE Journal of Oceanic Engineering*, Vol. **21**, No. 2, pp. 125–136, 1996.

[52] M. Stojanovic, "Underwater acoustic communications," in *Encyclopedia of Electrical and Electronics Engineering*, John G. Webster, ed. John Wiley and Sons, 1999, pp. 688–698.

[53] D. Green, "Acoustic modems, navigation aids, and networks for undersea operations," in *Proceedings of IEEE Oceans*, Sydney, Australia, pp. 1–6, 2010.

[54] S. Singh, S. Webster, L. Freitag, *et al.*, "Acoustic communication performance of the WHOI micro-modem in sea trials of the Nereus vehicle to 11000m depth," in *Proceedings of IEEE Oceans*, Biloxi, MS, pp. 1–6, 2009.

[55] K. Kredo, P. Djukic and P. Mohapatra, "STUMP: exploiting position diversity in the staggered TDMA underwater MAC protocol," in *Proceedings of IEEE INFOCOM Mini-Conference*, Rio de Janeiro, Brazil, pp. 2961–2965, 2009.

[56] G. Fan, H. Chen, L. Xie and K. Wang, "A hybrid reservation-based MAC protocol for underwater acoustic sensor networks," *Ad Hoc Networks*, Vol. **11**, No. 3, pp. 1178–1192, 2013.

[57] P. Karn, "MACA – a new channel access method for packet radio," in *Proceedings of the 9th ARRL Computer Networking Conference*, London, Ontario, Canada, 1990.

[58] M. Molins and M. Stojanovic, "Slotted FAMA: a MAC protocol for underwater acoustic networks," in *Proceedings of IEEE Oceans*, pp. 1–6, 2006.

[59] C. L. Fullmer and J. J. Garcia-Luna-Aceves, "Floor acquisition multiple access (FAMA) for packet-radio networks," in *Proceedings of SIGCOMM*, pp. 262–273, 1995.

[60] J. Partan, J. Kurose and B. N. Levine, "A survey of practical issues in underwater networks," *ACM SIGMOBILE Mobile Computing and Communications Review*, Vol. 11, No. 4, pp. 23–33, 2007.

[61] P. Xie, J.-H. Cui and L. Lao, "VBF: vector-based forwarding protocol for underwater sensor networks," in *Proceedings of IFIP Networking*, pp. 1216–1221, 2006.

[62] H. Yan, Z. J. Shi and J.-H. Cui, "DBR: depth-based routing for underwater sensor networks," in *Proceedings of the IFIP-TC6 Networking Conference on AdHoc and Sensor Networks, Wireless Networks, Next Generation Internet*, pp. 72–86, 2008.

[63] Y. Noh, U. Lee, P. Wang, B. S. C. Choi and M. Gerla, "VAPR: void-aware pressure routing for underwater sensor networks," *IEEE Transactions on Mobile Computing*, Vol. 12, No. 5, pp. 895–908, 2013.

[64] P. Xie and J.-H. Cui, "An FEC-based reliable data transport protocol for underwater sensor networks," in *Proceedings of International Conference on Computer Communications and Networks*, Honolulu, HI, pp. 747–753, 2007.

[65] P. Xie, Z. Zhou, Z. Peng, J.-H. Cui and Z. Shi, "SDRT: a reliable data transport protocol for underwater sensor networks," *Ad Hoc Networks*, Vol. 8, No. 7, pp. 708–722, 2010.

[66] M. Luby, M. Mitzenmacher, A. Shokrollahi, D. Spielman and V. Stemann, "Practical loss-resilient codes," in *Proceedings of ACM Symposium on Theory of Computing*, pp. 150–159, 1997.

[67] S. Misra and M. Khatua, "Cross-layer techniques and applications in wireless sensor networks," in *Using Cross-Layer Techniques for Communication Systems*, H. F. Rashvand and Y. S. Kavian, Ed. USA: IGI Global, 2012, pp. 94–119.

[68] M. C. Vuran, V. C. Gungor and O. B. Akan, "On the interdependence of congestion and contention in wireless sensor networks," in *Proceedings of the Third International Workshop on Measurement, Modeling, and Performance Analysis of Wireless Sensor Networks*, San Diego, CA, USA, 2005.

[69] L. D. P. Mendes and J. J. P. C. Rodrigues, "A survey on cross-layer solutions for wireless sensor networks," *Journal of Network and Computer Applications*, Vol. 34, No. 2, pp. 523–534, 2011.

[70] M. Erol-Kantarci, H. T. Mouftah and S. Oktug, "A survey of architectures and localization techniques for underwater acoustic sensor networks," *IEEE Communications Surveys and Tutorials*, Vol. 13, No. 3, pp. 487–502, 2011.

14 Wireless underground sensor networks

Wireless sensor networks (WSNs), due to their unique features such as fault tolerance, scalability, low production cost, and secured transmission, have found applications over widespread domains. Underground sensor networks have their constituent sensor nodes deployed underground, and are designed to operate underneath the earth's surface. Each of these nodes is equipped with a processor, a finite local storage memory, a radio, an antenna, and a power source unit. Underground sensor networks can be broadly classified into two categories, based on their data dissemination modes, namely, wired underground sensor networks, and wireless underground sensor networks (WUGSNs) [1–3]. Wired underground sensor networks [1–19] are composed of sensing devices, a small processor, limited memory, an antenna, and a power source. These devices are connected to data loggers on the surface by using wires. Data loggers act as transceivers between the sensors and the central data sink. At the receiver end, the data loggers receive and store data that are disseminated from the underground sensor devices. These data are then transmitted to a central data sink or network administrator site by the data loggers for further interpretation and analysis. On the other hand, WUGSNs consist of multiple wireless sensor nodes, each having a sensing unit, a local processing unit, a local storage memory, a radio, an antenna, and a constant source of power. Each of these nodes is capable of acting independently for transmitting the sensed data to the data sink. The nodes establish a multi-hop wireless communication network to ensure reliable and secured dissemination of sensed data.

The general applications of WUGSNs involve real-time soil monitoring in agricultural applications, measuring toxicity of soil for environment monitoring, infrastructure monitoring, underground mine detection, and border surveillance [1–3]. A WUGSN can effectively replace a traditional wired underground sensor network, as it overcomes most of the shortcomings of its wired counterpart. Unlike wired underground sensor networks [2], WUGSNs do not require concealment of sensing devices in order to protect them from environmental hazards, theft, and other issues of insecurity. WUGSNs are easy to deploy, and are cost effective in comparison to wired underground sensor networks. Moreover, WUGSNs are able to disseminate real-time data with higher reliability and security. Coverage density of WUGSN is also comparatively high, and any failure in a WUGSN can be detected with ease. Owing to these unequivocal advantages, WUGSNs can efficiently replace the wired underground sensor networks in most of their fields of applications. In this chapter, we focus on WUGSNs and their design, architecture, and application areas.

WUGSNs can either be buried underneath the soil or, alternatively, they can be positioned within open, but confined, areas such as tunnels and subways. In the first case, the sensor nodes communicate through the soil, whereas in the latter they transmit through the air medium. The primary difference between WUGSNs and terrestrial wireless sensor networks or WSNs is the medium of communication. Communication through solid substances such as soil, rock, dust, and mud proves to be more challenging in comparison with communication through air. The challenge lies in the establishment of efficient and reliable underground wireless links for data dissemination and multi-hop communication, in the context of WUGSNs. Moreover, WUGSNs are required to be unearthed for recharging purposes and, hence, it is very important to utilize the power feed in an optimal and efficient manner for each sensor node. Designing proper communication protocols for WUGSNs demands reconsideration of issues such as network topology, antenna design, and power conservation.

14.1   Applications

The applications of WUGSNs [1–3] cover a larger domain compared to the traditional terrestrial WSNs, as it involves monitoring underground soil, mineral, and water properties. WUGSNs are useful in several fields that require continuous or periodic monitoring of soil properties beneath the ground. These applications primarily involve soil property monitoring, environment monitoring, border surveillance, mining safety vigilance, infrastructure monitoring, and location determination. We discuss below the utility of WUGSNs in each of these domains [1–3].

14.1.1   Soil property monitoring

WUGSNs have found applications [1–3] in the field of irrigation. Technological advancements in agriculture demand periodic monitoring of different critical soil attributes [1, 4]. Owing to their efficient operability beneath the earth's surface, WUGSNs have proven to be prodigiously effective to enable intelligent irrigation. WUGSNs equipped with a special type of sensor node capable of measuring water or moisture levels, or mineral concentration in soil, continuously sense and report the soil property for centralized analyses of the data. These specific sensor nodes can be deployed over an agriculture field with ease, and a dense deployment of the nodes would ensure detailed local data reports. The concealment offered by the underground systems protects the sensor nodes from various environmental and physical damage. Accurate monitoring of water or moisture content, mineral concentration, salinity, and soil temperature is fundamental for systematic and proficient agriculture. Moreover, these underground sensor nodes can also be configured for determining the toxicity of soil that may be contaminated by the water source, or due to excess use of pesticides or fertilizers.

14.1.2 Environment monitoring

In addition to monitoring different soil attributes, WUGSNs can be configured to sense ground movements. These data are used to predict landslides and avalanches, and necessary evacuations and safeguarding could be initiated. Another useful application of these wireless underground sensor networks is monitoring and prediction of earthquakes [1–3]. A multi-hop communication network with its nodes deployed at different levels under the ground acts as an efficient system for monitoring the amplitude and frequency of the seismic waves, and, thus, predicts their future activity. Among other measures of environment monitoring, WUGSNs are used to keep track of glacier behavior and volcanic eruptions. Wireless sensor nodes are buried inside the glaciers to measure their movements and melting rate over time. A similar approach can also be used to predict and supervise the nature of eruptions of volcanoes.

14.1.3 Border surveillance

Underground wireless sensors can be deployed to sense pressure or seismic waves originating from the surface. This can be used for perimeter surveillance and border security maintenance [1–3]. Pressure waves generated as a result of impact on the ground surface caused by human activity or object movement can be captured by the wireless sensors deployed underneath the surface. This can be used for domestic surveillance purposes to detect intrusion, and thereby alert the house-keeper. It can also be used as a commercial product for different organizations that demand restricted entry to a secured zone. An advanced version, however, can be used for border surveillance and patrol applications. A dense deployment of wireless underground sensor nodes along the border perimeter, superficially under the surface, can activate an alarm following any unauthorized invasion. The basic working principle of these WUGSNs is conversion of pressure to displacement, which, in turn, is converted into electrical signals.

14.1.4 Mining safety vigilance

Safety in underground mining has always been a cause of concern [1–3]. Coal is an undisputed natural source of energy, and coal mining, consequently, has become even more relevant to modern civilization [6–19]. However, underground coal mining is inherently dangerous, as it involves multiple complex and critical processes. The main challenge in underground coal mining is to dig through unstable geological structures for coal harvesting. Growth of the cavity, over time, causes the underground tunnels to become prone to collapse. Unstable underground cavity structures often cause the tunnels to collapse, resulting in dangerous accidents in mines. An additional issue of concern is explosion of inflammable gases and leakage of poisonous gases inside the mines. Therefore, it is very important to continuously monitor different critical attributes inside the mines, and report them immediately. WUGSNs can be an effective, yet cheap, solution to these problems that may arise inside a mine. Different sensor nodes can be deployed inside the mines, under the soil for real-time monitoring of the tunnel structure, cavity growth rate, water leakage, and other critical mining attributes. The nodes can also detect and report the presence of poisonous and explosive gases inside the mines. Real-time analyses of these data at the data analytic center can be useful in avoiding mine accidents.

14.1.5   Infrastructure monitoring

WUGSNs are used to monitor the health and status of various infrastructures [5] that are located under the surface. Most infrastructures that span across cities or even states, such as plumbing and electrical wiring, are situated underneath the ground surface. WUGSNs can be deployed along these underground channels to detect leakages or malfunctioning of the infrastructures. Underground sensor networks are also useful in monitoring the health of various construction infrastructures such as buildings, roads, dams, and bridges. Specifically constructed wireless sensor nodes are capable of measuring stress, strain, and other parameters that dictate the health of a construction. A periodic assessment of these data can be very useful for much needed maintenance of these infrastructures.

14.1.6   Location determination

Wireless underground sensor nodes equipped with location-finding hardware such as global positioning system (GPS) can be used for traffic monitoring and navigation purposes [1–3]. A stationary GPS-enabled sensor device that is aware of its own location can help in providing navigation feeds to a vehicle moving on a road. A WUGSN can also report possible traffic congestion ahead, or the distance to the nearest fuel or car-fixing station, to a moving car. This may contribute towards a well-organized traffic management system.

14.2   Challenges in designing WUGSNs

Wireless underground sensor networks have emerged as a stimulating prospect due to their ability to operate in special environmental conditions, and have found widespread sphere of applications. The most intriguing feature of these sensor networks is their capability of sensing physical phenomena below the surface of the ground. The composite devices of a WUGSN are deployed beneath the surface, and are designed to sense various underground attributes. These devices are expected to communicate with one another through solid media such as soil, rock, dirt, and mud. Herein lies the primary challenge of designing such devices [1–3]. Wireless communication through solid media is far more complicated compared to the same through the air medium. Thus, it is very important to establish reliable and secured wireless links for efficient underground communication and data dissemination among the sensor nodes. Moreover, the fact that these sensor nodes are required to be unearthed for recharging purpose makes it crucial to design an underground network topology that curbs unnecessary power consumption. In this section, we discuss the factors that present challenges before the design of WUGSNs.

14.2.1   Underground communication channel design

In WUGSNs, dictating an efficient and reliable multi-hop underground transmission becomes very difficult. Communication through solid media experiences high levels of signal attenuation due to absorption by various underground particles. Another problem experienced in underground communications is path loss; this depends on the soil structure, which includes the water and mineral contents of the soil, density of minerals, and soil makeup. Electromagnetic (EM) waves are still the best alternative for wireless signal propagation in underground tunnels and mines. For shorter-range communications, magnetic induction (MI) waves can be used. It is very important, for both the cases, to design an underground wireless communication channel for sensor nodes that efficiently takes care of the typical underground communication problems. However, digital modes of communications in underground environments are yet to be explored [1–3].

14.2.2   Topology design

Topology design [1–3] for WUGSNs is vital to improve network reliability and network lifetime. Designing a suitable wireless underground network topology ensures optimal power consumption and thereby lengthens network lifetime. However, due to the hostile underground conditions [5], designing an appropriate topology becomes harder. Depending on the soil profile, moisture level, and other environmental conditions, signal attenuation takes place beneath the surface, and, therefore, it is very important to place the sensor nodes sufficiently close to the source of the event. Wireless underground sensor nodes are also required to be excavated from underneath the ground for recharging purposes. These are often placed at different depths for inter-node relay communications. The nodes that are close to the surface relay the data sensed by the deeper devices towards the sink. Therefore, it becomes a trade-off between the node deployment depth and the power supply to the node while designing a suitable network topology. It is necessary to structure a wireless network topology that minimizes the power usage and thereby extends the duration between consecutive device excavations. The deployment cost of these devices is also to be optimally reduced during topology construction.

14.2.3   Power consumption

Power is considered as a precious resource in the context of wireless sensor networks because each sensor node is equipped with a limited amount of power in the form of batteries prior to its deployment. Depending on the specific type of information that a sensor node acquires, its battery lifetime may vary. However, a sensor node is expected to have a reasonable lifetime of several years. A sensor node needs to be recharged periodically before its energy is completely depleted. For WUGSNs it becomes even more critical, as recharging of the battery following its discharge involves tedious actions such as excavating the sensor device, and redeploying it under the soil after it is recharged. Also, underground wireless communications demand greater power, as transmission through solid, lossy transmission channels formed by soil, rocks, and mud involves high levels of signal attenuation and absorption.

The sensor devices that are located near the surface can be recharged through induction techniques, but for devices that are buried deep in the soil, recharging is impossible without unearthing. Also, as the devices are deployed under the surface, solar energy harvesting becomes impossible. However, nodes can harvest energy from vibrational excitation, thermal gradients, and the background radio signal, although it remains to be explored if these harvesting techniques can provide sufficient amount of energy. Another straightforward approach to overcoming power scarcity is deploying the sensors with a larger built-in battery. Although it may increase the lifetime of the underground sensor devices, this would also increase the size of the device, and subsequently its cost. Therefore, it remains an overhead that is to be dealt with critically by using appropriate network topology, device hardware and energy source. A proper network topology may enhance the node lifetime and increase transmission reliability at the same time [1–3].

14.2.4   Antenna design

Selection of a suitable antenna size [1–3] that meets the hardware and transmission requirements plays a key role in designing WUGSNs. The transceiver hardware and the suitable antenna size selection are distinctively complex in WUGSNs in comparison to the terrestrial WSNs. Antennas in WUGSNs are designed to serve different special purposes. For a sensing device that is located near to the surface, the antenna needs to take care of reflection of the transmitted radiation at the soil–water interface. It also acts as a horizontal relaying hardware that assists signals transmitted from deep-buried sensors to reach the sink located at the surface. In the case of deeper sensor nodes, antennas help in the vertical relay of signals required for routing purposes. Therefore, the selection of suitable antenna size required for different actions needs to be made in a planned manner.

Size and directionality of antennas in wireless underground sensor devices are two crucial parameters that are to be chosen appropriately. For efficient long-range transmission and reception of lower-frequency signals (of MHz order), larger-sized antennas are required. However, larger antennas increase the device dimensions, which is highly undesired. So approaches that would serve long-range transmissions without increasing the antenna size are to be followed. Directionality of antennas corresponds to the direction of transmission and reception of signals. A wireless underground sensor needs to act in every direction to ensure uninterrupted communication. A wise and careful selection between the omnidirectional antenna and a set of unidirectional antennas is to be made to serve specific node functions. While the use of the omnidirectional antenna eliminates the need for multiple antennas in a sensor node, a proper combination of vertical and horizontal antennas can be used to avoid communication impairments that arise while using omnidirectional antennas. It remains an open research area to observe if other similar technologies can be used for improvement of underground communications.

14.2.5   Environmental hazards

The environmental conditions beneath the ground surface, in which the WUGSNs operate, are completely different from those for WSNs [1–3]. Several factors such as moisture, organic extracts, temperature extremities, rainfall, hailstorms, and even insects and plant roots pose threats for the design of WUGSNs. The devices should be manufactured to be resilient to these threats, and also hard enough to survive multiple excavations. Adequate protection is to be provided to these miniaturized devices so that they do not succumb to pressure caused by objects above the surface. Device overlays should be able to endure temperature extremities and water leakages. Thus, it is very important to study the environmental conditions of a region before deployment of underground wireless sensor devices.

14.3   Network architecture

The design of network architecture [1–3] by constructing suitable network topology is a cumbersome task for underground wireless communications. The interference of soil and other underground attributes makes the task difficult. As stated earlier, based on the positioning of the underground sensors, WUGSNs can be broadly classified into two categories: (a) topologies for WUGSNs buried underground, and (b) topologies for WUGSNs deployed in mines and tunnels. The primary difference between these two sensor positions is the modes of communication. Therefore it is necessary to have separate network topologies for each case. We discuss a few important network topologies in this section.

14.3.1   Topologies for WUGSNs buried underground

Network topologies for WUGSNs buried inside the soil can be classified into three categories [1–3]: underground topology, hybrid topology, and clustering chain topology.

14.3.1.1   Underground topology
In this network topology, the wireless sensor nodes are deployed underneath the surface, in the soil [2]. These sensors sense various underground attributes as a continuous function of time, and transmit the data sensed to a sink that is located above ground. Based on the location and depth of these underground sensors, there are two main network topologies: (i) uniform depth topology, and (ii) hierarchical depth topology [1–3].

Uniform depth topology
In uniform depth network topology, the underground sensor nodes are placed at the same level under the soil. These sensors are buried at a depth that may be covered by a single communication hop. Therefore this topology is designed for a typical single-hop

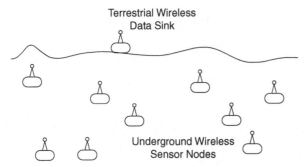

Terrestrial Wireless
Data Sink

Underground Wireless
Sensor Nodes

Figure 14.1   Underground topology (adopted from [2], with minor modifications).

communication network. The underground sensors disseminate the sensed data to a sink node that is located above the surface. The sink node can be designed either as a static node, or as a mobile one, as required. It can be noticed that whereas a static sink dictates a firm network topology, a mobile sink is able to gather data from a larger number of underground sensor nodes. Deployment of the sensor nodes may be done either in a planned way or in random fashion. In the random deployment of nodes, a mobile sink serves better than a static one, and also the deployment cost is less. On the contrary, a well-planned sensor deployment policy incurs higher cost, but guarantees higher reliability. A static sink node is the best fit in this case. Figure 14.1 depicts the uniform depth network topology for underground sensors [1–3].

Hierarchical depth topology
Hierarchical depth underground network topology is suitable for applications that involve the sensing of data at deeper regions under the soil. In cases where sensing of phenomena deep under the surface is necessary, single-hop communication is unsuitable. A multi-hop network topology is essential in such situations to relay the signal to the above-ground sink. Hierarchical depth network topology is made up of sensor nodes deployed at different strata inside the soil. Sensors at the bottom-most stratum sense the underground phenomena and disseminate the sensed data to the layer above. Sensors in the intermediate strata sense attributes and act as routers at the same time. These intermediate nodes may or may not collect data themselves, but essentially act as relay devices. Sensor nodes located near the surface commute with the above-ground sink and transmit the aggregated data. This multi-hop communication network is highly effective when sensors are deployed randomly at different depths over a larger area. The sink, in this case can also be static or mobile according to the application requirements. Figure 14.2 shows a hierarchical depth underground wireless network topology with three strata [1–3].

14.3.1.2   Hybrid topology
Hybrid network topology [2] is composed of a number of underground and terrestrial sensors. This mixed topology involves both underground communication through soil and above-ground communication through the air medium. The essence of hybrid

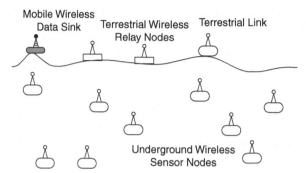

Figure 14.2   Hybrid topology (adopted from [2], with minor modifications).

topology is to increase the overall lifetime of the network by ensuring lesser power consumption, and data routing in a smaller number of hops. In this network topology, the underground sensor nodes sense different underground phenomena, and transmit the sensed data towards the superficially positioned sensor nodes, which, in turn, disseminate the data to the terrestrial sensor nodes. The terrestrial sensor nodes relay the data to a static or mobile above-ground sink node. This part of the network communication involves the air medium for transportation of data packets. Communication via radio signals through the air medium causes lower loss and consumes less power compared to underground communication.

Therefore, a hybrid network topology improves the network lifetime and the transmission reliability considerably. However, a major disadvantage of this topology is that the nodes are not strictly "underground" in nature, and, hence, are not completely concealed. Nevertheless, for underground applications that hardly require concealment of sensors, hybrid topology offers a power-efficient, and reliable mode of data transmission [1–3].

14.3.1.3   Clustering chain topology

Clustering chain topology [4] for WUGSNs, as proposed by Chen *et al.*, is a node-cluster-based design approach that reduces the overall power consumption for the underground network, and thus increases its lifetime. In this topology, the underground sensor nodes are organized into several clusters based on their spatial locations. Each of these clusters elects a cluster head to act as a local sink for the nodes in the cluster. Cluster heads are designed to transmit the received data packets to a base station. The sensor nodes are deployed randomly below the surface. Based on the location, connection quality, and residual energy level of the cluster heads, the common nodes choose to join a cluster. These cluster heads form a chain-like, uniformly distributed structure that ensures minimum power dissipation communication hops during data transmission. Clustering chain-type topology is found to be more efficient compared to line-type topology in some particular fields of application.

14.3.2   Topologies for WUGSNs deployed in mines and tunnels

For WUGSNs that are deployed underground but very close to, or on, the surface, such as in tunnels, subways, underground roads and railways or mines, the topology needs to be carefully designed. The sensor nodes are partially or completely revealed in situations like these, and yet should be designed to operate in underground environments. The signal propagation characteristics in such environments are significantly different from terrestrial signal transmission, or transmissions for completely buried WSNs. Based on the structure of the underground passage, appropriate wireless network topology is chosen. The tunnel model, the room–pillar model, and the shadow fading model are some of the popular ones that are used for topology design in such conditions. The selection of a particular model for a specific situation depends on the shape, depth, and position of the underground tunnel, subway, or mine [1–3].

14.4   Communication architecture

Communication architecture [1–3] for WUGSNs can be viewed as a classical five-layered protocol stack with an extension of cross-layered planes. In order to ensure efficient and reliable communications in underground WSNs, the protocol stack is required to be redesigned and modified. Cross-layered power management and task management planes enhance the protocol efficiency in underground environments. The main difficulty in designing an underground wireless network protocol stack is to overcome the different challenges proposed by the surrounding environment. This section thoroughly addresses the protocol stack, the cross-layer approach, and various research challenges encountered in the process.

14.4.1   Physical layer

The physical layer is the lowest in the protocol stack [1–3]. It is considered to be the most complex layer, as it deals with the hardware components, and therefore needs to be drafted appropriately. In the context of WUGSNs, it becomes even more complicated due to the miniaturized hardware components, and rough signal propagation media. The propagation medium for underground wireless sensor networks is solid channels made up of soil, rock, mud, and dust. The two different types of physical layer architectures that we will discuss in this section are: (a) communication using electromagnetic (EM) waves, and (b) communication using magnetic induction (MI) waves.

14.4.1.1   Communication using EM waves

A widely used physical layer architecture for wireless underground networks is communication through EM waves. However, EM waves suffer from high levels of signal loss and attenuation under the soil surface. The levels of loss and attenuation depend on various soil properties such as mineral content, mineral density, soil temperature, and volumetric water content. Moreover, due to the variable nature of these underground attributes, it is very difficult to design a uniform signal propagation channel. External environmental factors such as rain, hail storms, acid rain, and extreme heat often affect the soil properties, and thereby demand a significantly variation-tolerant underground communication architecture.

Another important aspect of physical layer architecture is to design an antenna that supports high-range transmission of low-frequency signals, and is yet small in size. Lower-frequency signals experience higher attenuation and loss compared to higher-frequency signals. Design of a suitable signal modulation scheme that consumes less power for the purpose also remains a research challenge [1–3].

14.4.1.2   Communication using MI waves

An alternative physical layer communication technology uses MI waves. MI-wave-based wireless underground communications overcome multiple problems of the traditional EM-wave-based communications. The magnetic permeability of soil and of water are almost the same as that of air [15]. Therefore, underground propagation mediums such as soil and water cause little deviation in the rate of attenuation of the magnetic field, and, hence, MI channel conditions remain almost constant in underground conditions [1–3]. Another constraint in EM-wave-based wireless transmissions is to miniaturize the antennas. Use of small coils for MI wave transmissions eliminates the need for antennas. However, one drawback of this technology is that the strength of the magnetic field drops much more quickly compared to that of the EM waves. High overall path loss may also be a problem for MI waves in soil.

Thus, a thorough study and analysis of the application field and its constraints, along with the applicability of EM and MI waves, should be performed before choosing the suitable physical layer technology. However, a combination of both technologies, properly distributed, may prove to be optimal for some environments.

14.4.2   Data link layer

The data link layer MAC protocols for terrestrial WSNs are broadly classified into two categories, namely, contention-based MAC protocols, and TDMA-based MAC protocols. However, each of these protocol categories is designed specifically for above-ground sensor networks, and hence would yield unsatisfactory results while performing in underground environment. The primary purpose of either protocol is to minimize the power consumption of the sensor nodes. Issues such as idle listening, overhearing, packet collision, and control packet overhead [16] need to be addressed appropriately in order to design an optimal power exhaustion protocol. Additionally, special considerations are required for WUGSNs, as underground wireless channel characteristics need to be accounted for [1–3]. In WSNs, energy is conserved by reducing the idle listening time, but in WUGSNs, in order to overcome path loss, radio transmissions would require higher power. The number of retransmissions, and choice of an optimal retransmission policy are also important in order to eliminate unnecessary power dissipation. An appropriate MAC layer protocol would also reduce the number of packet collisions during wireless transmissions.

Contention-based MAC layer schemes reduce the number of collisions over the network, but do so at the expense of additional power consumption and other undesired overheads. On the other hand, although TDMA-based MAC schemes are able to eliminate packet collisions by time-slot reservation policies, they introduce synchronization overhead. Since the devices operate on low duty cycle to save power, the device clock may drift by a considerable amount during these periods of sleep. As a consequence, the network may lose synchronization. Signals may also experience high bit error rate at the receiver end because of the lossy underground channel properties and limited availability of transmission energy. Therefore, it is crucial to choose between contention-based and TDMA-based MAC layer protocols, after thorough analysis of the trade-offs associated. Optimal packet size selection for WUGSNs is also important as it critically contributes towards power conservation for nodes, and improvement of quality of service.

14.4.3   Network layer

The routing protocols for terrestrial WSNs, as discussed before, are broadly classified into three categories – proactive, reactive, and geographical. However, these routing protocols are not suitable for underground environment due to several overheads involved. We discuss the working principle of each of the routing protocols, and examine the challenges faced in the context of WUGSNs. In proactive routing protocols, the nodes exchange routing information among themselves, periodically over time. Each node maintains a routing table that contains updated information about the optimal paths or routes to every other node. Prior to data transmission, a node consults its routing table, and takes the optimal route for data transmission. In reactive routing protocols, a node initiates a route discovery process in an on-demand basis, i.e. when the node requires transmitting data to a destination, it enquires about the optimal route towards the destination. The route is established and routing information is maintained until the completion of data transmission. However, after transmission, all routing information is erased. Both the routing policies discussed above suffer from signaling overhead, predetermined routes, and synchronization problems. In underground wireless signal propagation channels, changes in soil density, node depth, or water content cause the network to lose synchronization over sleep periods, or it may even become unreachable. Therefore, unlike terrestrial WSNs, the same data propagation routes cannot be used for long periods of time in WUGSNs.

Another well-established set of routing protocols is the geographical protocols. This set of protocols relies on the geographic location or global position coordinates of the wireless sensor nodes. Based on the position of its neighboring nodes, a node selects its immediate neighbors as the next hop. With each node acting in a similar manner, the overall routing within the network becomes optimized. However, a major limitation of this protocol is that each node must be aware of its global position. Thus, geographical routing protocols are applicable only for WUGSNs that have all of their nodes placed carefully in a preplanned manner. At the time of systematic deployment of nodes, the location information of nodes is recorded. However, for WUGSNs, where the devices are deployed randomly, this protocol falls short due to lack of proper location details of the underground wireless nodes. All these routing protocols treat all the underground sensing devices equally in selecting a route from a source to a sink. Some protocols take into account the current energy level of a sensor device while framing a route. In WUGSNs, the transmission power required for inter-node communications may vary greatly, and thus it is important to consider the power dissipation rate of a sensor along with its current residual power level during the establishment of a route. Issues such as low duty cycle, network topology, and applicability of multipath routing also need to be addressed before designing an appropriate routing protocol for WUGSNs [1–3].

14.4.4   Transport layer

The transport layer of the protocol stack is responsible for reliable transmission of data packets, data flow control, and network congestion control. A reliable transmission protocol guarantees proper identification and acknowledgement of data packets at the receiver end of the communication channel. Congestion control is needed to prevent the sink nodes from being overwhelmed by excessive amounts of data. Flow control is needed for the management of data transmission between network devices with limited memory, and to prevent a fast sender from overwhelming a slow receiver. Traditional transport-layer protocols for terrestrial sensor networks are inappropriate for underground environments due to the high and variable path loss rates of the underground communication links. The low data rates of WUGSNs also introduce a challenge in the design of a suitable transport-layer protocol. A cross-layered approach at the network layer to alleviate low data rates is to maximize terrestrial relays with higher data rates. Window-based transmission and retransmission policies are unsuitable for flow control in WUGSNs due to higher power consumption. An effective flow control strategy for such an environment requires the minimum number of retransmissions to preserve power. A reliable transport-layer architecture is able to distinguish between the different reasons of packet drop, such as packet drop due to high bit error rate of the underground communication channel, and packet loss due to network congestion. Also, different recovery strategies are to be adopted for such cases. For packet drops due to poor communication channel quality, the data transmission rate should be preserved to maintain a stable throughput. On the other hand, if packet drops occur due to network congestion, the data transmission rate should be checked and reduced to prevent the receiver from being overburdened. Therefore, it is highly important to consider certain issues such as lossy communication channels, network congestion, data transfer reliability, and power consumption while designing a suitable transport-layer protocol for WUGSNs [1–3].

14.4.5   Cross-layer design

The main challenges faced in cross-layering protocol design are utilization of sensor data for underground communication channels and soil property prediction, physical layer routing, and cross-layering between data link and transport layers. The underground channels are highly dependent on numerous soil properties, and therefore, the sensor nodes equipped with moisture sensors are capable of gathering information for the application layer. This information can also be used to adjust the cross-layering between the application layer and the lower layers in order to adjust the radio output, and appropriate route selection. On the other hand, in underground environments, the soil properties vary widely over short distances, and therefore to achieve power conservation, the use of cross-layer MAC and routing solution is essential.

In order to increase the overall network lifetime, routes should be chosen in such a way that the transmission power requirement and the chance of retransmission are minimized. This would maximize the channel utilization, and would also increase the data transmission reliability over the network. This routing information is collected at the physical-layer level, and is transferred to the network layer. Thus, a cross-layered approach is suggested for the physical-layer-based data routing protocols. Another cross-layered design is suggested for better understanding of the underground channel characteristics for the higher-level-protocol stack layers. The characteristic of the underground channel is primarily known only to the physical layer. Therefore, the higher-level layers need to be fed with the channel state information in order to gain maximum channel efficiency. A cross-layering between the transport layer and the data link layer could efficiently take care of this problem. Owing to the similarities in the functionality, cross-layering of these two layers is suggested for improvement of the network performance. An intelligently chosen cross-layered design effectively enhances the network performance for WUGSNs.

14.4.6   Extremely opportunistic routing

Extremely opportunistic routing or ExOR was originally proposed [17] to increase the throughput of unicast transmissions in ad-hoc networks. ExOR combines routing protocols with medium access control (MAC) to reduce loss rates in 802.11 mesh networks. Usual routing protocols are based on packet transmission via the optimal sequence of nodes. Another approach of cooperative diversity is based on broadcasts and relays. Although ExOR combines the advantages of cooperative diversity, on the other hand, only a single node is responsible for packet forwarding. The "best" receiver forwards the packet, thus avoiding unnecessary collisions and traffic congestion.

The key steps in ExOR are as follows.

(i)   Initially, the sender node broadcasts the packet.
(ii)   The receiver nodes analyze and decide the closest node to the destination. Such a node is termed the "best" node.
(iii)   The "best" node further transmits packets.
(iv)   After completion of every transmission of a particular batch of packets, steps (ii) through (iii) are repeated till the destination node is reached.

The states of nodes in ExOR are as follows.

(i)   Packet buffer: Such nodes are responsible for storing received packets.

(ii)   Local forwarder list: It contains a sequence of preferred nodes that are responsible for packet forwarding.

(iii) Forwarding timer: This timer denotes the time period for transmission. The timer is generally set to such a value so that packets can reach farther nodes.

(iv)   Transmission tracker: It computes the transmission rate from the sender.

(v)   Batch map: It shows the "best" node so that it can receive a copy of the packet.

ExOR greatly improvises the network throughput by considering the variation of the probability of delivery with distance. Moreover, as smaller number of packets are transmitted, compared to traditional routing techniques, ExOR also contributes to increase the lifetime of every node, which in turn positively affects the network lifetime.

14.4.7   Underground opportunistic routing protocol

The underground opportunistic routing protocol (UnOR) uses an ExOR-based underground scheme [17]. UnOR also contributes to the reliability of wireless communication. Initially, UnOR transmits packets from source to the sink. If the transmission fails, a very simple policy is followed. Adjacent nodes of each node decide and prepare a priority list table considering the cost of links while communicating. Thus every source node maintains such a table. During transmission, priority nodes of the table overhear the data packets from the sender node and store them in their buffer. This is a feature of radio frequency. These neighboring nodes retransmit the data and ensure a successful transmission. The same process is also applied for the packets containing acknowledgement. However, if the transmission is successful, such a procedure is ignored. The policy of overhearing and retransmission of the neighboring nodes improves the reliability to a great extent.

14.5   Wireless underground channels

Wireless underground channel design [1–3] is one of the key factors in the context of WUGSNs. The primary difference between the terrestrial WSNs and WUGSNs is the signal propagation medium. For WUGSNs, the medium of signal propagation is no longer air, but solid interface such as soil, rock, and mud. As a consequence, it is difficult to maintain the communication within the underground wireless links to maintain reliability and efficacy. Digital communications in WUGSNs are yet to be explored. Communications through EM [7, 8] waves may still fit for the underground environment. However, in underground conditions, EM waves may encounter critical issues such as high levels of attenuation, large antenna size, and dynamic channel conditions. EM waves are absorbed to a great extent by soil and rocks under the ground causing high signal attenuation or path loss. Path loss is also highly dependent on the soil makeup, i.e. mineral content of the soil, mineral density, moisture level, etc. Large antenna for underground sensing devices [9] are also a challenge for underground deployment, as discussed previously. These issues need to be taken care of before establishing of an efficient wireless communication channel for underground networks. An alternate mode of communication in under-soil conditions is through MI [8]. MI can be an efficient mode of communication over shorter ranges. In this section, the properties of the communication channel using EM waves are described along with the various properties of the wireless underground communication channel, and the effects of different underground factors on the channel. Also we discuss, in brief, the principle of underground communication using MI waves.

14.5.1 Wireless underground channel properties

Although comprehensive detail of underground signal propagation channel is yet to be established, in recent times, characteristics of the EM wave propagation through solid medium have been examined extensively [1–3]. A complete and thorough characterization of underground wireless channels is required in order to dictate an efficient communication channel. The major factors that are to be considered before designing an appropriate channel for communication are the high level of path loss, signal reflection, multipath fading, bit error rate (BER), reduced propagation velocity, and underground noise [18]. In this section, the different channel properties are analyzed from the under-soil–soil perspective, and the impact of different soil properties on those.

14.5.1.1 High path loss

Path loss in underground environments is highly associated with attenuation of the propagating signals [1–3]. EM waves suffer from high levels of attenuation while propagating under the surface through the solid medium. The attenuation is mainly caused by absorption of the wave energy by soil, rock, and water during propagation. The rate of loss of the signal strength is dependent on the original wave frequency, and the soil makeup. Water content, mineral content, and mineral density are some of the factors that determine the rate of signal attenuation under the ground. EM waves with lower frequencies have a lower attenuation rate than those with higher frequencies, over a given distance, under identical underground conditions.

14.5.1.2 Reflection from ground surface

Air and soil have different attenuation characteristics for a given signal frequency. These unequal attenuation characteristics introduce reflection and refraction of signals at the air–soil interface, i.e. at the ground surface. The wireless sensing devices buried at a superficial depth under the ground interact both with the deeply positioned underground sensors and the sensors located above ground. However, part of this communicating signal is reflected back into the ground from the soil–air interface. The actual path of communication between the two underground links is known as the *direct path*. The second path, which is known as the *reflection path* is generated by reflection of the propagating signal at the surface interface [3].

As shown in Figure 14.3, the direct path length for the inter-node underground communication is "d." The reflection path length, on the other hand is calculated to be "$(r_1 + r_2)$," which is, evidently, greater than the direct path length "d." The reflected path length can be ignored if the sensor nodes are buried deep in the soil, as the level of

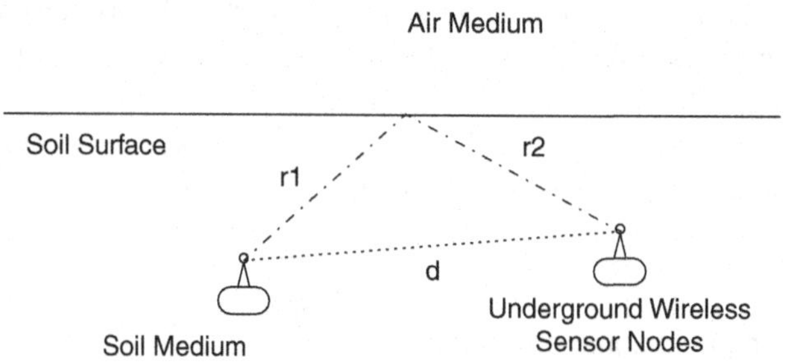

Air Medium

Figure 14.3   Reflection from ground surface [1–3].

attenuation is high. The channel can be considered as a single path channel in such cases. However, when the sensor nodes are deployed at shallow depths, near to the surface, the communication channel essentially becomes a two-path channel. The reflected signal may or may not improve the channel performance depending on certain communication criteria. The challenge is to identify the optimal burying depth for a particular operating frequency of the signal propagating. In a similar fashion, refraction also takes place at this interface causing the signals to deviate from their actual path.

14.5.1.3   Multipath fading

Apart from reflection from the ground surface, underground wireless channels exhibit an additional complication known as multipath fading [3, 9] of signals. The presence of rock, soil, and water within a solid underground wireless channel can disturb the homogeneity of the channel, and result in multipath fading. In above-ground wireless communications, EM waves become deviated and refracted from their original path due to variable air medium properties, and therefore the received signal exhibits the random behavior in wave amplitude and phase properties as a continuous function of time. In contrast, for underground wireless communication channels, variable wave characteristics appear due to the different node locations under the soil. Rather than time, the variable depth and random location of the sensor nodes trigger the anomaly for underground channels. This randomness is observed to follow Rayleigh probability distribution. The variable in this location is the node location, instead of time. Underground wireless signal propagation channels are, therefore, termed position dependent.

14.5.1.4   Bit error rate (BER)

Bit error is the number of bits that are altered during transmission over a communication channel due to noise, interference, and synchronization. The bit error rate (BER) of a communication channel can be defined as the ratio of the number of bit errors to the total number of bits transmitted during a unit time interval. The BER characteristics for an underground wireless communication channel are directly related to the various networking parameters. Hence, a thorough analysis of the BER characteristics proves to be useful while designing an efficient underground communication channel. The bit error rate of a communication channel for underground environments primarily depends on the channel model and the soil makeup – especially the volumetric water content of the soil, the signal modulation method, and the signal-to-noise ratio (SNR) for the channel. The depth at which the sensing devices are deployed significantly affects the path loss, which in turn results in fluctuations in the BER values. The more superficially the sensor nodes are positioned, the higher the fluctuation rate for the BER value. For sensor nodes deployed deep into the ground, the BER is stabilized. However, it is not feasible for an application-oriented design to have all its sensors deployed deep in the earth, as that would complicate the data-transfer route towards the above-ground data sinks. Therefore, it is highly recommended to choose an optimal depth and sensor deployment strategy for WUGSNs in order to obtain a stable BER value for the network [1–3].

14.5.1.5   Reduced signal propagation velocity

Signals in WUGSNs suffer from high levels of absorption by different under-soil components, such as rock, soil, mud, and water. The uneven channel characteristics absorb the wave energy by a greater extent, which in turn results in reduced wave propagation velocity. Those EM waves propagating through a channel made up of solid soil components will experience higher velocity loss compared to the same while propagating through air. The dielectric property of soil depends on the soil composition, the constituent particle size, water content of the soil, soil temperature, and several other factors. Most soil components have their dielectric constant values ranging from 1 to 80, resulting in a minimum of 10% degradation in the propagation velocity. The design of the underground wireless signal propagation channel, therefore, should be done with utmost care, keeping these factors in mind [1–3].

14.5.1.6   Noise

Similar to the air medium, underground communication channels also suffer from noise. Underground pipelines, electric motors, etc., are considered as noise sources. However, underground noise is generally of lower frequencies [1–3].

14.6   Effects of soil properties on wireless underground channels

Propagation of EM waves through soil is affected by the different intrinsic properties, such as: the volumetric water content of the soil, the composition of the soil, density of soil, soil particle size, soil temperature, and the operating frequency of the signal. The intensity of EM waves is governed by a dielectric constant of these waves. The parameters that are intrinsic to the soil decrease the dielectric constant of the propagating wave, and thus reduce the signal intensity. Below we discuss the effect of various soil properties on the wireless underground communication channel.

14.6.1   Volumetric water content

The presence of moisture within soil affects underground communication. Water content in soil is technically termed the volumetric content of water. The presence of water particles accounts for the lossy nature of the channel. As an EM wave propagates through soil with high water content, it results in path loss. This impacts on the attenuation of the communication signal and thus underground communication deteriorates.

14.6.2   Soil composition

It has been experimentally found that sandy soil attenuates signals less than clayey soil. One of the major reasons behind this is the volumetric content of water, because the water content in clayey soil is greater than in sandy soils. Also, the mineral content and mineral density of the soil affect the underground channel characteristics.

14.6.3   Density of soil

Soil density also significantly affects underground wireless communications through EM waves. An increase in soil density accounts for the higher path loss and subsequent signal attenuation under the surface.

14.6.4   Size of soil particles

Variation in particle size also varies signal intensity and amplitude. Based on the diameter, soil particles are generally of the following types: sand, silt, and clay. As the type of particle changes, so the water content changes, and thus the rate of attenuation also changes.

14.6.5   Soil temperature

Another factor that contributes to the attenuation of underground communication signals is the temperature of the soil. An increase in the temperature of the soil increases the dielectric value and thus reduces the intensity of the propagating signal. Temperature also modifies the water content of the soil.

14.6.6   Operating frequency

The frequency at which an EM signal operates is also responsible for its attenuation. The intensity of signals with a higher frequency is reduced more compared to those at a lower frequency at a fixed volumetric content of water.

14.7   Underground channel models

Communication channels for underground wireless signal propagation can be broadly classified into two categories: channels for WUGSNs buried underground, and channels for WUGSNs in mines and tunnels. In this section, we discuss these two underground channel models, and the challenges faced in designing them.

14.7.1   Communication channels for WUGSNs buried underground

For WUGSNs buried underground, the sensing devices are deployed at certain depths beneath the ground surface either randomly, or according to an outlined architecture. These sensor nodes may communicate with each other by using either EM waves or MI techniques. We discuss, in brief, the channel models for each of these wave-propagation techniques.

14.7.1.1   Communication using EM waves

Most research involving the suitable design of underground communication channels uses EM waves as the mode of signal transportation. EM waves are mostly used in multi-hop communications over long ranges. As discussed earlier, in order to develop an appropriate signal propagation channel for under-soil environments, the EM wave characteristics are needed to be learnt in detail along with the different challenges involved, such as reflection/refraction of the signal at the ground surface, multipath fading, sensor deployment depth, and soil makeup. There is a significant dependence on the operating frequency of the sensing devices as well. To reduce the path loss during transmission, the antenna size is needed to be increased, which, however, gives rise to another overhead, as the sensing devices for underground environments ought to be miniaturized. Therefore, although communication through EM waves is a suitable mode of communication in underground environments, the overheads involved are not reduced.

14.7.1.2   Communication using MI

The channel model for underground wireless communication through the solid soil medium using magnetic induction (MI) has been studied rigorously recently. MI has proven to be a strong alternative to EM-wave-based signal-propagation techniques as it has several additional benefits. The main advantage of using MI techniques [8] in under-soil communications is that the rate of attenuation of the magnetic fields is much less. This is because the magnetic permeability of the soil constituents is almost similar to that of air [15]. The channel conditions also remain constant for MI-based communications, and hence the channel dynamicity problem that was faced in EM-wave-based communications can be eliminated. Multipath fading is not an issue of concern for MI-based communications as the magnetic field generated is static in nature, and is non-propagating. The trade-off between lower signal frequency and large antenna size, which appears to be a big problem while communicating through EM waves, can be handled efficiently in MI-based communications. Transmission and reception of signals of even lower frequency ranges can be achieved by the use of a small coil of wire in induction-based signal propagation. The strength of the magnetic field produced by a coil depends on three different factors, namely, the number of turns in the coil, the cross-sectional area of the coil, and the magnetic permeability of the material placed in the core of the coil. However, as stated earlier, MI-based communications for WUGSNs are not suitable for long-range transmission of signals. Hence, a carefully designed hybrid of EM wave and MI-based propagation techniques may prove to be the best fit for underground wireless communications.

14.7.2   Communication channels for WUGSNs deployed in mines and tunnels

The design of the communication channel for underground mines, tunnels, and subways is different from the scenarios discussed previously, where the sensor nodes are buried deep underneath the ground. In mines, tunnels, or subways [7] the sensor nodes are deployed not at a superficial depth, but at considerable depths under the earth's crust. The two main models for communications in such conditions are: (a) the tunnel model, and (b) the room and pillar model. We describe each of these models in this section.

14.7.2.1   Tunnel model

The tunnel channel model [11] is adopted for signal propagation purposes in underground tunnel passageways and mining areas. There exist three main sub-models of this tunnel model, namely, the geometrical optical (GO) model [12], the wave-guide model, and the full wave model. In the GO model the waves are treated and modeled as optical rays, and are expected to follow certain characteristics of optical ray beams. The EM field, in this case, is computed as the weighted sum of all the contributing rays that are reflected from the tunnel walls. The GO model is capable of correctly predicting the path loss and signal delay at any position inside the tunnel during signal propagation. However, in order to do so, it requires a large amount of data bearing information related to the tunnel channel environment.

The wave-guide model [13] is designed to model the underground tunnel as a massive wave-guide with uneven, lossy channel walls. The wave-guide model fails to correctly predict the path loss and signal delay rate for a region that is close to the transmitter. However, unlike the GO model, this model does not require an enormous amount of information for the predictions. The full wave model [14], on the other hand, provides accurate results about the conventional tunnel geometries on a theoretical ground. Another tunnel model, developed in recent times, known as the multi-mode tunnel model, has solved most of the problems faced by the existing models. The multi-mode channel model can be viewed as a multi-mode wave-guide model. This model lays the ground for efficient and reliable communications in underground mines and tunnels.

14.7.2.2   Room and pillar model

The room and pillar model is applicable to underground cavities that are large in size, where the roof is supported by multiple pillars of random shapes. Experiments performed in the room and pillar environment suggest that the propagating signal experiences a high level of attenuation compared to that in a tunnel environment. Additionally, high multi-path fading is experienced throughout the cavity irrespective of its relative position to the transmitter. Owing to the vastness of the model, reflections and refractions from the side walls are minimal and negligible; however, reflections or refractions from the roof and ceiling of the cavity should always be taken into account. The room and pillar model can be considered as a planar wave-guide moving through a large underground cavity with randomly placed pillars. A simple multi-mode model combined with a shadow fading model can be used to describe the propagation of the EM waves in the planar wave-guide, and the slow fading of the signal during propagation at the same time.

14.8   Summary

In this chapter, we have presented the concept of wireless underground sensor networks (WUGSNs), and discussed their applications, advantages, and also the challenges encountered, in comparison with the traditional terrestrial WSNs. WUGSNs are deployed at various depths underneath the surface of the earth to monitor different underground parameters and infrastructures. We have examined the factors that affect the design of WUGSNs, and also discussed the different possible network topologies for WUGSNs. Following that, the communication architecture for WUGSNs was discussed along with various challenges faced at each layer, and the cross-layering approaches. Finally, we have illustrated the underground wireless communication channel design by including the various channel characteristics, effect of soil properties, and the channel models for communication.

References

[1] I. F. Akyildiz, W. Su, Y. Sankarasubramaniam and E. Cayirci, "Wireless sensor networks: a survey," *Computer Networks Journal (Elsevier)*, Vol. **38**, No. 4, pp. 393–422, March 2002.

[2] I. F. Akyildiz and E. P. Stuntebeck, "Wireless underground sensor networks: research challenges," *Ad-Hoc Networks Journal (Elsevier)*, Vol. **4**, pp. 669–686, July 2006.

[3] I. F. Akyildiz, Z. Sun and M. C. Vuran, "Signal propagation techniques for wireless underground communication networks," *Physical Communication Journal (Elsevier)*, Vol. **2**, No. 3, pp. 167–183, Sept. 2009.

[4] W. Chen, Y. Sun and H. Xu, "Clustering chain-type topology for wireless underground sensor networks," in *Proceedings of 8th World Congress on Intelligent Control and Automation*, Jinan, China, July 6–9 2010.

[5] M. Li and Y. Liu, "Underground structure monitoring with wireless sensor networks," in *6th International Symposium on Information Processing in Sensor Networks (ISPN)*, 2007.

[6] X. Niu, X. Huang, Z. Zhao, *et al.*, "The design and evaluation of a wireless sensor network for mine safety monitoring," in *Global Telecommunications Conference, 2007, GLOBECOM'07 IEEE*, 2007.

[7] Z. Sun and I. F. Akyildiz, "Channel modeling and analysis for wireless networks in underground mines and road tunnels," *IEEE Transactions on Communications*, Vol. **58**, No. 6, June 2010.

[8] Z. Sun and I. F. Akyildiz, "Magnetic induction communications for wireless underground sensor networks," *IEEE Transactions on Antennas and Propagation*, Vol. **58**, No.7, July 2010.

[9] Z. Sun and I. F. Akyildiz, "Underground wireless communication using magnetic induction," in *IEEE International Conference on Communications 2009, ICC 09*, pp. 1–5, 2009.

[10] S. Yarkan and H. Arslan, "Statistical wireless channel propagation characteristics in underground mines at 900MHz," in *IEEE Military Communications Conference, 2007, MILCOM 2007*, pp. 1–7, 2007.

[11] C. Cerasoli, "RF propagation in tunnel environments," in *Proceedings IEEE Military Communications Conference, MILCOM'04*, Vol. **1**, pp. 363–369, Nov. 2004.

[12] D. Porrat, "Radio propagation in hallways and streets for UHF communications," Ph.D. thesis, Stanford University, 2002.

[13] A. G. Emslie, R. L. Lagace and P. F. Strong, "Theory of the propagation of UHF radio waves in coal mine tunnels," *IEEE Transactions on Antennas and Propagation*, Vol. **23**, No. 2, pp. 192–205, 1975.

[14] Y. P. Zhang, G. X. Zheng and J. H. Sheng, "Excitation of UHF radio waves in tunnels," *Microwave and Optical Technology Letters,* Vol. **22**, No. 6, pp. 408–410, 1999.

[15] J. Sojdehei, P. Wrathal and D. Dinn, "Magneto-inductive (MI) communications," in *OCEANS, 2001. MTS/IEEE Conference and Exhibition*, Vol. **1**, pp. 513–519, 2001.

[16] K. Kredo and P. Mohapatra, "Medium access control in wireless sensor networks," *Computer Networks (Elsevier)*, Vol. **51**, No. 4, March 2007.

[17] S. Biswas and R. Momis, *ExOR: Opportunistic Multihop Routing for Wireless Networks, SIGCOMM'05*, Pennsylvania, USA, 2005.

[18] L. L. Mehmet, C. Vurany and I. F. Akyildiz "Characteristics of underground channel for wireless underground sensor networks," *Physical Communication*, Vol. **2**, No. 3, pp. 167–183, Sept. 2009.

[19] S. Bhattacharjee, P. Roy, S. Ghosh, S. Misra and M. S Obaidat, "Wireless sensor network-based fire detection, alarming, monitoring and prevention system for Bord-and-Pillar coal mines," *Journal of Systems and Software*, Vol. **85** (3), pp. 571–581, March 2012.

References

R.G. Aghaei, M.A. Rahman, W. Gueaieb and A.E. Saddik, "Ant colony-based reinforcement learning algorithm for routing in wireless sensor networks," in *Instrumentation and Measurement Technology Conference (IMTC)*, Warsaw, Poland, 2007.

N. Ahmed, Y. Dong, T. Bokareva *et al.*, "Detection and tracking using wireless sensor networks," in *Proceedings of the 2007 International Conference on Embedded Networked Sensor Systems*, pp. 425–426, Sydney, Australia, November 2007.

N. Ahmed, S.S. Kanhere and S. Jha, "Ensuring area coverage in hybrid wireless sensor networks," in *Proceedings of the 3rd international Conference on Mobile ad-hoc and Sensor Networks (MSN'07)*, H. Zhang, S. Olariu, J. Cao and D.B. Johnson, Ed. Berlin, Heidelberg: Springer-Verlag, pp. 548–560, 2007.

J. Ai, J. Kong and D. Turgut, "An adaptive coordinated medium access control for wireless sensor networks," in *Proceedings of the International Symposium on Computers and Communications*, Vol. **1**, pp. 214–219, July 2004.

M.A. Ainslie, *Principles of Sonar Performance Modelling*, Springer, ISBN: 9783540876625, 2010.

K. Akkaya and M. Younis, "An energy-aware QoS routing protocol for wireless sensor network," in *Proceedings of the 23rd International Conference on Distributed Computing Systems*, pp. 710–715, May 2003.

K. Akkaya and M. Younis, "A survey on routing protocols for wireless sensor networks," *Ad hoc Networks*, Vol. **3**, No. 3, pp. 325–349, 2005.

I.F. Akyildiz, D. Pompili and T. Melodia, "Underwater acoustic sensor networks: research challenges," *Journal of Ad Hoc Networks, Elsevier*, Vol. **3**, No. 3, pp. 257–279, March 2005.

I.F. Akyildiz, W. Su, Y. Sankarasubramaniam and E. Cayirci, "A survey on sensor networks," *IEEE Communications Magazine*, pp. 102–114, August 2002.

I. Akyildiz, W. Su, Y. Sankarasubramaniam and E. Cayirci, "A survey on sensor networks," *IEEE Communications Magazine*, Vol. **40**, No. 8, pp. 102–114, August 2002.

I.F. Akyildiz, W. Su, Y. Sankarasubramaniam and E. Cayirci, "A survey on sensor networks," *IEEE Communications Magazine*, Vol. **40**, No. 8, pp. 102–114, Aug. 2002.

I.F. Akyildiz, W. Su, Y. Sankarasubramaniam and E. Cayirci, "A survey on sensor networks," *IEEE Communication Magazine*, Vol. **40**, No. 8, pp. 102–114, 2002.

I.F. Akyildiz, D. Pompili and T. Melodia, "Underwater acoustic sensor networks: research challenges," *Ad Hoc Networks*, Vol. **3**, No. 3, pp. 257–279, May 2005.

I.F. Akyildiz, T. Melodia and K.R. Chowdhury, "A survey on wireless multimedia sensor networks," *Computer Networks*, Vol. **51**, No. 4, p. 921, 2007.

I.F. Akyildiz, T. Melodia and K.R. Chowdhury, "Wireless multimedia sensor networks: applications and testbeds," *Proceedings of the IEEE*, Vol. **96**, No. 10, pp. 1588–1605, Oct. 2008.

I. F. Akyildiz, W. Su, Y. Sankarasubramaniam and E. Cayirci, "Wireless sensor networks: a survey," *Computer Networks*, Vol. **38**, pp. 393–422, March 2002.

I. F. Akyildiz, D. Pompili and T. Melodia, "Underwater acoustic sensor networks: research challenges," *Ad Hoc Networks*, Vol. **2**, No. 3, pp. 257–279, 2005.

I. F. Akyildiz, W. Su, Y. Sankarasubramaniam and E. Cayirci, "Wireless sensor networks: a survey," *Computer Networks*, Vol. **38**, No. 4, pp. 393–422, 2002.

I. F. Akyildiz, W. Su, Y. Sankarasubramaniam and E. Cayirci, "A survey on sensor networks," *IEEE Communications Magazine*, Vol. **40**, No. 8, pp. 102–114, 2002.

I. F. Akyildiz, D. Pompili and T. Melodia, "State of the art in protocol research for underwater acoustic sensor networks," *Mobile Computing and Communications Review*, Vol. **11**, No. 4, 2007.

I. F. Akyildiz, W. Su, Y. Sankarasubramaniam and E. Cayirci, "Wireless sensor networks: a survey," *Computer Networks Journal (Elsevier)*, Vol. **38**, No. 4, pp. 393–422, March 2002.

I. F. Akyildiz and E. P. Stuntebeck, "Wireless underground sensor networks: research challenges," *Ad-Hoc Networks Journal (Elsevier)*, Vol. **4**, pp. 669–686, July 2006.

I. F. Akyildiz, Z. Sun and M. C. Vuran, "Signal propagation techniques for wireless underground communication networks," *Physical Communication Journal (Elsevier)*, Vol. **2**, No. 3, pp. 167–183, Sept. 2009.

A. Al-Dhalaan and I. Lambadaris, "Semidefinite programming for wireless sensor localization with lognormal shadowing," in *Proceedings of the 2009 IEEE International Conference on Sensor Technologies and Applications*, pp. 187–193, 2009.

J. N. Al-Karaki, R. Ul-Mustafa and A. Kamal, "Data aggregation in wireless sensor networks – exact and approximate algorithms," in *Proceedings of IEEE Workshop on High Performance Switching and Routing (HPSR)*, pp. 241–245, Phoenix, Arizona, April 2004.

I. T. Almalkawi, M. G. Zapata, J. N. Al-Karaki and J. M. Pozo, "Wireless multimedia sensor networks: current trends and future directions," *Sensor*, Vol. **10**, No. 7, pp. 6662–6717, Jul. 2010

N. A. Alsindi, K. Pahlavan and B. Alavi, "An error propagation aware algorithm for precise cooperative indoor localization," in *Proceedings of IEEE Military Communications Conference (MILCOM)*, pp. 1–7, Washington, DC, USA, 2006.

E. Altman, T. Basar, T. Jimenez and N. Shimkin, "Competitive routing in networks with polynomial costs," *IEEE Transactions on Automatic Control*, Vol. **47**, No. 1, pp. 92–96, 2002.

G. Anastasi, M. Conti and M. Di Francesco, "Data collection in sensor networks with data mules: an integrated simulation analysis," in *Proceedings of IEEE Symposium on Computers and Communications, 2008 (ISCC 2008)*, Marrakech, Morocco, 2008, pp. 1096–1102.

G. Anastasi, M. Conti, M. Di Francesco and A. Passarella, "Energy conservation in wireless sensor networks: a survey," *Ad Hoc Networks*, Vol. **7**, No. 3, pp. 537–568, 2009.

A. Arora, P. Dutta, S. Bapat *et al.*, "A Line in the Sand: a wireless sensor network for target detection, classification, and tracking," *Computer Networks*, Vol. **46**, No. 5, pp. 605–634, December 2004.

W. L. Au Whitlow, *The Sonar of Dolphins*, Springer, ISBN: 9780387978352, 1993.

M. Awad, K. Wong and Z. Li, "An integrated overview of the open literature's empirical data on the indoor radiowave channel's delay properties," *Proceedings of the IEEE Transactions on Antennas and Propagation*, Vol. **56**, No. 5, May 2008.

D. O. Awduche, "MPLS and traffic engineering in IP networks," *IEEE Communications Magazine*, Vol. **37**, No. 12, pp. 42–47, Dec. 1999.

M. Ayaz and A. Abdullah, "Underwater wireless sensor networks: routing issues and future challenges," in *Proceedings of the 7th ACM International Conference on Advances in Mobile Computing & Multimedia*, Kuala Lumpur, Malaysia, pp. 370–375, 2009.

J. Bachrach and C. Taylor, "Localization in sensor networks," *Handbook of Sensor Networks: Algorithms and Architectures*, ISBN: 978-0-471-68472-5, 2005.

A. Bagtzoglou and A. Novikov, "Chaotic behavior and pollution dispersion characteristics in engineered tidal embayments: a numerical investigation," *Journal of American Water Resources Association*, Vol. **43**, No. 1, pp. 207–219, 2007.

A. Bagtzoglou and A. Novikov, "Chaotic behavior and pollution dispersion characteristics in engineered tidal embayments: a numerical investigation," *Journal of American Water Resources Association*, Vol. **43**, No. 1, pp. 207–219, 2007.

X. Bai, Z. Yun, D. Xuan, W. Jia and W. Zhao, "Pattern mutation in wireless sensor deployment," in *Proceedings IEEE International Conference on Computer Communications*, 2010, DOI: 10.1109/INFCOM.2010.5462076.

P. Balister, B. Bollobas, A. Sarkar and S. Kumar, "Reliable density estimates for coverage and connectivity in thin strips of finite length," in *Proceedings of the ACM International Conference on Mobile Computer Networks*, pp. 75–86, 2007.

F. Bao, R. Hu, R. Li and L. Cheng, "Principles and implementation challenges of cooperative transmission for realistic WSNs," *Proceedings of the Second International Conference on Signal Processing Systems (ICSPS)*, Vol. **2**, pp.V2-334–V2-338, 5–7 July 2010.

R. Barr, J.C. Bicket, D.S. Dantas *et al.*, "On the need for system-level support for ad hoc and sensor networks," *Operating System Review*, Vol. **36**, No. 2, pp. 1–5, 2002.

M. Batalin and G. Sukhatme, "Coverage, exploration and deployment by a mobile robot and communication network," *Telecommunication Systems*, Vol. **26**, No. 2–4, pp. 181–196, 2004.

L. Benini and G. DeMicheli, *Dynamic Power Management: Design Techniques and CAD Tools*, Norwell, MA: Springer, 1997.

M. Berg, O. Cheong, M. Kreveld and M. Overmars, *Computational Geometry: Algorithms and Applications*, 3rd ed. Berlin: Springer-Verlag, 2008.

P. Bergamo and G. Mazzini, "Localization in sensor networks with fading and mobility," in *Proceedings of International Symposium on Personal, Indoor and Mobile Radio Networks*, pp. 750–754, 2002.

S. Bhattacharjee, P. Roy, S. Ghosh, S. Misra and M.S. Obaidat, "Wireless sensor network-based fire detection, alarming, monitoring and prevention system for bord-and-pillar coal mines," *Journal of Systems and Software*, Vol. **83**, No. 3, pp. 571–581, March 2012.

S. Bhattacharjee, P. Roy, S. Ghosh, S. Misra and M.S Obaiclat, "Wireless sensor network-based fire detection, alarming, monitoring and prevention system for Bord-and-Pillar coal mines," *Journal of systems and software*, Vol. **85** (3), pp. 571–581, March 2012.

S. Bhatti and J. Xu, "Survey of target tracking protocols using wireless sensor network," in *Proceedings of International Conference on Wireless and Mobile Communications (ICWMC)*, Cannes, La Bocca, pp. 110–115, 2009.

Y. Bi, L. Sun, J. Ma, *et al.*, "HUMS: an autonomous moving strategy for mobile sinks in data-gathering sensor networks," in *EURASIP Journal on Wireless Communications and Networking*, Vol. **2007**, 2007.

S. Biswas and R. Momis, *ExOR: Opportunistic Multihop Routing for Wireless Networks*, *SIGCOMM'05*, Pennsylvania, USA, 2005.

D. Blough, M. Leoncini, G. Resta and P. Santi, "On the symmetric range assignment problem in wireless ad hoc networks," in *Proceedings IFIP 17th World Computer Congress/TC1 Stream/ 2nd IFIP International Conference Theor. Comput. Sci., Found. Inf. Technol. Era Netw. Mobile Comput.*, pp. 71–82, 2002.

Z.S. Bojkovic, B.M. Bakmaz and M.R. Bakmaz, "Security issues in wireless sensor networks," *International Journal of Communications*, Issue 1, Vol. **2**, pp. 108–110, 2008.

T. Bokareva, W. Hu, S. Kanhere *et al.*, "Wireless sensor networks for battlefield surveillance," in *Proceedings of Land Warfare Conference*, Brisbane, October 2006.

I. Borg and P. Groenen, *Modern Multidimensional Scaling, Theory and Applications*, 2nd ed. New York: Springer, 2005.

J.-Y.L. Boudec and P. Thiran, "Network calculus," in *LNCS*, Vol. **2050**, New York: Springer-Verlag, 2001.

N. Boudriga and M.S. Obaidat, "Mobility and security issues in wireless ad-hoc sensor networks," in *Proceedings of IEEE Globcom 2005 Conference*, 2005.

A. Boukerche, J. Feng, R. Werner, Y. Du and Y. Huang, "Reconstructing the plenoptic function from wireless multimedia sensor networks," in *Proceedings of 33rd IEEE Conference on Local Computer Networks (LCN)*, Montreal, Quebec, Canada, pp. 74–81, Oct. 2008.

A. Boukerche, Y. Du, J. Feng and R. Pazzi, "A reliable synchronous transport protocol for wireless image sensor networks," in *Proceedings of IEEE Symposium on Computers and Communications, ISCC 2008*, Marrakech, Morocco, pp. 1083–1089, July 2008.

A. Boulis and M. Srivastava, "Node-level energy management for sensor networks in the presence of multiple applications," in *Proceedings of IEEE International Conference on Pervasive Computing and Communications (PerCom)*, Dallas, USA, pp. 41–49, 2003.

D. Braginsky and D. Estrin, "Rumor routing algorithm for sensor networks," in *Proceedings of the First Workshop on Sensor Networks and Applications (WSNA)*, Atlanta, GA, October 2002.

S. Bravn and C.J. Screenen, "A new model for updating software in wireless sensor networks," *IEEE Network*, Vol. **20**, No. 6, pp. 42–47, Nov./Dec. 2006.

L.M. Brekhovskikh, *Fundamentals of Ocean Acoustics*, 3rd ed. Springer, ISBN: 9780387954677, 2003.

N. Bulusu, J. Heidemann and D. Estrin, "GPS-less low cost outdoor localization for very small devices," Computer Science Department, University of Southern California, Technical Report 00–729, Apr. 2000.

N. Bulusu, J. Heidemann and D. Estrin, "Adaptive beacon placement," in *Proceedings of the 21st International Conference on Distributed Computing Systems*, Phoenix, Arizona, April 2001.

N. Bulusu, V. Bychkovskiy, D. Estrin and J. Heidemann, "Scalable, ad hoc deployable RF-based localization," in *Proceedings of Grace Hopper Celebration of Women in Computing Conference*, Vancouver, British Columbia, Canada, 2002.

D.E. Burgner and L.A. Wahsheh, "Security of wireless sensor networks," in *Proceedings of the 2011 Eighth International Conference on Information Technology: New Generations*, pp. 315–320, 2011.

M. Busse, T. Haenselmann and W. Effelsberg, "TECA: a topology and energy control algorithm for wireless sensor networks," in *International Symposium on Modeling, Analysis and Simulation of Wireless and Mobile Systems (MSWiM '06)*, Torremolinos, Malaga, Spain, ACM, Oct. 2–6, 2006.

M. Caccamo, L. Zhang, S. Lui and G. Buttazzo, "An implicit prioritized access protocol for wireless sensor networks," in *Proceedings of 23rd IEEE Real-Time Systems Symposium*, pp. 39–48, Dec. 2002.

E. Callaway, Jr., *Wireless Sensor Networks: Architectures and Protocols*, Auerbach Publications, 2003.

T. Camp, J. Boleng and V. Davies, "A survey of mobility models for ad hoc network research," *Wireless Communication and Mobile Computing*, Vol. **2**, No. 5, pp. 483–502, 2002.

J. Campbell, P.B. Gibbons, S. Nath, *et al.*, "IrisNet: an internet-scale architecture for multimedia sensors," in *Proceedings of the 13th Annual ACM International Conference on Multimedia*, New York, USA, pp. 81–88, 2005.

H. Cao, K. Parker and A. Arora, "O-MAC: a receiver centric power management protocol," in *Proceedings of the 2006 IEEE International Conference on Network Protocols*, pp. 311–332, 2006.

S. Capkun, M. Hamdi and J. Hubaux, "GPS-free positioning in mobile ad-hoc networks," in *Proceedings of the 34th Annual Hawaii International Conference on System Sciences*, pp. 3481–3490, 2001.

S. Capkun, M. Cagalj and M. Srivastava, "Secure localization with hidden and mobile base stations," in *Proceedings of the 2006 IEEE INFOCOM*, 2006.

N. M. Carbone and W. S. Hodgkiss, "Effects of tidally driven temperature fluctuations on shallow-water acoustic communications at 18 kHz," *IEEE Journal of Oceanic Engineering*, Vol. **25**, No. 1, pp. 84–94, 2000.

M. Cardei, J. Wu and S. Yang, "Topology control in ad hoc wireless networks with hitch-hiking," in *Proceedings of the IEEE Communications Society Conference on Sensory Ad Hoc Communication Networks*, pp. 480–488, 2005.

D. W. Carman, P. S. Kruus and B. J. Matt, "Constraints and approaches for distributed sensor network security," NAI Labs Tech. Rep. No. 00–010, 2000.

W. Carman, P. S. Kruus and B. J. Matt, "Constraints and approaches for distributed sensor network security," Tech. Rep. 00–010, NAI Labs, Network Associates, Inc., Glenwood, MD, 2000.

M. J. Caruso and L. S. Withanawasam, "Vehicle detection and compass applications using AMR magnetic sensors, AMR sensor documentation," http://www.magneticsensors.com/datasheets/amr.pdf

C. Cerasoli, "RF propagation in tunnel environments," in *Proceedings IEEE Military Communications Conference, MILCOM'04*, Vol. **1**, pp. 363–369, Nov. 2004.

A. Chakrabarti, A. Sabharwal and B. Aazhang, "Using predictable observer mobility for power efficient design of sensor networks," in *Proceedings of the 2nd International Conference on Information Processing in Sensor Networks (IPSN'03)*, F. Zhao and L. Guibas, Ed. Berlin, Heidelberg: Springer-Verlag, pp. 129–145, 2003.

A. Chalak, S. Misra and M. S. Obaidat, "A cluster-head selection algorithm for wireless sensor networks," in *Proceedings of the 2010 IEEE International Conference on Electronics, Circuits and Systems, ICECS 2010*, pp. 130–133, Athens, Greece, Dec. 2010.

A. Chandrakasan and R. Brodersen, *Low Power CMOS Digital Design*, Norwell, MA: Springer, 1996.

A. Chandrakasan, R. Min, M. Bhardawi, S.-H. Cho and A. Wang, "Power aware wireless micro-sensor systems," in *Proceedings of the 2002 International European Solid-State Device Research Conference, ESSDERC'02*, pp. 47–54, Florence, Italy, 2002.

R. Chandrasekar, M. S. Obaidat, S. Misra and F. Peña-Mora, "A secure and energy-efficient scheme for group-based routing in heterogeneous ad-hoc sensor networks and its simulation analysis," *SIMULATION: Transactions of the Society for Modeling and Simulation International*, Vol. **84**, No. 2/3, pp. 131–146, Feb. 2008.

J. Chang and L. Tassiulas, "Energy conserving routing in wireless ad hoc networks," in *Proceedings of the Nineteenth International Annual Joint Conference of the IEEE Computer and Communications Societies (INFOCOM)*, 2000.

J.-H. Chang and L. Tassiulas, "Maximum lifetime routing in wireless sensor networks," *IEEE/ACM Transactions on Networking*, Vol. **12**, No. 4, pp. 609–619, Aug. 2004.

W.-R. Chang, H.-T. Lin and Z.-Z. Cheng, "CODA: a continuous object detection and tracking algorithm for wireless ad hoc sensor networks," in *Proceedings of IEEE Consumer Communications and Networking Conference*, Las Vegas, NV, USA, pp. 168–174, 2008.

X. Chang, "Network simulations with OPNET," in *Proceedings of the 1999 Winter Simulation Conference*, pp. 307–314, 1999.

S. Cheekiralla and D.W. Engels, "A functional taxonomy of wireless sensor network devices," *Proceedings of the 2005 International Conference on Broadband Networks Conference, BroadNets 2005*, Vol. **2**, pp. 949–956, 2005.

B. Chen, K. Jamieson, H. Balakrishnan and R. Morris, "SPAN: an energy-efficient coordination algorithm for topology maintenance in ad-hoc wireless networks," *Wireless Networks*, Vol. **8**, No. 5, pp. 481–494, Sep. 2002.

W.-P. Chen, J.C. Hou and L. Sha, "Dynamic clustering for acoustic target tracking in wireless sensor networks," *IEEE Transactions on Mobile Computing*, Vol. **3**, No. 3, pp. 258–271, 2004.

Y.-S. Chen and Y.-J. Liao, "HVE-mobicast: a hierarchical-variant-egg-based mobicast routing protocol for wireless sensornets," in *Proceedings of Wireless Communications and Networking Conference (WCNC)*, Las Vegas, NV, pp. 697–702, 2006.

B. Chen, K. Jamieson, H. Balakrishnan and R. Morris, "Span: an energy efficient coordination algorithm for topology maintenance in ad-hoc wireless networks," Mobicom, Rome, Italy, pp. 70–84, July 2001.

A. Chen, S. Kumar and T. Lai, "Designing localized algorithms for barrier coverage," in *Proceedings ACM International Conference on Mobile Computer Networks*, pp. 63–74, 2007.

S. Chen, G. Yang and S. Chen, "A security routing mechanism against Sybil attack for wireless sensor networks," in *Proceedings of the 2010 International Conference on Communications and Mobile Computing*, Vol. **1**, pp. 142–146, 2010.

C.T. Chen and F.J. Millero, "Speed of sound in seawater at high pressures," *Journal of the Acoustical Society of America*, Vol. **62**, No. 5, pp. 1129–1135, 1977.

W. Chen, Y. Sun and H. Xu, "Clustering chain-type topology for wireless underground sensor networks," in *Proceedings of 8th World Congress on Intelligent Control and Automation*, Jinan, China, July 6–9 2010.

K.-Y. Cheng, K.-S. Lui and V. Tam, "Localization in sensor networks with limited number of anchors and clustered placement," in *Proceedings of Wireless Communications and Networking Conference (WCNC)*, Kowloon, pp. 4425–4429, 2007.

S. Cheng, J. Li and G. Horng, "An adaptive cluster-based routing mechanism for energy conservation in mobile ad hoc networks," Personal Communication, June 2012.

W. Cheng, M. Li, K. Liu, *et al.*, "Sweep coverage with mobile sensors," in *Proceedings of the IEEE International Symposium on Parallel Distribution Processes*, 2008, DOI: 10.1109/IPDPS. 2008.4536245.

E. Cheong, J. Liebman, J. Liu and F. Zhao, "TinyGALS: a programming model for event-driven embedded systems," in *Proceedings of the 2003 Annual ACM Symposium on Applied Computing, SAC 2003*, pp. 698–704, Melbourne, FL, March 2003.

C. Chiasserini and E. Magli, "Energy consumption and image quality in wireless video-surveillance networks," in *Proceedings of 13th IEEE International Symposium on Personal, Indoor and Mobile Radio Communications (PIMRC)*, pp. 2357–2361, Sept. 2002.

Y. Cho, L. Bao and T. Michael, "LAAC: a location-aware access control protocol," in *Proceedings of the 2006 Annual International Conference on Mobile and Ubiquitous Systems: Networking & Services*, pp. 1–7, July 2006.

C.-Y. Chong and S.P. Kumar, "Sensor networks: evolution, opportunities, and challenges," *Proceedings of IEEE*, Vol. **91**, No. 8, pp. 1247–1256, 2006.

C.-Y. Chong and S.P. Kumar, "Sensor networks: evolution, opportunities, and challenges," *Proceedings of the IEEE*, Vol. **91**, No. 8, pp. 1247–1256, August 2003.

Y. Chraibi, "Localization in wireless sensor networks," Master's Degree Project, Stockholm, Sweden, 2005.

M. Chu, H. Haussecker and F. Zhao, "Scalable information-driven sensor querying and routing for ad hoc heterogeneous sensor networks," in *The International Journal of High Performance Computing Applications*, Vol. **16**, No. 3, pp. 293–313, August 2002.

http://citeseer.ist.psu.edu/article/ye02energyefficient.html

C.S. Clay and H. Medwin, *Acoustical Oceanography: Principles and Applications*. New York: John Wiley & Sons, ISBN: 978-0471160410, 1977, pp. 88 and 98–99.

S. Climent, J.V. Capella, N. Meratnia and J.J. Serrano, "Underwater sensor networks: a new energy efficient and robust architecture," *SENSORS*, Vol. **12**, pp. 704–731, 2012.

R.F.W. Coates, *Underwater Acoustic Systems*, John Wiley & Sons, ISBN: 9780470215449, 1989.

http://www.consensus.tudelft.nl/documents_papers/vanDam03.pdf

M. Conti, R. Di Pietro, L.V. Mancini and A. Mei, "Requirements and open issues in distributed detection of node identity replicas in WSN," in *Proceedings of the 2006 IEEE International Conference on Systems, Man and Cybernetics*, pp. 1468–1473, Oct. 2006.

Crossbow, Wireless Sensor Platform, website: http://www.xbow.com/Products/Wireless SensorNetworks.htm

R. Cruz, "A calculus for network delay. I. Network elementsin isolation," *IEEE Transactions on Information Theory*, Vol. **37**, pp. 114–131, Jan. 1991.

www.cs.nyu.edu/~nikos/personal/pubs/sow2002.pdf

http://www.cse.unsw.edu.au/~sensar/hardware/hardware_survey.html

J.-H. Cui, J. Kong, M. Gerla and S. Zhou, "Challenges: building scalable mobile underwater wireless sensor networks for aquatic applications," University of Connecticut, USA, UCONN CSE technical report UbiNET-TR05–02; last update, September 2005.

J.-H. Cui et al., "Challenges: building scalable mobile underwater wireless sensor networks for aquatic applications," UCONN CSE technical report UbiNET-TR05–02; last update, Sept. 2005.

J.-H. Cui, J. Kong, M. Gerla and S. Zhou, "Challenges: building scalable mobile underwater wireless sensor networks for aquatic applications," *IEEE Network*, Vol. **20**, No. 3, pp. 12–18, 2006.

T.V. Dam and K. Langendoen, "An adaptive energy-efficient MAC protocol for wireless sensor networks," in *Proceedings of International Conference on Embedded Networked Sensor Systems (SenSys)*, pp. 171–180, Nov. 2003.

W. Dargie and C. Poellabauer, *Fundamentals of Wireless Sensor Networks: Theory and Practice*, John Wiley & Sons, 2010.

W. Dargie and C. Poellabauer, *Fundamentals of Wireless Sensor Networks*, John Wiley & Sons, 2010.

http://db.csail.mit.edu/labdata/labdata.html, 29 Feb. 2012.

B. Deb, S. Bhatnagar and B. Nath, "A topology discovery algorithm for sensor networks with applications to network management," Technical Report, Rutgers University, May 2001.

B. Deb, S. Bhatnagar and B. Nath, "Multi-resolution state retrieval in sensor networks," in *Proceedings of the First IEEE International Workshop on Sensor Network Protocols and Applications, 2003*, pp. 19–29, 11 May 2003.

I. Demirkol, C. Ersoy and F. Alagöz, "MAC protocols for wireless sensor networks: a survey," *IEEE Communications Magazine*, Vol. **44**, No. 4, pp. 115–121, April 2006.

J. Deng, Y. Han, W. Heinzelman and P. Varshney, "Balanced-energy sleep scheduling scheme for high density cluster-based sensor networks," in *Proceedings of the 4th Workshop on Applications and Services in Wireless Networks (ASWN)*, 2004.

J. Deng, R. Han and S. Mishra, "Defending against path-based DoS attacks in wireless sensor networks," in *Proceedings of the 3rd ACM Workshop on Security of Ad Hoc and Sensor Networks*, 2005.

J. Deng, R. Han and S. Mishra, "Decorrelating wireless sensor network traffic to inhibit traffic analysis attacks," *Pervasive and Mobile Computing Journal, Elsevier*, Vol. **2**, No. 2, pp. 159–186, 2006.

M. W. Denny, *How the Ocean Works: an Introduction to Oceanography*, Princeton University Press, ISBN: 9780691126470, 2008.

S. Dhurandher, S. Misra, M. S. Obaidat and S. Khairwal, "UWSim: a simulator for underwater wireless sensor networks," *Simulation: Transactions of the Society for Modeling and Simulation International, SCS*, Vol. **84**, No. 7, pp. 327–338, July 2008.

S. K. Dhurandher, S. Misra, M. S. Obaidat and N. Gupta, "QDV: a quality-based distance vector routing for wireless sensor networks using ant colony optimization," in *Proceedings of the 4th IEEE International Conference on Wireless and Mobile Computing, Networking and Communications: The First International Workshop in Wireless and Mobile Computing, Networking and Communications (IEEE SecPriWiMob'08)*, pp. 598–602, Avignon, France, October 12–14, 2008.

S. Dhurandher, S. Khairwal, M. S. Obaidat and S. Misra, "Efficient data acquisition in underwater wireless sensor ad hoc networks," *IEEE Wireless Communications*, Vol. **16**, No. 6, pp. 70–78, December 2009.

S. K. Dhurandher, S. Misra, M. S. Obaidat and S. Khairwal, "UWSim: a simulator for underwater wireless sensor networks," *Simulation: Transactions of the Society for Modeling and Simulation International, SCS*, Vol. **84**, No. 7, pp. 327–338, July 2008.

S. Dhurandher, M. S. Obaidat, G. Jain, I. Mani Ganesh and V. Shashidhar, "An efficient and secure routing protocol for wireless sensor networks using multicasting," in *Proceedings of the IEEE/ACM International Conference on Green Computing and Communications, Green Com 2010*, pp. 374–379, Hangzhou, China, December 2010.

S. K. Dhurandher, M. S. Obaidat, D. Gupta, N. Gupta and A. Asthana, "Energy efficient routing for wireless sensor networks in urban environments," in *Proceedings of the 2010 Workshop on Web and Pervasive Security (WPS)-GlobCom 2010*, Miami, FL, 2010.

S. K. Dhurandher, S. Khairwal, M. S. Obaidat and S. Misra, "Efficient data acquisition in underwater wireless sensor ad hoc networks," *IEEE Wireless Communications*, Vol. **16**, No. 6, pp. 70–78, Dec. 2009.

S. K. Dhurandher, S. Misra, H. Mittal, A. Agarwal and I. Woungang, "Using ant-based agents for congestion control in ad-hoc wireless sensor networks," *Cluster Computing (Springer)*, Vol. **14**, pp. 41–53, 2011.

S. Dhurandher, S. Misra, M. S. Obaidat and S. Khairwal, "UWSim: a simulator for underwater wireless sensor networks," *Simulation: Transactions of the Society for Modeling and Simulation International, SCS*, Vol. **84**, No. 7, pp. 327–338, July 2008.

S. K. Dhurandher, M. S. Obaidat and M. Gupta, "An efficient technique for geocast region holes in underwater sensor networks and its performance evaluation," *Simulation: Modeling Practice and Theory, Elsevier*, Vol. **19**, No. 9, pp. 2102–2116, Sep. 2011.

S. Dhurandher, M. S. Obaidat, S. Misra and S. Khairwal, "Efficient data acquisition in underwater wireless sensor ad-hoc networks," *IEEE Wireless Communications*, Vol. **16**, No. 6, pp. 70–78, Dec. 2009.

S. Dhurandher, S. Misra, M.S. Obaidat and N. Gupta, "An ant colony optimization approach for reputation and quality-of-service-based security in wireless sensor networks," *Security and Communications Networks, Wiley*, Vol. **2**, No. 2, pp. 215–224, March/April, 2009.

S. Dhurandher, M.S. Obaidat, G. Jain, I. Mani Ganesh and V. Shashidhar, "An efficient and secure routing protocol for wireless sensor networks using multicasting," in *Proceedings of the IEEE/ACM International Conference on Green Computing and Communications, Green Com 2010*, pp. 374–379, Hangzhou, China, Dec. 2010.

S. Dhurandher, M.S. Obaidat, D. Gupta, N. Gupta and A. Asthana, "Network layer based secure routing protocol for wireless ad hoc sensor networks in urban environments," in *Proceedings of the IEEE ICETE 2010-International Conference on Wireless Information Networks and Systems, WINSYS 2010*, pp. 23–30, Athens, Greece, 2010.

A. Dimitrievski, V. Pejovska and D. Davcev, "Security issues and approaches in WSN," *International Journal of Peer to Peer Networks (IJP2P)*, Vol. **2**, No. 2, pp. 24–42, April 2011.

R. Dobrescu *et al.*, "Embedding wireless sensors in UPnP services networks," *NAUN International Journal of Communications*, Vol. **1**, No. 2, pp. 62–67, 2007.

M. Dorigo, A. Colorni and V. Maniezzo, "The ant system: optimization by a colony of cooperating agents," *IEEE Transactions on System, Man, and Cybernetics–Part B*, Vol. **26**, No. 1, pp. 1–13, 1996.

M. Dorigo, D.G. Caro and L.M. Gambardella, "Ant algorithms for discrete optimization," *Artificial Life*, Vol. **5**(2), pp. 137–172, 1999.

DS-UWB, website: http://www.radio-electronics.com/info/wireless/uwb/ds-uwb.php

DS-UWB, website: http://www.networkworld.com/news/tech/2004/0614techupdate.html

X. Du and H.-H. Chen, "Security in wireless sensor networks," *IEEE Wireless Communications*, Vol. **15**, No. 4, pp. 60–66, Aug. 2008.

M. Duarte and Y. Hu, "Vehicle classification in distributed sensor networks," *Journal of Parallel and Distributed Computing*, Vol. **64**, No. 7, pp. 826–838, July 2004.

S. Dulman, T. Nieberg, J. Wu and P. Havinga, "Trade-off between traffic overhead and reliability in multipath routing for wireless sensor networks," in *WCNC Workshop*, Vol. **3**, pp. 1918–1922, New Orleans, LA, March 2003.

A. Dunkels, T. Voigt, J. Alonso and H. Ritter, "Distributed TCP caching for wireless sensor networks," in *Proceedings of the Mediterranean Ad Hoc Networking Workshop (MedHoc-Net)*, June 2004.

A. Durresia, V. Paruchuri and L. Barolli, "Clustering protocol for sensor networks," in *20th International Conference, AINA 2006*, Vol. **2**, pp. 18–20, April 2006.

B.D. Dushaw, P.F. Worcester, B.D. Cornuelle and B.M. Howe, "On equations for the speed of sound in seawater," *Journal of the Acoustical Society of America*, Vol. **93**, No. 1, pp. 255–275, 1993.

P.K. Dutta and A.K. Arora, "Sensing Civilians, Soldiers, and Cars," The Ohio State University Department of Computer and Information Science Technical Report OSU-CISRC-12/03-TR66, 2003.

W. Eddy, "A new convex hull algorithm for planar sets," *ACM Transactions on Mathematical Software*, Vol. **3**, No. 4, pp. 398–403, 1977.

C.T. Ee and R. Bajcsy, "Congestion control and fairness for many-to-one routing in sensor networks," in *Proceedings of ACM Sensys'04*, Baltimore, MD, Nov. 3–5, 2004, pp. 148–161.

A.G. Emslie, R.L. Lagace and P.F. Strong, "Theory of the propagation of UHF radio waves in coal mine tunnels," *IEEE Transactions on Antennas and Propagation*, Vol. **23**, No. 2, pp. 192–205, 1975.

http://en.wikipedia.org/wiki/Wireless_sensor_network

C. Enz, N. Scolari and U. Yodprasit, "Ultra low-power radio design for wireless sensor networks," in *Proceedings of the 2005 IEEE International Workshop on Radio-Frequency Integration Technology: Integrated Circuits for Wideband Communication and Wireless Sensor Networks*, pp. 1–17, 30 Nov./Dec. 2005.

M. Erol, L.F.M. Vieira and M. Gerla, "Localization with DiveNRise (DNR) beacons for underwater acoustic sensor networks," in *Proceedings of ACM WUWNet*, pp. 97–100, 2007.

M. Erol-Kantarci, H.T. Mouftah and S. Oktug, "A survey of architectures and localization techniques for underwater acoustic sensor networks," *IEEE Communications Surveys and Tutorials*, Vol. **13**, No. 3, pp. 487–502, 2011.

L. Eschenauer and V.D. Gligor, "A key management scheme for distributed sensor networks," in *Proceedings of the 9th ACM Conference on Computer and Communications Security*, pp. 41–47, 2002.

L. Eschenauer and V.D. Giligor, "A key management scheme for distributed sensor networks," in *Proceedings of the Adaptive Random Key Distribution Schemes for Wireless Sensor Networks*, 2002.

M.F. Fallon, G. Papadopoulos, J.J. Leonard and N.M. Patrikalakis, "Cooperative AUV navigation using a single maneuvering surface craft," *International Journal of Robotics Research*, Vol. **29**, pp. 1461–1474, 2010.

G. Fan, H. Chen, L. Xie and K. Wang, "A hybrid reservation-based MAC protocol for underwater acoustic sensor networks," *Ad Hoc Networks*, Vol. **11**, No. 3, pp. 1178–1192, 2013.

Q. Fang, F. Zhao and L. Guibas, "Lightweight sensing and communication protocols for target enumeration and aggregation," in *Proceedings of the 4th ACM International Symposium on Mobile Ad-hoc Networking and Computing (MOBIHOC)*, pp. 165–176, 2003.

W. Feng, B. Code, E. Kaiser, M. Shea and W. Feng, "Panoptes: scalable low-power video sensor networking technologies," in *Proceedings of the Eleventh ACM International Conference on Multimedia*, pp. 90–91, Nov. 2003.

W.C. Feng, E. Kaiser and M. LeBaillif, "Panoptes: scalable low-power video sensor networking technologies," *ACM Transactions on Multimedia Computing, Communications, and Applications*, Vol. **1**, No. 2, pp. 151–167, May 2005.

S. Fitzpatrick and L. Meertens, "Diffusion based localization," private communication, 2004.

R. Frank, *Understanding Smart Sensors*, 2nd ed. Norwood, MA: Artech House, 2000.

A.E. Franke, T.-J. King and R.T. Howe, "Integrated MEMS technologies," *MRS Bulletin*, Vol. **26**, No. 4, pp. 291–295, 2001.

C.L. Fullmer and J.J. Garcia-Luna-Aceves, "Floor acquisition multiple access (FAMA) for packet-radio networks," in *Proceedings of SIGCOMM*, pp. 262–273, 1995.

S. Ganeriwal, A. Kansal and M. Srivastava, "Self aware actuation for fault repair in sensor networks," in *Proceedings IEEE International Conference Robotics and Automation*, Vol. **5**, pp. 5244–5249, 2004.

S. Ganeriwal, S. Capkun, C. Han and M. Srivastava, "Secure time synchronization service for sensor networks," in *Proceedings of the ACM Workshop on Wireless Security (WSNA'05)*, 2005.

C.F. García-Hernández, P.H. Ibargüengoytia-González, J. García-Hernández and J.A. Pérez-Díaz, "Wireless sensor networks and applications: a survey," *International Journal of Computer Science and Network Security*, Vol. **7**, No. 3, pp. 264–273, March 2007.

M. Gerla and K. Xu, "Multimedia streaming in large-scale sensor networks with mobile swarms," *Special Interest Group on Management of Data*, Vol. **32**, pp. 72–76, 2003.

M.C. Ghanbari, J. Hughes, M.C. Sinclair and J.P. Eade, *Principles of Performance Engineering for Telecommunication and Information Systems*, Hertfordshire, United Kingdom: IEE, 1997.

A. Ghosh, "Estimating coverage holes and enhancing coverage in mixed sensor networks," in *Proceedings of the 29th Annual IEEE International Conference on Local Computer Networks (LCN '04)*, Washington, DC, USA, 2004, pp. 68–76.

E. Giancoli, F. Jabour and A. Pedroza, "CTCP: reliable transport control protocol for sensor networks," in *Proceedings of the International Conference on Intelligent Sensors, Sensor Networks and Information Processing (ISSNIP)*, Sydney, Australia, December 2008, pp. 493–498.

B. Girod, A. Aaron, S. Rane and D. Monedero, "Distributed video coding," *Proceedings of the IEEE*, Vol. **93**, pp. 71–83, 2005.

B.P. Godfrey and D. Ratajczak, "Naps: scalable, robust topology management in wireless ad-hoc networks," *ISPN'04*, Berkeley, CA, ACM, April 26–27, 2004.

R.L. Graham, "An efficient algorithm for determining the convex hull of a finite planar set," *Information Processing Letters*, Vol. **1**, pp. 132–133, 1972.

J. Granjal, R. Silva and J. Silva, "Security in wireless sensor networks", *CISUC UC*, June 2008.

D. Green, "Acoustic modems, navigation aids, and networks for undersea operations," in *Proceedings of IEEE Oceans*, Sydney, Australia, pp. 1–6, 2010.

P. Guangyu, M. Gerla and X. Hong, "LANMAR: landmark routing for large scale wireless ad hoc networks with group mobility," in *Mobile and Ad Hoc Networking and Computing (MobiHOC)*, Boston, USA, pp. 11–18, Aug. 2000.

P. Guo, G. Zhu and L. Fang, "An adaptive coverage algorithm for large-scale mobile sensor networks," *Ubiquitous Intelligence and Computing, Lecture Notes in Computer Science*, Vol. **4159**/2006, pp. 468–477, 2006.

M. Gupta, M.S. Obaidat and S. Dhurandher, "Energy-efficient wireless sensor networks," in *Green Information and Communication Systems*, M.S. Obaidat, A. Alagan and I. Woungang, Eds. Elsevier, 2013.

S. Hadim and N. Mohamed, "Middleware challenges and approaches for wireless sensor networks," *IEEE Distributed Systems Online*, Vol. **7**, No. 3, pp. 1–13, March 2006.

R. Hall, "An improved geocast for mobile ad hoc networks," *IEEE Transactions on Mobile Computing*, Vol. **10**, No. 2, pp. 254–266, Feb. 2011.

R.N. Handcock, D.L. Swain, G.J. Bishop-Hurley, *et al.*, "Monitoring animal behaviour and environmental interactions using wireless sensor networks, GPS collars and satellite remote sensing," *Sensors*, Vol. **9**, No. 5, pp. 3586–3603, 2009.

T. He, J.A. Stankovic, L. Chenyang and T. Abdelzaher, "SPEED: a stateless protocol for real-time communication in sensor networks," in *Proceedings of 23rd International Conference on Distributed Computing Systems*, pp. 46–55, Providence, RI, May 2003.

T. He, C. Huang, B.M. Blum, J.A. Stankovic and T. Abdelzaher, "Range-free localization schemes for large scale sensor networks," in *Proceedings of the 9th Annual International Conference on Mobile Computing and Networking*, San Diego, California, USA, pp. 81–95, 2003.

J. Heidemann *et al.*, "Underwater sensor networking: research challenges and potential applications," USC/ISI technical report ISI-TR-2005–603.

J. Heidemann, M. Stojanovic and M. Zorzi, "Underwater sensor networks: applications, advances and challenges," *Philosophical Transactions of the Royal Society A*, Vol. **370**, pp. 158–175, 2012.

J. Heidemann, W. Ye, J. Wills, A. Syed and Y. Li, "Research challenges and applications for underwater sensor networking," in *Proceedings of IEEE Wireless Communication and Networking Conference*, Las Vegas, NV, USA, pp. 228–235, 2006.

W. Heinzelman, A. Chandrakasan and H. Balakrishnan, "Energy-efficient communication protocols for wireless microsensor networks," in *Proceedings of the Hawaii International Conference on Systems Sciences*, Vol. **8**, Jan. 2000.

W. Heinzelman, A. Chandrakasan and H. Balakrishnan, "Energy-efficient communication protocol for wireless microsensor networks," in *Proceedings of the 33rd Hawaii International Conference on System Sciences*, Vol. **8**, p. 8020, January 2000.

W. Heinzelman, A. Chandrakasan and H. Balakrishnan, "An application-specific protocol architecture for wireless microsensor networks," *IEEE Transactions on Wireless Communications*, Vol. **1**, No. 4, pp. 660–670, Oct. 2002.

W. Heinzelman, J. Kulik and H. Balakrishnan, "Adaptive protocols for information dissemination in wireless sensor networks," in *Proceedings of the ACM/IEEE Mobicom Conference (MobiCom '99)*, pp. 174–185, Seattle, WA, August, 1999.

W. Heinzelman, A. Chandrakasan and H. Balakrishnan, "Energy-efficient communication protocol for wireless microsensor networks," in *Proceedings of the 33rd Hawaii International Conference on System Sciences (HICSS '00)*, Vol. **2**, p. 10, January 2000.

W.R. Heinzelman, A. Chandrakasan and H. Balakrishnan, "Energy efficient communication protocol for wireless microsensor networks," in *Proceedings of the 33rd Annual Hawaii International Conference on System Sciences*, Jan. 4–7, 2000.

W.B. Heinzelman *et al.*, "Application specific protocol architecture for wireless microsensor networks," *IEEE Transactions on Wireless Communications*, pp. 660–670, 2002.

W.R. Heinzelman, A. Chandrakasan and H. Balakrishnan, "Energy-efficient communication protocol for wireless microsensor networks," in *Proceedings of the Hawaii International Conference on System Sciences*, Vol. **2**, p. 10, Jan. 2000.

W.R. Heinzelman, J. Kulik and H. Balakrishnan, "Adaptive protocols for information dissemination in wireless sensor networks," in *Proceedings of the ACM/IEEE International Conference on Mobile Computing and Networking*, pp. 174–185, 1999.

S. Hengstler and H. Aghajan, "WiSNAP: a wireless image sensor network application platform," in *Proceedings of 2nd International Conference on Testbeds and Research Infrastructures for the Development of Networks and Communities*, Barcelona, Spain, pp. 6–12, Mar. 2006.

J. Hightower and G. Borriello, "Location systems for ubiquitous computing," in *IEEE Computer*, Vol. **34**, No. 8, pp. 57–66, Aug. 2001.

J. Hill, R. Szewczyk, A. Woo *et al.*, "System architecture directions for networked sensors," in *Proceedings of the 9th ACM International Conference on Architectural Support for Programming Languages and Operating Systems*, pp. 93–104, Cambridge, MA, USA, Nov. 2000.

J. Hill, R. Szewczyk, A. Woo *et al.*, "System architecture directions for network sensors", in *Proceedings of the International Conference on Architectural Support for Programming Languages and Operating Systems (ASPLOS), ASPLOS 2000*, Cambridge, November 2000.

J. Hill and D. Culler, "Mica: a wireless platform for deeply embedded networks," *IEEE Micro*, Vol. **22**, No. 6, pp. 12–24, Nov.–Dec. 2002.

X. Hong and Q. Liang, "An access based energy efficient communication protocol for wireless microsensor networks," in *15th IEEE International Symposium on Personal, Indoor and Mobile Radio Communications (PIMRC 2004)*, Vol. **2**, pp. 1022–1026, Sept. 5–8, 2004.

A. Howard, M. Mataric and G. Sukhatme, "Mobile sensor network deployment using potential fields: a distributed, scalable solution to the area coverage problem," *Distrib. Autonom. Robot. Syst.*, Vol. **5**, pp. 299–308, 2002.

A. Howard and M. Mataric, "Cover me! A self-deployment algorithm for mobile sensor networks," in *Proceedings of the International Conference on Robotics Automation*, Washington, DC, USA, 2002, DOI: 10.1.1.16.1394.

A. Howard, M.J. Mataric and G.S. Sukhatme, "Mobile sensor network deployment using potential fields: a distributed, scalable solution to the area coverage problem," in *Proceedings of the 6th*

International Symposium on Distributed Autonomous Robotics Systems (DARS02), Fukuoka, Japan, June 2002.

Y.-C. Hu, A. Perrig and D.B. Johnson, "Packet leashes: a defense against wormhole attacks in wireless networks," in *Proceedings of IEEE Infocom*, 2003.

R. Huang, G.V. Zaruba and M. Huber, "Complexity and error propagation of localization using interferometric ranging," in *Proceedings of IEEE International Conference on Communications (ICC)*, Glasgow, Scotland, pp. 3063–3069, 2007.

C. Huang and Y. Tseng, "The coverage problem in a wireless sensor network," *Mobile Network Applications*, Vol. **10**, pp. 519–528, 2005.

J. Hui and D. Culler, "The dynamic behavior of a data dissemination protocol for network programming at scale," in *Proceedings of the 2nd International Conference on Embedded Networked Sensor Systems (SenSys)*, pp. 81–94, 2004.

B. Hull, K. Jamieson and H. Balakrishnan, "Mitigating congestion in wireless sensor networks," in *Proceedings of ACM Sensys'04*, Baltimore, MD, Nov. 3–5, 2004.

M. Ikeda, E. Kulla, L. Barolli and M. Takizawa, "Wireless ad-hoc networks performance evaluation using NS-2 and NS-3 network simulators," in *Proceedings of the IEEE 2011 International Conference on Complex, Intelligent, and Software Intensive Systems*, pp. 40–45, 2011.

M. Ilyas and I. Mahgoub, *Handbook of Sensor Networks: Compact Wireless and Wired Sensing Systems*, Boca Raton, FL: CRC Press, 2004.

C. Intanagonwiwat, R. Govindan and D. Estrin, "Directed diffusion: a scalable and robust communication paradigm for sensor networks," in *Proceedings of ACM MobiCom 2000*, pp. 56–67, Boston, MA, 2000.

C. Intanagonwiwat, R. Govindan, D. Estrin, J. Heidemann and F. Silva, "Directed diffusion for wireless sensor networking," *IEEE/ACM Transactions on Networking*, Vol. **11**, No. 1, pp. 2–16, Feb. 2003.

C. Intanagonwiwat, R. Govindan and D. Estrin, "Directed diffusion: a scalable and robust communication paradigm for sensor networks," in *Proceedings of the ACM MobiCom'00*, Boston, MA, pp. 56–67, 2000.

S. Intille, "Designing a home of the future," *IEEE Pervasive Computing*, Vol. **1**, No. 2, pp. 76–82, April 2002.

M.T. Isik and O.B. Akan, "A three dimensional localization algorithm for underwater acoustic sensor networks," *IEEE Transactions on Wireless Communications*, Vol. **8**, No. 9, pp. 4457–4463, 2009.

R. Iyer and L. Kleinrock, "QoS control for sensor networks," in *Proceedings of the 2003 IEEE International Conference on Communications*, Vol. **1**, May 2003.

Y.G. Iyer, S. Gandham and S. Venkatesan, "STCP: a generic transport layer protocol for wireless sensor networks," in *Proceedings of 14th International Conference on Computer Communications and Networks (ICCCN 05)*, San Diego, California USA, Oct. 17–19, 2005, pp. 449–454.

Y.G. Iyer, S. Gandham and S. Venkatesan, "STCP: a generic transport layer protocol for wireless sensor networks," in *Proceedings of IEEE International Conference on Computer Communications and Networks (ICCCN)*, USA, pp. 449–454, 2005.

V. Jacobson, "Congestion avoidance and control," in *Proceedings on Communications Architectures and Protocols (SIGCOMM '88)*, V. Cerf (Ed.), ACM, New York, NY, USA, pp. 314–329, 1988.

R. Jain, *The Art of Computer Systems Performance Analysis*, New York: Wiley, 1991.

D. Johnson and D. Maltz, "Dynamic source routing in ad hoc wireless networks," in *Mobile Computing*, T. Imelinsky and H. Korth, Eds., Kluwer Academic Publishers, pp. 153–181, 1996.

J. Jones and M. Atiquzzaman, "Transport protocols for wireless sensor networks: state-of-the-art and future directions," *International Journal of Distributed Sensor Networks*, Vol. 3, No. 1, pp. 119–133, 2007.

J.H. Jun, B. Xie and D.P. Agrawal, "Wireless mobile sensor networks: protocols and mobility strategies," in *Guide to Wireless Sensor Networks*, S. Misra, I. Woungang and S.C. Misra, Ed. Springer, 2009, pp. 607–634.

D. Juneja, N. Arora and S. Bansal, "An ant-based routing algorithm for detecting attacks in wireless sensor networks," *International Journal of Computational Intelligence Research*, Vol. 6, No. 2, pp. 311–330, 2010.

R. Jurdak, C.V. Lopes and P. Baldi, "Software acoustic modems for short range mote-based underwater sensor networks," in *Proceedings of IEEE Oceans*, Singapore, pp. 1–7, May 2006.

R. Jurdak, P.M.Q. Aguiar, P. Baldi and C.V. Lopes, "Software modems for underwater sensor networks," in *Proceedings of Oceans*, pp. 1–6, June 2007.

R. Jurdak, A.G. Ruzzelli, G.M.P. O'Hare and C.V. Lopes, "Mote-based underwater sensor networks: opportunities, challenges, and guidelines," *Telecommunication System*, Vol. 37, No. 1–3, pp. 37–47, 2008.

H.K. Kalita and A. Kar, "Simulator based performance analysis of wireless sensor network – a new approach," in *First IEEE International Workshop on Wireless Communication and Networking Technologies for Rural Enrichment*, pp. 461–465, 2011.

V. Kannan, and S. Ahmed, "A resource perspective to wireless sensor network security," in *Proceedings of the 2011 Fifth International Conference on Innovative Mobile and Internet Services in Ubiquitous Computing*, pp. 94–99, 2011.

K. Kant, *Introduction to Computer System Performance Evaluation*, New York: McGraw-Hill, 1992.

J.N. Karaki and A.E. Kamal, "Routing techniques in wireless sensor networks: a survey," *IEEE Wireless Communications*, Vol. 11, No. 6, pp. 6–28, Dec. 2004.

C. Karlof and D. Wagner, "Secure routing in wireless sensor networks: attacks and counter-measures," *AdHoc Networks (Elsevier)*, Vol. 1, pp. 299–302, 2003.

C. Karlof, N. Sastry and D. Wagner, "Tinysec: a link layer security architecture for wireless sensor networks," in *Second ACM Conference on Embedded Networked Sensor Systems (SensSys 2004)*, Nov. 2004.

C. Karlof and D. Wagner, "Secure routing in wireless sensor networks: attacks and counter-measures," *Elsevier's AdHoc Networks Journal, Special Issue on Sensor Network Applications and Protocols*, in *First IEEE International Workshop on Sensor Network Protocols and Applications*, University of California at Berkeley, Berkeley, USA, 2003.

P. Karn, "MACA – a new channel access method for packet radio," in *Proceedings of the 9th ARRL Computer Networking Conference*, London, Ontario, Canada, 1990.

B. Karp and H.T. Kung, "GPSR: greedy perimeter stateless routing for wireless sensor networks," in *Proceedings of the 6th Annual ACM/IEEE International Conference on Mobile Computing and Networking (MobiCom '00)*, Boston, MA, Aug. 2000.

T. Kavitha and D. Sridharan, "Security vulnerabilities in wireless sensor networks: a survey," *Journal of Information Assurance and Security*, Vol. 5, No. 1, pp. 31–44, 2010.

S.M. Kay, *Fundamentals of Statistical Signal Processing: Estimation Theory*. Upper Saddle River, New Jersey: Prentice-Hall, 1993.

S. Kim, S. Pakzad, D. Culler, *et al.*, "Health monitoring of civil infrastructures using wireless sensor networks," in *Proceedings of Information Processing in Sensor Networks (IPSN)*, pp. 254–263, 2007.

J. Kim, S. Kim, D. Kim and W. Lee, "Low-energy efficient clustering protocol for ad-hoc wireless sensor network," in *15th International Symposium on Personal, Indoor and Mobile Radio Communications, 2004 (PIMRC 2004)*, Vol. **2**, pp. 1022–1026, Sept. 5–8, 2004.

L. Kirousis, E. Kranakis, D. Krizanc and A. Pelc, *Power Consumption in Packet Radio Networks STACS 97*, vol. **1200**. Berlin: Springer-Verlag, 1997, pp. 363–374.

J. Kong *et al.*, "Building underwater ad-hoc networks and sensor networks for large scale real-time aquatic application," *Proceedings IEEE MILCOM*, 2005.

J. Kong, J.-H. Cui, D. Wu and M. Gerla, "Building underwater ad-hoc networks and sensor networks for large scale real-time aquatic applications," in *Proceedings of IEEE Military Communication Conference*, Atlantic City, NJ, USA, pp. 1535–1541, 2005.

C. Kraub, M. Schneider and C. Eckert, "Defending against false endorsement-based DoS attacks in wireless sensor networks," in *Proceedings of the 1st ACM Conference on Wireless Network Security*, 2008.

C. Kraub, M. Schneider and C. Eckert, "An enhanced scheme to defend against false-endorsement-based DoS attacks in WSNs," in *Proceedings of the IEEE International Conference on Wireless & Mobile Computing, Networking & Communication*, 2008.

K. Kredo II and P. Mohapatra, "Medium access control in wireless sensor networks," in *Computer Networks Journal*, Vol. **51**, pp. 961–994, 2007.

K. Kredo, P. Djukic and P. Mohapatra, "STUMP: exploiting position diversity in the staggered TDMA underwater MAC protocol," in *Proceedings of IEEE INFOCOM Mini-Conference*, Rio de Janeiro, Brazil, pp. 2961–2965, 2009.

K. Kredo and P. Mohapatra, "Medium access control in wireless sensor networks," *Computer Networks (Elsevier)*, Vol. **51**, No. 4, March 2007.

N. Krishnamurthi and S.J. Yang, "Feasibility and performance analysis of sensor modeling in OPNET," in *Online Proceedings of OPNETWORK 2007*, Washington, DC, Aug. 2007.

F. Kuhn, R. Wattenhofer and A. Zollinger, "Worst-case optimal and average-case efficient geometric ad-hoc routing," in *Proceedings of the 4th ACM International Conference on Mobile Computing and Networking*, pp. 267–278, 2003.

J. Kulik, W.R. Heinzelman and H. Balakrishnan, "Negotiation-based protocols for disseminating information in wireless sensor networks," *Wireless Networks*, Vol. **8**, No. 2–3, pp. 169–185, March/May 2002.

J. Kulik, W. Heinzelman and H. Balakrishnan, "Negotiation-based protocols for disseminating information in wireless sensor networks," *Wireless Networks*, Vol. **8**, pp. 169–185, 2002.

P. Kulkarni, D. Ganesan, P. Shenoy and Q. Lu, "SensEye: a multi-tier camera sensor network," in *Proceedings of the 13th Annual ACM International Conference on Multimedia*, New York, USA, pp. 229–238, 2005.

D. Kumar, T.C. Aseri and R.B. Patel, "EEHC: energy efficient heterogeneous clustered scheme for wireless sensor networks," *Computer Communications*, Vol. **32**, issue 4, pp. 662–667, March 2009.

S. Kumar, T. Lai and A. Arora, "Barrier coverage with wireless sensors," in *Proceedings of the ACM International Conference on Mobile Computer Networks*, pp. 284–298, 2005.

T.P. Lambrou and C.G. Panayiotou, "Collaborative area monitoring using wireless sensor networks with stationary and mobile nodes," *EURASIP Journal on Advances in Signal Processing*, Vol. **2009**, 2009.

K. Langendoen and N. Reijers, "Distributed localization in wireless sensor networks: a quantitative comparison," *Computer Networks*, Vol. **43**, pp. 499–518, 2003.

L. Lazos, S. Capkun and R. Poovendran, "ROPE: robust position estimation in wireless sensor network," in *Proceedings of the Fourth International Conference on Information Processing in Sensor Networks (IPSN' 05)*, 2005.

V. Lecuire, C.D. Faundez and N. Krommenacker, "Energy-efficient image transmission in sensor networks," *International Journal of Sensor Networks*, Vol. **4**, pp. 37–47, 2008.

J.C. Lee *et al.*, "Key management issues in wireless sensor networks: current proposals and future developments," *IEEE Wireless Communications*, Vol. **14**, No. 5, pp. 76–84, Oct. 2007.

H.J. Lee *et al.*, "Centroid-based movement assisted sensor deployment schemes in wireless sensor networks," in *Proceedings of the 70th IEEE Vehicular Technology Conference Fall (VTC 2009-Fall)*, Anchorage, Alaska, USA, Sep. 2009.

S. Leung, W. Gomez and J.J. Kim, "Zigbee mesh network simulation using Opnet and study of routing selection," ENSC 427: Communication Networks, final project, 2009.

P. Levis and D. Culler, "Mote: a tiny virtual machine for sensor networks," in *Proceedings of the 2002 International Conference on Architectural Support for Programming Languages and Operating Systems, ASLOS'2002*, pp. 85–95, San Jose, CA, 2002.

P. Levis, N. Patel, D. Culler and S. Shenker, "Trickle: a self regulating algorithm for code propagation and maintenance in wireless sensor networks," in *Proceedings of the 1st USENIX/ACM Symposium on Networked Systems Design and Implementation*, pp. 15–28, 2004.

P. Levis, N. Patel, D. Culler and S. Shenker, "Trickle: a self regulating algorithm for code propagation and maintenance in wireless sensor networks," in *Proceedings of 1st Symposium on Networked Systems Design and Implementation*, San Francisco, CA, Mar. 29–31, 2004.

L. Li and J.Y. Halpern, "Minimum-energy mobile wireless networks revisited," in *IEEE International Conference on Communications (ICC) 2001*, Vol. **1**, pp. 278–283, 2001.

Q. Li, J. Aslam and D. Rus, "Hierarchical power-aware routing in sensor networks," in *Proceedings of the DIMACS Workshop on Pervasive Networking*, May 2001.

B.M. Li, Z. Li and A.V. Vasilakos, "A survey on topology control in wireless sensor networks: taxonomy, comparative study, and open issues," *Proceedings of the IEEE*, Vol. **25**, No. 10, pp. 2367–2380, 2013.

N. Li, J. Hou and L. Sha, "Design and analysis of an MST-based topology control algorithm," *IEEE Transactions on Wireless Communications*, Vol. **4**, No. 3, pp. 1195–1206, May 2005.

Z. Li, W. Dehaene and G. Gielen, "A 3-tier UWB-based indoor localization system for ultra-low-power sensor networks," *IEEE Transactions on Wireless Communications*, Vol. **8**, No. 6, pp. 2813–2818, June 2009.

C. Li, P. Wang, H.-H. Chen and M. Guizani, "A cluster based on-demand multi-channel MAC protocol for wireless multimedia sensor networks," in *Proceedings of IEEE International Conference on Communications*, Beijing, China, pp. 2371–2376, May 2008.

S. Li, R. Neelisetti, C. Liu and A. Lim, "Delay-constrained high throughput protocol for multi-path transmission over wireless multimedia sensor networks," in *International Symposium on a World of Wireless, Mobile and Multimedia Networks (WoWMoM)*, Newport Beach, USA, pp. 1–8, June 2008.

S. Li, A. Lim, S. Kulkarni and C. Liu, "EDGE: a routing algorithm for maximizing throughput and minimizing delay in wireless sensor networks," in *Proceedings of the 26th Military Communications Conference (MILCOM'07)*, Oct. 2007.

M. Li and Y. Liu, "Underground structure monitoring with wireless sensor networks," in *6th International Symposium on Information Processing in Sensor Networks (ISPN)*, 2007.

P.-K. Liao, M.-K. Chang and C. Kuo, "Contour line extraction with wireless sensor networks," *Proceedings of the 2005 IEEE International Conference on Communications, ICC 2005*, Vol. **5**, pp. 3202–3206, May 2005.

D.J. Lilja, *Measuring Computer Performance*, Cambridge: Cambridge University Press, 2000.

H. Lim and J.C. Hou, "Localization for anisotropic sensor networks," in *Proceedings of INFOCOM*, Vol. **1**, pp. 138–149, 2005.

S. Lin, J. Zhang, G. Zhou *et al.*, "ATPC: adaptive transmission power control for wireless sensor networks," in *Proceedings of the ACM International Conference on Embedded Network Sensor Systems*, 2006, DOI: 10.1145/1182807.1182830.

P. Lin, C. Qiao and X. Wang, "Medium access control with a dynamic duty cycle for sensor networks," in *Proceedings of the IEEE Wireless Communications and Networking Conference (WCNC)*, Vol. **3**, pp. 1534–1539, March 2004.

W. Lin, D. Li, Y. Tan, J. Chen and T. Sun, "Architecture of underwater acoustic sensor networks: a survey," in *Proceedings of Intelligent Networks and Intelligent Systems*, pp. 155–159, 2008.

S. Lindsey and C. Raghavendra, "PEGASIS: power-efficient gathering in sensor information systems," *IEEE Aerospace Conference Proceedings*, Vol. **3**, pp. 1125–1130, 2002.

S. Lindsey and C.S. Raghavendra, "PEGASIS: power-efficient gathering in sensor information systems," in *IEEE Aerospace Conference Proceedings*, 2002.

Y. Liu, I. Elhanany and H. Qi, "An energy-efficient QoS-aware media access control protocol for wireless sensor networks," in *Proceedings of IEEE MASS 2005*.

B. Liu and D. Towsley, "A study of the coverage of large-scale sensor networks," in *Proceedings of the IEEE International Conference on Mobile Ad-Hoc Sensor Systems*, pp. 475–483, 2004.

B. Liu, O. Dousse, J. Wang and A. Saipulla, "Strong barrier coverage of wireless sensor networks," in *Proceedings ACM International Symposium on Mobile Ad Hoc Network Computers*, pp. 411–420, 2008.

J. Liu and B. Li, "Mobilegrid: capacity-aware topology control in mobile ad hoc networks," in *Proceedings of the 11th International Conference Computer Communications Networks*, pp. 570–574, 2002.

X. Liu, Q. Wang, L. Sha and W. He, "Optimal QoS sampling frequency assignment for real-time wireless sensor networks," in *Proceedings of 24th IEEE Real-Time Systems Symposium*, pp. 308–319, Dec. 2003.

G. Lu, B. Krishnamachari and C. Raghavendra, "An adaptive energy-efficient and low-latency MAC for data gathering in sensor networks," in *Proceedings of the Fourth International Workshop on Algorithms for Wireless, Mobile, Ad Hoc and Sensor Networks (WMAN)*, 2004.

K. Lu *et al.*, "A framework for a distributed key management scheme in heterogeneous wireless sensor networks," *IEEE Transactions on Wireless Communications*, Vol. **7**, No. 2, pp. 639–647, Feb. 2008.

S. Lu, V. Bharghavan and R. Srikant, "Fair scheduling in wireless packet networks," *IEEE/ACM Trans. Network*, Vol. **7**, pp. 473–489, Aug. 1999.

M. Luby, M. Mitzenmacher, A. Shokrollahi, D. Spielman and V. Stemann, "Practical loss-resilient codes," in *Proceedings of ACM Symposium on Theory of Computing*, pp. 150–159, 1997.

J. Luo and J.P. Hubaux, "Joint mobility and routing for lifetime elongation in wireless sensor networks," in *Proceedings of the 24th Annual Joint Conference of the IEEE Computer and Communications Societies (INFOCOM, 2005)*, Miami, FL, USA, Vol. **3**, 2005, pp. 1735–1746.

K.V. Mackenzie, "Discussion of sea water sound-speed determinations," *Journal of the Acoustical Society of America*, Vol. **70**, No. 3, pp. 801–806, 1981.

R. MacRuairi, M.T. Keane and G. Coleman, "A wireless sensor network application requirements taxonomy," in *2008 IEEE International Conference on Sensor Technologies and Applications, IEEE Computer Society*, pp. 209–216, 2008.

M. Maimour, C. Pham and J. Amelot, "Load repartition for congestion control in multimedia wireless sensor networks with multipath routing," in *Proceedings of 3rd International Symposium on Wireless Pervasive Computing, ISWPC*, pp. 11–15, Oct. 2008.

A. Mainwaring, J. Polastre, R. Szewczyk, D. Culler, and J. Anderson, "Wireless sensor networks for habitat monitoring," in *Proceedings of the 2nd Annual International Conference on Mobile and Ubiquitous Systems: Networking and Services, MobiQuitous'05*, pp. 479–481, July 2005.

J. Mainwaring, R. Polastre, R. Szewczyk, D. Culler and J. Anderson, "Wireless sensor networks for habitat monitoring," in *Proceedings of 2002 ACM International Workshop on Wireless Sensor Networks and Applications*, pp. 88–97, Sept. 2002.

D. Malan, T. Fulford-Jones, M. Welsh and S. Moulton, "Codeblue: an ad hoc sensor network infrastructure for emergency medical care," in *Proceedings of the International Workshop on Wearable and Implantable Body Sensor Networks*, 2004.

Q. Mamun, S. Ramakrishnan and B. Srinivasan, "Multi-chain oriented logical topology for wireless sensor networks," in *2nd International Conference, Computer Engineering and Technology (ICCET)*, April 16–18, 2010.

A. Manjeshwar and D.P. Agarwal, "TEEN: a routing protocol for enhanced efficiency in wireless sensor networks," in *Proceedings of the 15th International Workshop on Parallel and Distributed Computing: Issues in Wireless Networks and Mobile Computing*, pp. 2009–2015, April 2001.

A. Manjeshwar and D.P. Agarwal, "APTEEN: a hybrid protocol for efficient routing and comprehensive information retrieval in wireless sensor networks," in *Proceedings of the 2002 International Parallel and Distributed Processing Symposium, IPDPS*, pp. 195–202, 2002.

V. Mansouri, M. MohammadNia-Awal, Y. Ghiassi-Farrokhfal and B. Khalaj, "Dynamic scheduling MAC protocol for large scale sensor networks," in *Proceedings of the 2005 IEEE Mobile Adhoc and Sensor Systems Conference*, Nov. 2005.

G. Mao, B. Fidan and B.D.O. Anderson, "Wireless sensor network localization techniques," *Computer Networks*, Vol. **51**, No. 10, pp. 2529–2553, 2007.

A. Marco, R. Casas, J.L. Sevillano *et al.*, "Synchronization of multi-hop wireless sensor networks at the application layer," *IEEE Wireless Communications*, Vol. **18**, No. 1, pp. 82–88, Feb. 2011.

C. Marghescu, M. Pantazica, A. Brodeala and P. Svasta, "Simulation of a wireless sensor network using OPNET," in *Proceedings of the 2011 IEEE International Symposium for Design and Technology in Electronic Packaging (SIITME 2011)*, pp. 249–252, 2011.

M. Marina and S. Das, "Routing performance in the presence of unidirectional links in multihop wireless networks," in *Proceedings of the ACM International Symposium on Mobile Ad Hoc Network Computing*, pp. 12–23, 2002.

M. Maroti, B. Kusy, G. Balogh, *et al.*, "Radio interferometric geolocation," in *Proceedings of International Conference on Embedded Networked Sensor Systems (SenSys)*, pp. 1–12, 2005.

MBOFDM-UWB, website: http://www.radio-electronics.com/info/wireless/uwb/mb-ofdm-uwb.php

L.L. Mehmet, C. Vurany and I.F. Akyildiz "Characteristics of underground channel for wireless underground sensor networks," *Physical Communication*, Vol. **2**, No. 3, pp. 167–183, Sept. 2009.

T. Melodia, H. Kulhandjian, L.C. Kuo and E. Demirors, "Advances in underwater acoustic networking," in *Mobile Ad Hoc Networking: Cutting Edge Directions*, Second Edition, S. Basangni *et al.*, Ed. John Wiley & Sons, 2013, pp. 804–852.

L.D.P. Mendes and J.J.P.C. Rodrigues, "A survey on cross-layer solutions for wireless sensor networks," *Journal of Network and Computer Applications*, Vol. **34**, No. 2, pp. 523–534, 2011.

S. Methley, *Essentials of Wireless Mesh Networking*, Cambridge University Press, 2009.

J. Michel, "Assessment and recovery of submerged oil: current state analysis," Research & Development Center, U.S. Coast Guard, 2006; http://www.crrc.unh.edu/workshops/submerged_oil/submerged_oil_workshop_report.pdf (accessed May 10, 2008).

S. Misra, K. Abraham, M.S. Obaidat and P. Krishna, "LAID: a learning automata based scheme for intrusion detection in wireless sensor networks," *Security and Communications Networks, Wiley*, Vol. **2**, No. 2, pp. 105–115, March/April 2009.

S. Misra, M.S. Obaidat, S. Sanchita and D. Mohanta, "An energy-efficient, and secured routing protocol for wireless sensor networks," in *Proceedings of the 2009 SCS/IEEE International Symposium on Performance Evaluation of Computer and Telecommunication Systems, SPECTS 2009*, pp. 185–192, Istanbul, Turkey, July 2009.

S. Misra, K.I. Abraham, M.S. Obaidat and P.V. Krishna, "Intrusion detection in wireless sensor networks: the S-model learning automata approach," in *Proceedings of the 4th IEEE International Conference on Wireless and Mobile Computing, Networking and Communications: The First International Workshop in Wireless and Mobile Computing, Networking and Communications (IEEE SecPriWiMob'08)*, pp. 603–607, Avignon, France, October 12–14, 2008.

S. Misra, V. Tiwari and M.S. Obaidat, "Adaptive learning solution for congestion avoidance in wireless sensor networks," in *Proceedings of the IEEE/ACS International Conference on Computer Systems and Applications, AICCSA 2009*, pp. 478–484, Rabat, Morocco, May 2009.

S. Misra, V. Tiwari and M.S. Obaidat, "LACAS: learning automata-based congestion avoidance scheme for healthcare wireless sensor networks," *IEEE Journal on Selected Area on Communications (JSAC)*, Vol. **27**, No. 4, pp. 466–479, May 2009.

S. Misra, M.P. Kumar and M.S. Obaidat, "Connectivity preserving localized coverage algorithm for area monitoring using wireless sensor networks," *Computer Communication Journal, Elsevier*, Vol. **34**, No. 12, pp. 1484–1496, 2011.

S. Misra, V. Tiwari and M.S. Obaidat, "LACAS: learning automata-based congestion avoidance scheme for healthcare wireless sensor networks," *IEEE Journal on Selected Areas in Communications*, Vol. **27**, No. 4, pp. 466–479, May 2009.

S. Misra and M. Khatua, "Cross-layer techniques and applications in wireless sensor networks," in H. Rashvand and Y. Kavian (Eds.), *Using Cross-Layer Techniques for Communication Systems*, pp. 94–119, 2012, Hershey, PA: Information Science Reference. doi: 10.4018/978-1-4666-0960-0.ch004.

S. Misra and S. Singh, "Localized policy-based target tracking using wireless sensor networks," *ACM Transactions on Sensor Networks*, Vol. **8**, No. 3, pp. 27:1–27:30, 2012.

S. Misra, I. Woungang and S.C. Misra, *Guide to Wireless Sensor Networks*, London: Springer-Verlag, 2009.

S. Misra, V. Tiwari and M.S. Obaidat, "LACAS: learning automata-based congestion avoidance scheme for healthcare wireless sensor networks," *IEEE Journal on Selected Area on Communications (JSAC)*, Vol. **27**, No. 4, pp. 466–479, May 2009.

S. Misra and M.S. Obaidat, "Fire monitoring and alarm system for underground coal mines bord-and-pillar panel using wireless sensor networks," *Journal of Systems and Software*, Vol. **85**, No. 3, pp. 571–581, 2012.

S. Misra, K. Abraham, M.S. Obaidat and P. Krishna, "LAID: a learning automata based scheme for intrusion detection in wireless sensor networks," *Security and Communications Networks, Wiley*, Vol. **2**, No. 2, pp. 105–115, March/April, 2009.

S. Misra, S. Dash, M. Khatua, A. V. Vasilakos and M. S. Obaidat, "Jamming in underwater sensor networks: detection and mitigation," *IET Communications*, Vol. **6**, No. 14, pp. 2178–2188, 2012.

S. Misra, A. Ghosh, A. Sagar and M. S. Obaidat, "Detection of identity-based attacks in wireless sensor networks using signalprints," in *2010 IEEE GlobCom 2010 Workshop on Web and Pervasive Security (WPS)*, Miami, FL, Dec. 2010.

S. Misra, M. S. Obaidat, S. Sanchita and D. Mohanta, "An energy-efficient, and secured routing protocol for wireless sensor networks," in *Proceedings of the 2009 SCS/IEEE International Symposium on Performance Evaluation of Computer and Telecommunication Systems, SPECTS 2009*, pp. 185–192, Istanbul, Turkey, July 2009.

S. Misra, K. I. Abraham, M. S. Obaidat and P. V. Krishna, "Intrusion detection in wireless sensor networks: the S-model learning automata approach," in *Proceedings of the 4th IEEE International Conference on Wireless and Mobile Computing, Networking and Communications: The First International Workshop in Wireless and Mobile Computing, Networking and Communications (IEEE SecPriWiMob'08)*, pp. 603–607, Avignon, France, Oct. 12–14, 2008.

S. Misra and A. A. Ghosh, "The effects of variable sound speed on localization in underwater sensor networks," in *Proceedings of Australasian Telecommunication Networks and Applications Conference*, pp. 1–4, 2011.

S. Misra and M. Khatua, "Cross-layer techniques and applications in wireless sensor networks," in *Using Cross-Layer Techniques for Communication Systems*, H. F. Rashvand and Y. S. Kavian, Ed. USA: IGI Global, 2012, pp. 94–119.

M. Molins and M. Stojanovic, "Slotted FAMA: a MAC protocol for underwater acoustic networks," in *Proceedings of IEEE Oceans*, pp. 1–6, 2006.

D. Moore, J. Leonard, D. Rus and S. Teller, "Robust distributed network localization with noisy range measurements," in *Proceedings of the 2nd International Conference on Embedded Networked Sensor Systems*, Baltimore, MD, USA, pp. 50–61, 2004.

B. Nancharaiah, G. F. Sudha and M. B. R. Murthy, "A scheme for efficient topology management of wireless ad hoc networks using the MARI algorithm," in *16th IEEE International Conference on Networks, ICON 2008*, Dec. 12–14, 2008.

S. Narayanaswamy, V. Kawadia, R. Sreenivas and P. Kumar, "Power control in ad-hoc networks: theory, architecture, algorithm and implementation of the COMPOW protocol," in *Proceedings of the European Wireless Conference*, 2002. [Online.] Available: http://citeseerx.ist.psu.edu/viewdoc/summary?doi=10.1.1.23.5186.

S. Nath, Y. Ke, P. B. Gibbons, B. Karp and S. Seshan, "A distributed filtering architecture for multimedia sensors technical report IRP-TR-04-16," in *First Workshop on Broadband Advanced Sensor Networks (BaseNets)*, Aug. 2004.

P. Nicopolitidis, M. S. Obaidat, G. I. Papadimitriou and A. S. Pomportsis, *Wireless Networks*, John Wiley & Sons, 2003.

P. Nicopolitidis, M. S. Obaidat and G. Papadimitriou, *Wireless Networks*, Wiley, 2003.

P. Nicopolitidis, M. S. Obaidat and G. I. Papadimitriou, *Wireless Networks*, John Wiley & Sons, 2003.

D. Niculescu and B. Nath, "Trajectory-based forwarding and its applications," in *Proceedings of the Ninth Annual International Conference on Mobile Computing and Networking (MobiCom)*, 2003.

D. Niculescu and B. Nath, "Ad hoc positioning system (APS) using AoA," in *Proceedings of INFOCOM*, pp. 1734–1743, 2003.

T. Nieberg, S. Dulman, P. Havinga, L. van Hoesel and J. Wu, "Collaborative algorithms for communication in wireless sensor networks," in *Workshop on European Research on Middleware and Architectures for Complex and Embedded Systems*, Pisa, Italy, April 2003.

X. Niu, X. Huang, Z. Zhao, *et al.*, "The design and evaluation of a wireless sensor network for mine safety monitoring," in *Global Telecommunications Conference, 2007, GLOBECOM'07 IEEE*, 2007.

Y. Noh, U. Lee, P. Wang, B.S.C. Choi and M. Gerla, "VAPR: void-aware pressure routing for underwater sensor networks," *IEEE Transactions on Mobile Computing*, Vol. **12**, No. 5, pp. 895–908, 2013.

A. Novikov and A. Bagtzoglou, "Hydrodynamic model of the lower Hudson River estuarine system and its application for water quality management," *Water Resources Management*, Vol. **20**, No. 2, pp. 257–276, 2006.

A. Novikov and A. Bagtzoglou, "Hydrodynamic model of the lower Hudson River estuarine system and its application for water quality management," *Water Resources Management*, Vol. **20**, No. 2, pp. 257–276, 2006.

M.S. Obaidat, P. Nicopolitidis and J.-S. Li, "Security in wireless sensor networks," *Security and Communications Networks, Wiley*, Vol. **2**, No. 2, pp. 101–103, March/April 2009.

M.S. Obaidat, "On the characterization of ultrasonic transducers using pattern recognition techniques," *IEEE Transactions on Systems, Man, and Cybernetics*, Vol. **23**, No. 5, pp. 1443–1450, Sep./Oct. 1993.

M.S. Obaidat and J.W. Ekis, "An automated system for characterizing ultrasonic transducers using pattern recognition," *IEEE Transactions on Instrumentation and Measurement*, Vol. **40**, No. 5, pp. 847–850, October 1991.

M.S. Obaidat and D.S. Abu-Saymeh, "Methodologies for characterizing ultrasonic transducers using neural network paradigms," *IEEE Transactions on Industrial Electronics*, Vol. **39**, No. 6, pp. 529–536, Dec. 1992.

M.S. Obaidat, H. Khalid and B. Sadoun, "Ultrasonic transducers characterization by neural networks," *Information Sciences Journal, Elsevier*, Vol. **107**, No. 1–4, pp. 195–215, June 1998.

M.S. Obaidat and H. Khalid, "Performance evaluation of neural network paradigms for ultrasonic transducers characterization," in *Proceedings of the IEEE International Conference on Electronics, Circuits and Systems*, pp. 370–376, Dec. 1995.

M.S. Obaidat and D.S. Abu-Saymeh, "Performance comparison of neural networks and pattern recognition techniques for classifying ultrasonic transducers," in *Proceedings of the 1992 ACM Symposium on Applied Computing*, pp. 1234–1242, Kansas City, MO, March 1992.

M.S. Obaidat and D.S. Abu-Saymeh, "Neural network and pattern recognition techniques for characterizing ultrasonic transducers," in *Proceedings of the 1992 IEEE Phoenix Conference on Computers and Communications*, pp. 729–735, April 1992.

M.S. Obaidat and N. Boudriga, *Security of e-Systems and Computer Networks*, Cambridge University Press, 2007.

M.S. Obaidat, *Fundamentals of Performance Evaluation of Computer and Telecommunication Systems*, Wiley, 2010.

M.S. Obaidat and G.I. Papadimitriou, Eds., *Applied System Simulation: Methodologies and Applications*, Springer, 2003.

M.S. Obaidat, "Advances in performance evaluation of computer and telecommunications networking," *Computer Communication Journal, Elsevier*, Vol. **25**, Nos. 11–12, pp. 993–996, 2002.

M.S. Obaidat, "ATM systems and networks: basics issues, and performance modeling and simulation," *Simulation: Transactions of the Society for Modeling and Simulation International, SCS*, Vol. **78**, No. 3, pp. 127–138, 2003.

M.S. Obaidat, "Performance evaluation of telecommunication systems: models issues and applications," *Computer Communications Journal, Elsevier*, Vol. **34**, No. 9, pp. 753–756, 2003.

M.S. Obaidat and N. Boudriga, *Security of e-Systems and Computer Networks*, Cambridge University Press, 2007.

M.S. Obaidat and N. Boudriga, *Fundamentals of Performance Evaluation of Computer and Telecommunication Systems*, Wiley, 2010.

M.S. Obaidat, A. Alagan and I. Woungang, Eds., *Handbook of Green Information and Communications Systems*, Elsevier, 2013.

M.S. Obaidat, P. Nicopolitidis and J.-S. Li, "Security in wireless sensor networks," *Security and Communications Networks, Wiley*, Vol. **2**, No. 2, pp. 101–103, March/April, 2009.

M.S. Obaidat and N. Boudriga, *Security of e-Systems and Computer Networks*, Cambridge University Press, 2007.

T. Ojha and S. Misra, "HASL: high-speed AUV-based silent localization for underwater sensor networks," in *Proceedings of the International Conference on Heterogeneous Networking for Quality, Reliability, Security and Robustness*, LNICST 115, pp. 128–140, Greater Noida, India, 2013.

T. Ojha and S. Misra, "MobiL: a 3-dimensional localization scheme for mobile underwater sensor networks," in *Proceedings of National Conference on Communications (NCC)*, pp. 1–5, New Delhi, India, 2013.

T. Ojha, M. Khatua and S. Misra, "Tic-tac-toe-arch: a self-organizing virtual architecture for underwater sensor networks," *IET Wireless Sensor Systems*, Vol. **3**, No. 4, pp. 307–316, Dec. 2013.

T. Ojha and S. Misra, "HASL: high-speed AUV-based silent localization for underwater sensor networks," in *Proceedings of the International Conference on Heterogeneous Networking for Quality, Reliability, Security and Robustness, LNICST 115*, Greater Noida, India, pp. 128–140, 2013.

T. Ojha and S. Misra, "MobiL: a 3-dimensional localization scheme for mobile underwater sensor networks," in *Proceedings of National Conference on Communications*, New Delhi, India, pp. 1–5, 2013.

G. Padmavathi and D. Shanmugapriya, "A survey of attacks, security mechanisms and challenges in wireless sensor networks," *International Journal of Computer Science and Information Security (IJCSIS)*, Vol. **4**, Nos. 1 & 2, pp. 1–9, 2009.

G. Padmavathi and D. Shanmugapriya, "A survey of attacks, security mechanisms and challenges in wireless sensor networks," *International Journal of Computer Science and Information Security*, Vol. **4**, Nos. 1 & 2, 2009.

J.B. Paduan, D.W. Caress, D.A. Clague, C.K. Paull and H. Thomas, "High-resolution mapping of mass wasting, tectonic, and volcanic hazards using the MBARI mapping AUV," in *International Conference on Seafloor Mapping for Geohazard Assessment*, Forio d'Ischia, Italy, May 11–13, 2009.

A. Pal, "Localization algorithms in wireless sensor networks: current approaches and future challenges," *Network Protocols and Algorithms*, Vol. **2**, No. 1, 2010.

G.I. Papadimition, A.S. Pomportsis, P. Nicopolitidis and M.S. Obaidat, *Wireless Network*, Wiley, 2002.

S.-J. Park, R. Vedantham, R. Sivakumar and I.F. Akyildiz, "GARUDA: achieving effective reliability for downstream communication in wireless sensor networks," *IEEE Transactions on Mobile Computing*, Vol. **7**, No. 2, pp. 214–230, Feb. 2008.

J. Partan, J. Kurose and B.N. Levine, "A survey of practical issues in underwater networks," *ACM SIGMOBILE Mobile Computing and Communications Review*, Vol. **11**, No. 4, pp. 23–33, 2007.

S. Pattem, B. Krishnamachari and R. Govindan, "The impact of spatial correlation on routing with compression in wireless sensor networks," in *Proceedings of the Third International Symposium on Information Processing in Sensor Networks (IPSN)*, 2004.

K. Paul, R.R. Choudhuri and S. Bandyopadhyay, "Survivability analysis of ad hoc wireless network architecture," in *Proceedings of the IFTP-TC6/European Commission International Workshop on Mobile and Wireless Communication Networks*, Springer, Vol. **1818**, pp. 31–46, 2000.

M. Perillo and W. Heinzelman, "DAPR: a protocol for wireless sensor networks utilizing an application-based routing cost," in *Proceedings of the 2004 IEEE Wireless Communications and Networking Conference (WCNC)*, Vol. **3**, pp. 1540–1545, March 2004.

M. Perillo and W. Heinzelman, "Sensor management policies to provide application QoS," *Ad Hoc Networks (Elsevier)*, Vol. **1**, No. 2–3, pp. 235–246, 2003.

T. Pering, T. Burd and R. Brodersen, "The simulation and evaluation of dynamic voltage scaling algorithm," in *Proceedings of the 1998 International Symposium on Low Power Electronics and Design, ISLPED '98*, pp. 76–81, Monterey, CA, 1998.

A. Perrig, J. Stankovic and D. Wagner, "Security in wireless sensor networks," *Communications of the ACM*, Vol. **47**, No. 6, pp. 53–57, 2004.

A. Perrig, R. Szewzyk, J.D. Tygar, V. Wen and D.E. Culler, "SPINS: security protocols for sensor networks," *Wireless Networks*, Vol. **8**, No. 5, pp. 521–534, 2002.

A. Perrig, J. Stankovic and D. Wagner, "Security in wireless sensor networks," *Communications of the ACM*, Vol. **47**, No. 6, pp. 53–57, June 2004.

A. Perrig, "Secure routing in sensor networks." http://www.cylab.cmu.edu/default.aspx?id=1985.

A. Perrig, J. Stankovic and D. Wagner, "Security in wireless sensor networks," *Communications of the ACM*, Vol. **47**, No. 6, pp. 53–57, June 2004.

A. Perrig, R. Szewczyk, J.D. Tygar, V. Wen and D.E. Culler, "SPINS: security protocols for sensor networks," *Wireless Networks Journal*, Vol. **8**, pp. 521–534, 2002.

H. Pham and S. Jha, "An adaptive mobility-aware MAC protocol for sensor networks (MS-MAC)," in *Proceedings of the 2004 IEEE International Conference on Mobile Ad-hoc and Sensor Systems*, pp. 558–560, October 2004.

H. Pham and S. Jha, "An adaptive mobility-aware MAC protocol for sensor networks (MS-MAC)," in *Proceedings of IEEE International Conference on Mobile Ad-hoc and Sensor Systems (MASS)*, pp. 214–226, Oct. 2004.

R.F. Pierret, *Introduction to Microelectronics Fabrication*, Menlo Park, CA: Addison-Wesley, 1990.

P.R. Pinet, *Invitation to Oceanography*, 5th Edition, Jones & Bartlett Learning, ISBN: 9780763759933, 2008.

J. Polastre, R. Szewczyk and D. Culler, "Telos: enabling ultra-low power wireless research," in *Proceedings of the 2005 International Symposium on Information Processing, PSN 2005*, pp. 364–369, April 2005.

J. Polastre, J. Hill and D. Culler, "Versatile low power media access for wireless sensor networks," in *Proceedings of the International Conference on Embedded Network Sensor Systems*, pp. 95–107, 2004.

http://www.polastre.com/papers/sensys04-bmac.pdf

D. Pompili, T. Melodia and I.F. Akyildiz, "Deployment analysis in underwater acoustic wireless sensor networks," in *Proceedings of ACM Workshop on Underwater Networks*, Los Angeles, California, USA, pp. 48–55, September 2006.

D. Pompili, T. Melodia and I.F. Akyildiz, "Three-dimensional and two-dimensional deployment analysis for underwater acoustic sensor networks," *Ad Hoc Networks*, Vol. **7**, No. 4, pp. 778–790, 2009.

D. Pompili and T. Melodia, "An architecture for ocean bottom underwater acoustic sensor networks (UWASN)," in *Proceedings of Mediterranean Ad Hoc Networking Workshop*, Bodrum, Turkey, 2004.

D. Porrat, "Radio propagation in hallways and streets for UHF communications," Ph.D. thesis, Stanford University, 2002.

G.J. Pottie and W.J. Kaiser, "Wireless integrated network sensors," *Communications of the ACM*, Vol. **43**, No. 5, pp. 51–58, May 2000.

G.J. Pottie and W.J. Kaiser, "Embedding the internet: wireless integrated network sensors," *Communications of the ACM*, Vol. **43**, No. 5, pp. 51–58, May 2000.

J. Preisig, "Acoustic propagation considerations for underwater acoustics communications network development," *ACM SIGMOBILE Mobile Computing and Communications Review*, Vol. **11**, No. 4, Oct. 2007.

N.B. Priyantha, H. Balakrishnan, E. Demaine, and S. Teller, "Anchor-free distributed localization in sensor networks," in Proceedings *of the 1st International Conference on Embedded Networked Sensor Systems*, Los Angeles, California, USA, p. 340, 2003.

J. Proakis, E. Sozer, J. Rice and M. Stojanovic, "Shallow water acoustic networks," *Proceedings of IEEE Communication Magazine*, Vol. **39**, pp. 114–119, 2001.

R. Puri, A. Majumdar, P. Ishwar and K. Ramchandran, "Distributed video coding in wireless sensor networks," *IEEE Signal Processing Magazine*, Vol. **23**, pp. 94–106, 2006.

K. Römer and F. Mattern, "The design space of wireless sensor networks," *IEEE Wireless Communications*, Vol. **11**, No. 6, pp. 54–61, Dec. 2004.

J. Rabaey, J. Ammer, J. da Silva, D. Patel and S. Roundy, "Picoradio supports ad hoc ultra-low power wireless networking," *IEEE Computer Magazine*, pp. 42–48, July 2002.

H. Radha, M. Schaar and Y. Chen, "The MPEG-4 fine-grained scalable video coding method for multimedia streaming over IP," *IEEE Transactions Multimedia*, Vol. **3**, pp. 53–68, 2001.

C.S. Raghavendra, C. Meesookho and S. Narayanan, "Collaborative classification applications in sensor networks," in *Proceedings of 2002 IEEE Sensor Array and Multichannel Signal Processing Workshop*, pp. 370–374, August 2002.

V. Raghunathan, C. Schurgers, S. Park and M. Srivastava, "Energy-aware wireless microsensor networks," *IEEE Signal Processing Magazine*, Vol. **19**, No. 2, pp. 40–50, March 2002.

M. Rahimi, R. Baer, J. Warrior, D. Estrin and M. Srivastava, "Cyclops: in situ image sensing and interpretation in wireless sensor networks," in *Proceedings of ACM Conference on Embedded Networked Sensor Systems (SENSYS)*, 2005.

M.A. Rahman, A. El Saddik and W. Gueaieb, "Wireless sensor network transport layer: state of the art," in *Sensors*, LNEE 21, S.C. Mukhopadhyay and R.Y.M. Huang (Eds.), Springer, 2008, pp. 221–245.

M. Rahman, R.G. Aghaei, A.E. Saddik and W. Gueaieb, "M-IAR: biologically inspired routing protocol for wireless multimedia sensor networks," in *Proceedings Instrumentation and Measurement Technology Conference (IMTC)*, Victoria, British Columbia, Canada, pp. 1823–1827, October 2008.

C. Rahul and J. Rabaey, "Energy aware routing for low energy ad hoc sensor networks," in *IEEE Wireless Communications and Networking Conference (WCNC)*, Vol. **1**, pp. 350–355, Orlando, FL, March 2002.

V. Rajendran, J. Garcia-Luna-Aceves and K. Obraczka, "Energy-efficient, application-aware medium access for sensor networks," in *Proceedings of the 2005 IEEE International Conference on Mobile Ad-Hoc and Sensor Systems*, Nov. 2005.

V. Rajendran, K. Obraczka and J. Garcia-Luna-Aceves, "Energy-efficient collision-free medium access control for wireless sensor networks," in *Proceedings of the 2003 International Conference on Embedded Networked Sensor Systems*, Nov. 2003.

C. Ramachandran, M.S. Obaidat, S. Misra and F. Pena-Mora, "A secure, and energy-efficient scheme for group-based routing in heterogeneous ad-hoc sensor networks and its simulation analysis," *Simulation: Transactions of the Society for Modeling and Simulation International,* Vol. **84**, No. 2–3, pp. 131–146, Feb./March 2008.

C. Ramachandran, M.S. Obaidat, S. Misra and F. Pena-Mora, "A secure, and energy-efficient scheme for group-based routing in heterogeneous ad-hoc sensor networks and its simulation analysis," *Simulation: Transactions of the Society for Modeling and Simulation International,* Vol. **84**, No. 2–3, pp. 131–146, Feb./March 2008.

C. Ramachandran, M.S. Obaidat, S. Misra and F. Pena-Mora, "A secure, and energy-efficient scheme for group-based routing in heterogeneous ad-hoc sensor networks and its simulation analysis," *Simulation: Transactions of the Society for Modeling and Simulation International,* Vol. **84**, No. 2–3, pp. 131–146, Feb./March 2008.

V. Rathod and M. Mehta, "Security in wireless sensor network: a survey," *GANPAT University Journal of Engineering & Technology,* Vol. **1**, No. 1, pp. 24–43, 2011.

S. Ratnasamy and B. Karp, "GHT: a geographic hash table for data-centric storage," in *Proceedings of the First ACM International Workshop on Wireless Sensor Networks and Applications (WSNA),* 2002.

P. Raviraj, H. Sharif, M. Hempel and S. Ci, "An energy efficient MAC approach for mobile wireless sensor networks systems communications," in *Proceedings of the 2005 IEEE Computer Systems and Applications,* pp. 370–375, 2005.

D.R. Raymond and S.F. Midkiff, "Denial-of-service in wireless sensor networks: attacks and defenses," *IEEE Pervasive Computing,* Vol. **7**, pp. 74–81, March 2008.

D. Raymond *et al.*, "Effects of denial of sleep attacks on wireless sensor network MAC protocols," in *Proceedings of the 7th Annual IEEE Systems, Man, and Cybernetics (SMC) Information Assurance Workshop (IAW),* IEEE Press, pp. 297–304, 2006.

P. Rentala, R. Musunuri, S. Gandham and U. Saxena, "Survey on sensor networks," in *Proceedings of International Conference on Mobile Computing and Networking,* 2001.

http://www.rfm.com/products/data/tr1000.pdf

V. Rodoplu and T.H. Meng, "Minimum energy mobile wireless networks," *IEEE Journal Selected Areas in Communications,* Vol. **17**, No. 8, pp. 1333–1344, Aug. 1999.

V. Rodoplu and T. Meng, "Minimum energy mobile wireless networks," *IEEE Journal on Selected Areas in Communication,* Vol. **17**, No. 8, pp. 1333–1344, Aug. 1999.

J. O'Rourke, "Convex hulls in 2D," Chapter 3 in *Computational Geometry in C,* 2nd ed. Cambridge University Press, 1998.

N. Sadagopan, B. Krishnamachari and A. Helmy, "The ACQUIRE mechanism for efficient querying in sensor networks," in *Proceedings of the First International Workshop on Sensor Network Protocol and Applications,* pp. 149–155, May 2003.

N. Sah, N. Prakash, A. Kumar, Da Kumar and De Kumar, "Optimizing the path loss of wireless indoor propagation models using CSP algorithms," in *Second International Conference on Computer and Network Technology,* pp. 324–328, 2010.

A. Saipulla, C. Westphal, B. Liu and J. Wang, "Barrier coverage of line-based deployed wireless sensor networks," in *Proceedings IEEE International Conference on Computer Communications,* pp. 127–135, 2009.

Y. Sankarasubramaniam, O.B. Akan and I.F. Akyildiz, "ESRT: event-to-sink reliable transport in wireless sensor networks," in *Proceedings of the 4th ACM International Symposium on Mobile Ad Hoc Networking & Computing (MobiHoc '03),* ACM, New York, NY, USA, 2003, pp. 177–188.

P. Santi and D. Blough, "The critical transmitting range for connectivity in sparse wireless ad hoc networks," *IEEE Transactions Mobile Computing*, Vol. **2**, No. 1, pp. 25–39, Jan.–Mar. 2003.

P. Sarisaray, G. Gur, S. Baydere and E. Harmanc, "Performance comparison of error compensation techniques with multipath transmission in wireless multimedia sensor networks," in *15th International Symposium on Modeling, Analysis, and Simulation of Computer and Telecommunication Systems*, Istanbul, Turkey, pp. 73–86, Oct. 2007.

A. Savvides and M. Srivastava, "A distributed computation platform for wireless embedded sensing," in *Proceedings of the 2002 International Conference on Computer Design, ICCD02*, pp. 220–225, Sep. 2002.

A. Savvides, C.-C. Han and M. Srivastava, "Dynamic fine-grained localization in ad-hoc networks of sensors," in *Proceedings of the Seventh ACM Annual International Conference on Mobile Computing and Networking (MobiCom)*, pp. 166–179, July 2001.

A. Savvides, C.C. Han and M.B. Srivastava, "Dynamic fine grained localization in ad hoc networks of sensors," in *Proceedings of the 7th Annual International Conference on Mobile Computing and Networking*, Rome, Italy, pp. 166–179, 2001.

N. Saxena, A. Roy and J. Shin, "Dynamic duty cycle and adaptive contention window based Qos-MAC protocol for wireless multimedia sensor networks," *Computer Networks*, Vol. **52**, pp. 2532–2542, 2008.

M. Schulkin and H.W. Marsh, "Sound absorption in sea water," *Journal of the Acoustical Society of America*, Vol. **34**, pp. 864–865, 1962.

C. Schurgers, V. Tsiatsis, S. Ganeriwal and M. Srivastava, "Optimizing sensor networks in the energy-latency-density design space," *IEEE Transactions on Mobile Computing*, Vol. **1**, No. 1, pp. 70–80, January/March 2002.

C. Schurgers and M.B. Srivastava, "Energy efficient routing in wireless sensor networks," in *Proceedings on Communications for Network-Centric Operations: Creating the Information Force-MILCOM*, McLean, VA, 2001.

C. Schurgers, V. Tsiatsis and M.B. Srivastava, "STEM: topology management for energy efficient sensor networks," in *Aerospace Conference Proceedings, IEEE*, 2002.

C. Schurgers, V. Tsiatsis, S. Ganeriwal and M.B. Srivastava, "Topology management for sensor networks: exploiting latency and density," *MOBIHOC'02*, Lausanne, Switzerland, ACM, June 9–11, 2002.

K. Seada and A. Helmy, "Efficient and robust geocasting protocols for sensor networks," *Computer Communications*, Vol. **29**, No. 2, pp. 151–161, 2006.

W.K. Seah, Z.A. Eu and H-P. Tan, "Wireless sensor networks powered by ambient energy harvesting (WSN-HEAP) – survey and challenges," in *Proceedings of the 1st International Conference on Wireless Communication, Vehicular Technology, Information Theory and Aerospace & Electronics Systems Technology, Wireless VITAE 2009*, pp. 1–5, May 2009.

W.K.G. Seah and H.-X. Tan, "Multipath virtual sink architecture for underwater sensor networks," in *Proceedings of IEEE Oceans*, Singapore, pp. 1–6, 2006.

A. Seema and M. Reisslein, "Towards efficient wireless video sensor networks: a survey of existing node architectures and proposal for a flexi-WVSNP design," *Communications Surveys & Tutorials, IEEE*, Vol. **13**, Issue 3, pp. 462–486, 2011.

http://www.sensorsportal.com/HTML/DIGEST/E_24.htm

http://www.sensorwaresystems.com/

S.D. Senturia, *Microsystems Design*, Norwell, MA: Springer, 2001.

R.C. Shah and J. Rabaey, "Energy aware routing for low energy ad hoc sensor networks," in *IEEE Wireless Communications and Networking Conference (WCNC)*, Vol. **1**, pp. 350–355, Orlando, FL, March 2002.

R.C. Shah, S. Roy, S. Jain and W. Brunette, "Data MULEs: modeling a three-tier architecture for sparse sensor networks," in *Proceedings of the First IEEE International Workshop on Sensor Network Protocols and Applications*, Anchorage, Alaska, USA, 2003, pp. 30–41.

Y. Shang, W. Ruml, Y. Zhang and M.P.J. Fromherz, "Localization from mere connectivity," in *Proceedings of the 4th ACM International Symposium on Mobile ad hoc Networking & Computing*, Annapolis, Maryland, USA, pp. 201–212, 2003.

A. Sharif, V. Potdar and E. Chang, "Wireless multimedia sensor network technology: a survey," in *7th IEEE International Conference on Industrial Informatics*, pp. 606–613, June 2009.

K. Sharma and M.K. Ghose, "Wireless sensor networks: an overview on its security threats," *International Journal of Computer Applications (IJCS), Special Issue on Mobile Ad-hoc Networks*, pp. 42–45, 2010.

C.-C. Shen, C. Srisathapornphat and C. Jaikaeo, "Sensor information networking architecture and applications," *IEEE Personal Communications*, pp. 52–59, August 2001.

E. Shi and A. Perrig, "Designing secure sensor networks," *IEEE Wireless Communications*, pp. 38–43, December 2004.

E. Shi and A. Perrig, "Designing secure sensor networks," *IEEE Communications Magazine*, Vol. **44**, No. 4, 2004.

E. Shi and A. Perrig, "Designing secure sensor networks," *IEEE Wireless Communications Magazine*, pp. 38–43, Dec. 2004.

L. Shu, Y. Zhang, L.T. Yang, Y. Wang and M. Hauswirth, "Geographic routing in wireless multimedia sensor networks," in *Proceedings of International Conference on Future Generation Communication and Networking*, pp. 68–73, 2008.

S.N. Simic and S. Sastry, "Distributed environmental monitoring using random sensor networks," in *Proceedings of Information Processing in Sensor Networks, IPSN 2003*, pp. 582–592, 2003.

S. Singh, M. Woo, and C. Raghavendra, "Power-aware routing in mobile ad hoc networks," in *Proceedings of the Fourth ACM/IEEE International Conference on Mobile Computing*, 2005.

S. Singh, S. Webster, L. Freitag, *et al.*, "Acoustic communication performance of the WHOI micro-modem in sea trials of the Nereus vehicle to 11000m depth," in *Proceedings of IEEE Oceans*, Biloxi, MS, pp. 1–6, 2009.

B. Sinopoli, C. Sharp, L. Schenato, S. Schaffert and S. Sastry, "Distributed control applications within sensor networks," *Proceedings of the IEEE*, Vol. **91**, No. 8, pp. 1235–1246, Aug. 2003.

R. Sivakumar, P. Sinha and V. Bharghavan, "CEDAR: a core-extraction distributed ad hoc routing algorithm," *IEEE Journal of Selected Areas in Communications*, Vol. **17**, No. 8, pp. 1454–1465, 1999.

A. Skordas, C. Chirstopoulos and T. Ebrahimi, "The JPEG 2000 still image compression standard," *IEEE Signal Processing Magazine*, Vol. **18**, No. 5, pp. 36–58, Sept. 2001.

S. Slijepcevic, M. Potkonjak, V. Tsiatsis, S. Zimbeck and M.B. Srivastava, "On communication security in wireless ad-hoc sensor networks," in *Proceedings of 11th IEEE International Workshop on Enabling Technologies: Infrastructure for Collaborative Enterprises, WETICE2002*, June 2002.

K. Sohrabi and G. Pottie, "Performance of a novel self organization protocol for wireless ad hoc sensor networks," in *Proceedings of the IEEE 50th Vehicular Technology Conference*, pp. 1222–1226, 1999.

K. Sohrabi, and J. Pottie, "Protocols for self-organization of a wireless sensor networks," *IEEE Personal Communications*, Vol. **7**, No. 5, pp. 16–27, Oct. 2000.

K. Sohraby, D. Minoli and T. Znati, *Wireless Sensor Networks: Technology, Protocols, and Applications*, John Wiley & Sons, 2007.

J. Sojdehei, P. Wrathal and D. Dinn, "Magneto-inductive (MI) communications," in *OCEANS, 2001. MTS/IEEE Conference and Exhibition*, Vol. **1**, pp. 513–519, 2001.

B. Son, Y.-S. Her and J. Kim, "A design and implementation of forest-fires surveillance system based on wireless sensor networks for South Korea mountains," *International Journal of Computer Science and Network Security (IJCSNS)*, Vol. **6**, No. 9, pp. 124–130, 2006.

F. Stann, and J. Heidemann, "RMST: reliable data transport in sensor networks," in *Proceedings of the First IEEE International Workshop on Sensor Network Protocols and Applications*, May 2003, pp. 102–112.

Stargate Platform, Crossbox, website: http://www.xbow.com/Products/XScale.htm

T. Stathopoulos, J. Heidemann and D. Estrin, "A remote code update mechanism for wireless sensor networks," Tech. Rep. CENS-TR-30, University of California, Los Angeles, Center for Embedded Networked Computing, Nov. 2003. http://www.isi.edu/~johnh/PAPERS/Stathopoulos03b.pdf.

M. Stemm and R.H. Katz, "Measuring and reducing energy consumption of network interfaces in hand-held devices," *IEICE Transactions on Communications*, Vol. **E80**-B, No. 8, pp. 1125–1131, Aug. 1997.

M. Stojanovic, "Underwater acoustic communication," *Encyclopedia of Electrical and Electronics Engineering*, J.G. Webster, Ed., Wiley, 1999, Vol. **22**, pp. 688–98.

M. Stojanovic, "Recent advances in high speed under water acoustic communication," *IEEE Journal of Oceanic Engineering*, Vol. **21**, No. 2, pp. 125–36, 1996.

M. Stojanovic, "On the relationship between capacity and distance in an underwater acoustic channel," in *Proceedings of ACM Workshop on Underwater Networks*, pp. 41–47, 2006.

M. Stojanovic, "Recent advances in high-speed underwater acoustic communications," *IEEE Journal of Oceanic Engineering*, Vol. **21**, No. 2, pp. 125–136, 1996.

M. Stojanovic, "Underwater acoustic communications," in *Encyclopedia of Electrical and Electronics Engineering*, John G. Webster, ed. John Wiley and Sons, 1999, pp. 688–698.

I. Stojmenovic and X. Lin, "GEDIR: loop-free location based routing in wireless networks," in *Proceedings of the 1999 International Conference on Parallel and Distributed Computing and Systems, Boston, MA, USA*, Nov. 3–6, 1999.

L. Su, C. Liu, H. Song and G. Cao, "Routing in intermittently connected sensor networks," in *Proceedings of the 2008 IEEE International Conference on Network Protocols (ICNP)*, 2008.

L. Subramanian and R.H. Katz, "An architecture for building self configurable systems," in *Proceedings of IEEE/ACM Workshop on Mobile Ad Hoc Networking and Computing*, pp. 63–73, Boston, MA, August 2000.

Y. Sun, H. Ma, L. Liu and Y. Zheng, "ASAR: an ant-based service-aware routing algorithm for multimedia sensor networks," *Frontiers of Electrical and Electronic Engineering, China*, Vol. **3**, pp. 25–33, 2008.

Z. Sun and I.F. Akyildiz, "Channel modeling and analysis for wireless networks in underground mines and road tunnels," *IEEE Transactions on Communications*, Vol. **58**, No. 6, June 2010.

Z. Sun and I.F. Akyildiz, "Magnetic induction communications for wireless underground sensor networks," *IEEE Transactions on Antennas and Propagation*, Vol. **58**, No.7, July 2010.

Z. Sun and I.F. Akyildiz, "Underground wireless communication using magnetic induction," in *IEEE International Conference on Communications 2009, ICC 09*, pp. 1–5, 2009.

H. Sundani, H. Li, V. Devabhaktuni, M. Alam and P. Bhattacharya, "A survey and comparisons on wireless sensor network simulators," *International Journal of Computer Networks (IJCN)*, Vol. **2**, No. 5, pp. 249–261, 2010.

S. Sundresh, K. Wooyoung and G. Agha, "SENS: a sensor environment and network simulator", in *Proceedings of 37th Annual Simulation Symposium*, pp. 221–228, Apr. 2004.

A. Swami, Q. Zhao, Y.-W. Hong and L. Tong (Eds.), *Wireless Sensor Networks: Signal Processing and Communication Perspectives*, John Wiley & Sons, 2007.

Technical Report: International Association for Oil and Gas Producers, "Fundamentals of underwater sound," Report No. 406, May 2008.

L. Teng and Y. Zhang, "SeRA: a secure routing algorithm against sinkhole attacks for mobile wireless sensor networks," in *Proceedings of the 2010 Second International Conference on Computer Modeling and Simulation*, Vol. **4**, pp. 79–82, 2010.

N. Tezcan and W. Wang, "ART: an asymmetric and reliable transport mechanism for wireless sensor networks," *International Journal of Sensor Networks*, Vol. **2**, Nos. 3–4, pp. 188–200, 2007.

W.H. Thorp and D.G. Browning, "Attenuation of low frequency sound in the ocean," *Journal of Sound and Vibration*, Vol. **26**, pp. 576–578, 1973.

S. Tilak, N.B. Abu-Ghazaleh and W. Heinzelman, "A taxonomy of sensor network communication models," *Mobile Computing and Communication Review*, Vol. **6**, No. 2, pp. 28–36, April 2002.

A. Timm-Giel, K. Murray, M. Becker, *et al.*, "Comparative simulations of WSN," *ICT-MobileSummit*, 2008.

Y.-C. Tseng, M.-S. Pan and Y.-Y. Tsai, "Wireless sensor networks for emergency navigation," *IEEE Computer*, Vol. **39**, No. 7, pp. 55–62, 2006.

Y. Tseng, C. Hsu and T. Hsieh, "Power-saving protocols for IEEE 802.11-based multi-hop ad hoc networks," *Computer Networks*, Vol. **43**, No. 3, pp. 317–337, 2003.

T. Van Dam and K. Langendoen, "An adaptive energy-efficient MAC protocol for wireless sensor networks," in *Proceedings of the ACM International Conference on Embedded Network Sensor Systems*, pp. 171–180, 2003.

L. van Hoesel and P. Havinga, "A lightweight medium access protocol (LMAC) for wireless sensor networks: reducing preamble transmissions and transceiver state switch," in *Proceedings of the 1st International Workshop on Networked Sensing Systems*, pp. 205–208, Tokyo, Japan, 2004.

I. Vasilescu, K. Kotay, D. Rus, M. Dunbabin and P. Corke, "Data collection, storage, and retrieval with an underwater sensor network," in *Proceedings of ACM SenSys*, San Diego, CA, pp. 154–165, Nov. 2005.

V.V. Veeravalli and J.F. Chamberland, "Detection in sensor networks," in *Wireless Sensor Networks: Signal Processing and Communications Perspectives*, A. Swami, Q. Zhao, Y.W. Hong and L. Tong (Eds.), John Wiley & Sons, 2007, pp. 119–148.

N. Vlajic and D. Stevanovic, "Sink mobility in wireless sensor networks: when theory meets reality," *Proceedings of the 32nd International Conference on Sarnoff Symposium (SARNOFF'09)*, NJ, USA, 2009.

H. Vogt, "Exploring message authentication in sensor networks," in *Security in Ad-hoc and Sensor Networks (ESAS), First European Workshop, Vol. 3313 of Lecture Notes in Computer Science*, pp. 19–30. Springer, 2004.

M.C. Vuran, V.C. Gungor and O.B. Akan, "On the interdependence of congestion and contention in wireless sensor networks," in *Proceedings of the Third International Workshop on Measurement, Modeling, and Performance Analysis of Wireless Sensor Networks*, San Diego, CA, USA, 2005.

Intrusion detection in wireless networks – Micheal Krishnan. http://walrandpc.eecs.berkeley.edu/228S06/Projects/KrishnanProject.pdf.

C.-Y. Wan, S.B. Eisenman and A.T. Campbell, "CODA: congestion detection and avoidance in sensor networks," in *Proceedings of ACM Sensys '03*, Los Angeles, CA, Nov. 5–7, 2003.

C.-Y. Wan, S.B. Eisenman, A.T. Campbell and J. Crowcroft, "Siphon: overload traffic management using multi radio virtual sinks in sensor networks," in *Proceedings of ACM Sensys '05*, San Diego, CA, Nov. 2–4, 2005.

C.-Y. Wan, A.T. Campbell and L. Krishnamurthy, "PSFQ: a reliable transport protocol for wireless sensor networks," in *Proceedings of the First ACM International Workshop on Wireless Sensor Networks and Applications (WSNA'02)*, Atlanta, Georgia, USA, Sept. 2002.

C. Wang, K. Sohraby, V. Lawrence, L. Bo and Y. Hu, "Priority–based congestion control in wireless sensor networks," *IEEE International Conference on Sensor Networks, Ubiquitous, and Trustworthy Computing (SUTC'06)*, 2006.

C. Wang, B. Li, K. Sohraby, M. Daneshmand and Y. Hu, "Upstream congestion control in wireless sensor networks through cross-layer optimization," *IEEE Journal on Selected Areas in Communications*, Vol. 25, No. 4, pp. 786–795, 2007.

X. Wang, G. Xing, Y. Zhang, *et al.*, "Integrated coverage and connectivity configuration in wireless sensor networks," in *Proceedings ACM International Conference Embedded Network Sensor Systems*, pp. 28–39, 2003.

G. Wang, G. Cao and T. La Porta, "Movement-assisted sensor deployment," *IEEE Transactions on Mobile Computing*, Vol. 5, No. 6, pp. 640–652, June 2006.

G. Wang, G. Cao and T. LaPorta, "A bidding protocol for deploying mobile sensors," in *Proceedings of the IEEE International Conference on Networks protocols*, pp. 315–324, 2003.

Q. Wang and W. Yang, "Energy consumption model for power management in wireless sensor networks," in *Proceedings of the 4th Annual IEEE Communications Society Conference on Sensor, Mesh and Ad Hoc Communications and Networks, SECON '2007*, pp.142–151, June 2007.

W. Wang, "Traffic analysis, modeling and their applications in energy-constrained wireless sensor networks – on network optimization and anomaly detection," PhD Thesis, No. 78, Dept of Information Technology and Media, Mid Sweden University, Sundsvall, Sweden, 2010.

Y. Wang, G. Attebury and B. Ramamurthy, "A survey of security issues in wireless sensor networks," *IEEE Communication Surveys*, Vol. 8, No. 2, pp. 2–23. 2006.

B. Wang, H.B. Lim and D. Ma, "A survey of movement strategies for improving network coverage in wireless sensor networks," *Computer Communications*, Vol. 32, Nos. 13–14, pp. 1427–1436, 2009.

G. Wang, G. Cao and T.L. Porta, "Movement-assisted sensor deployment," *IEEE Transactions on Mobile Computing*, Vol. 5, No. 6, pp. 640–652, 2006.

G. Wang, G. Cao, P. Berman and T.F. La Porta, "Bidding protocols for deploying mobile sensors," *IEEE Transactions on Mobile Computing*, Vol. 6, No. 5, pp. 563–576, 2007.

Y.C. Wang, W.C. Peng, M.H. Chang and Y.C. Tseng, "Exploring load-balance to dispatch mobile sensors in wireless sensor networks," in *Proceedings of 16th International Conference on Computer Communications and Networks,(ICCCN 2007)*, Honolulu, HI, 2007, pp. 669–674.

G. Wang, T. Wang, W. Jia, M. Guo, and J. Li, "Adaptive location updates for mobile sinks in wireless sensor networks," *The Journal of Supercomputing*, Vol. 47, No. 2, pp. 127–145, 2009.

W. Wang, V. Srinvasan and K. Chua, "Using mobile relays to prolong the lifetime of wireless sensor networks," in *Proceedings of the 11th Annual International Conference on Mobile Computing and Networking (MobiCom '05)*, ACM, New York, NY, USA, 2005, pp. 270–283.

W. Wang, D. Peng, H. Wang, H. Sharif and H.H. Chen, "Energy-constrained distortion reduction optimization for wavelet-based coded image transmission in wireless sensor networks," *Multimedia*, Vol. **10**, pp. 1169–1180, 2008.

Y. Wang, A. Reibman and S. Lin, "Multiple description coding for video delivery," *Proceedings of the IEEE*, Vol. **93**, pp. 57–70, 2005.

J. Wang, D. Li, M. Zhou and D. Ghosal, "Data collection with multiple mobile actors in underwater sensor networks," in *Proceedings of IEEE Workshop on Delay/Disruption-Tolerant Mobile Networks*, pp. 216–221, 2008.

http://webhosting.devshed.com/c/a/Web-Hosting-Articles/Wireless-Sensor-Networks-part2-Limitations/1/

http://webs.cs.berkeley.edu/800demo/

S. Weinmann, M. Kochhal and L. Schwiebert, "Power efficient topologies for wireless sensor networks," in *Proceedings of the 2001 International Conference on Parallel Processing, ICPP 2001*, pp. 156–163, Valencia, Spain.

E.W. Weisstein, "Cassini ovals", From *MathWorld* – A Wolfram Web Resource [Online] http://mathworld.wolfram.com/CassiniOvals.html.

S. Wielens, M. Galetzka and P. Schneider, "Design support for wireless sensor networks based on the IEEE 802.15.4 standard," in *Proceedings of the 2008 IEEE International Symposium on Personal, Indoor and Mobile Radio Communications, PIMRC 2008*, pp. 1–5, Sept. 2008.

J.E. Wieselthier, J. Nguyen and A. Ephremides, "On the construction of energy-efficient broadcast and multicast trees in wireless networks," in *Proceedings of the IEEE International Conference on Computer Communications*, Vol. **2**, pp. 585–594, 2000.

Wireless LAN Medium Access Control (MAC) and Physical Layer (PHY) Specification, IEEE Std. 802.11.

WISEBED, website: http://wisebed.eu/site/

H. Woithe, D. Boehm and U. Kremer, "Improving Slocum glider dead reckoning using a Doppler velocity log," in *Proceedings of MTS/IEEE OCEANS*, Waikoloa, HI, pp. 1–5, 2011.

S. Wong and B. MacDonald, "A topological coverage algorithm for mobile robots," in *Proceedings of the IEEE/RSJ International Conference on Intelligent Robots Systems*, Vol. **2**, pp. 1685–1690, 2003.

A. Woo and D. Culler, "A transmission control scheme for media access in sensor networks," in *Proceedings of the 7th Annual International Conference on Mobile Computing and Networking*, Vol. **6**, No. 5, pp. 221–235, Rome, Italy, 2001.

A.D. Wood and J.A. Stankovic, "Denial of service attacks in sensor networks," *IEEE Computer Magazine*, pp. 54–62, 2002.

G. Wu, X. Chen and M.S. Obaidat, "A high efficient node capture attack algorithm in wireless sensor network based on route minimum key set," *Security and Communication Networks, Wiley*, Vol. **6**, No. 2, pp. 230–238, 2013.

J. Wu and S. Yang, "SMART: a scan-based movement-assisted sensor deployment method in wireless sensor networks," in *Proceedings of 24th Annual Joint Conference of the IEEE Computer and Communications Societies (INFOCOM 2005)*, Miami, USA, Vol. **4**, 2005, pp. 2313–2324.

P. Xie, J.-H. Cui and L. Lao, "VBF: vector-based forwarding protocol for underwater sensor networks," in *Proceedings of IFIP Networking*, pp. 1216–1221, 2006.

P. Xie and J.-H. Cui, "An FEC-based reliable data transport protocol for underwater sensor networks," in *Proceedings of International Conference on Computer Communications and Networks*, Honolulu, HI, pp. 747–753, 2007.

P. Xie, Z. Zhou, Z. Peng, J.-H. Cui and Z. Shi, "SDRT: a reliable data transport protocol for underwater sensor networks," *Ad Hoc Networks*, Vol. **8**, No. 7, pp. 708–722, 2010.

K. Xing, S. Sundhar, R. Srinivasan, *et al.*, "Attacks and countermeasures in sensor networks: a survey," in *Network Security*, S. Huang, D. MacCallum and D.-Z. Du, Ed. Springer, 2005, pp. 1–28.

Y. Xu, J. Heidemann and D. Estrin, "Geography-informed energy conservation for ad-hoc routing," in *Proceedings of the Seventh Annual ACM/IEEE International Conference on Mobile Computing and Networking*, pp. 70–84, 2001.

Y. Xu, J. Winter and W.-C. Lee, "Prediction-based strategies for energy saving in object tracking sensor networks," in *Proceedings of the IEEE International Conference on Mobile Data Management (MDM)*, pp. 346–357, 2004.

Y. Xu, S. Bien, Y. Mori, J. Heidemann and D. Estrin, "Topology control protocols to conserve energy in wireless ad-hoc networks," Technical Report 6, University of California, Los Angeles, Center for Embedded Networked Computing, January 2003.

W. Xu, K. Ma, W. Trappe and Y. Zhang, "Jamming sensor networks: attack and defense strategies," *IEEE Network*, Vol. **20**, No. 3, pp. 41–47, Spring, 2006.

M. Yaghmaee and D. Adjeroh, "A model for differentiated service support in wireless multimedia sensor networks," in *Proceedings of 17th International Conference on Computer Communications and Networks*, Virgin Islands, USA, pp. 1–6, Aug. 2008.

H. Yan, Z.J. Shi and J.-H. Cui, "DBR: depth-based routing for underwater sensor networks," in *Proceedings of the IFIP-TC6 Networking Conference on AdHoc and Sensor Networks, Wireless Networks, Next Generation Internet*, pp. 72–86, 2008.

Z. Yang and Y. Liu, "Quality of trilateration: confidence-based iterative localization," *IEEE Transactions on Parallel and Distributed Systems*, Vol. **21**, No. 5, pp. 631–640, 2010.

H. Yang and B. Sikdar, "A protocol for tracking mobile targets using sensor networks," in *Proceedings of IEEE International Workshop on Sensor Network Protocols and Applications*, pp. 71–81, 2003.

T. Yang, G. Mino, E. Spaho, *et al.*, "A simulation system for multi mobile events in wireless sensor networks," in *Proceedings of the IEEE 2011 International Conference on Advanced Information Networking and Applications*, pp. 411–418, 2011.

Y. Yang, M.I. Fonoage and M. Cardei, "Improving network lifetime with mobile wireless sensor networks," *Computer Communications*, Vol. **33**, No. 4, pp. 409–419, 2010.

F. Yao, A. Demers and S. Shenker, "A scheduling model for reduced CPU energy," in *Proceedings of the 1995 Annual Symposium on Foundations of Computer Science, FOCS'95*, pp. 374–382, October 1995.

Y. Yao and J. Gehrke, "The Cougar approach to in-network query processing in sensor networks", *SIGMOD Record*, Vol. **31**, No. 3, Sep. 2002.

S. Yarkan and H. Arslan, "Statistical wireless channel propagation characteristics in underground mines at 900MHz," in *IEEE Military Communications Conference, 2007, MILCOM 2007*, pp. 1–7, 2007.

W. Ye, J. Heidemann and D. Estrin, "An energy-efficient MAC protocol for wireless sensor networks", in *21st Annual Joint Conference of the IEEE INFOCOM*, Vol. **3**, pp. 1567–1576, June 2002.

W. Ye, J. Heidemann and D. Estrin, "An energy-efficient MAC protocol for wireless sensor networks," in *Proceedings of the 2200 IEEE INFOCOM*, pp. 1567–1567, June 2002.

F. Ye, H. Luo, J. Cheng, S. Lu and L. Zhang, "A two-tier data dissemination model for large-scale wireless sensor networks," in *Proceedings of ACM/IEEE MOBICOM*, 2002.

M. Ye, C. Li, G. Chen and J. Wu, "EECS, an energy efficient clustering scheme in wireless sensor networks," in *24th IEEE International Performance, Computing and Communication Conference* IPCCC, pp. 535–554, April 7–9, 2005.

W. Ye, J. Heidemann and D. Estrin, "An energy-efficient MAC protocol for wireless sensor networks," in *Proceedings of the IEEE International Conference on Computer Communications*, Vol. 3, pp. 1567–1576, 2002.

W. Ye, F. Silva and J. Heidemann, "Ultra-low duty cycle MAC with scheduled channel polling," in *Proceedings 4th International Conference on Embedded Network Sensor Systems*, pp. 321–334, 2006.

W.-L. Yeow, C.-K. Tham and W.-C. Wong, "Energy efficient multiple target tracking in wireless sensor networks," *IEEE Transactions on Vehicular Technology*, Vol. 56, No. 2, pp. 918–928, 2007.

W.-L. Yeow, C.-K. Tham, and W.-C. Wong, "A novel target movement model and energy efficient target tracking in sensor networks," in *Proceedings of IEEE VTC – Spring*, pp. 2825–2829, 2005.

J. Yick, B. Mukherjee and D. Ghosal, "Wireless sensor network survey," *Computer Networks (Elsevier)*, Vol. 52, No. 12, pp. 2292–2330, 2008.

J. Yick, B. Mukherjee and D. Ghosal, "Wireless sensor network survey," *Elsevier's Computer Networks Journal*, Vol. 52, No. 12, pp. 2292–2330, 2008.

J. Yick, B. Mukherjee and D. Ghosal, "Wireless sensor network survey," *Elsevier's Computer Networks Journal*, Vol. 52, No. 12, pp. 2292–2330, Aug. 2008.

O. Younis and S. Fahmy, "Distributed clustering in ad-hoc sensor networks: a hybrid, energy-efficient approach," in *Proceedings of the Twenty-Third Annual Joint Conference of the IEEE Computer and Communications Societies (INFOCOM)*, 2004.

Y. Yu, D. Estrin and R. Govindan, "Geographical and energy-aware routing: a recursive data dissemination protocol for wireless sensor networks," UCLA Computer Science Department Technical Report, UCLA-CSD TR-01–0023, May 2001.

M. Zamalloa, K. Seada, B. Krishnamachari and A. Helmy, "Efficient geographic routing over lossy links in wireless sensor networks," *ACM Transactions on Sensor Networks*, Vol. 4, No. 3, pp. 1–33, 2008.

H. Zhang, A. Arora, Y. Choi and M.G. Gouda, "Reliable bursty convergecast in wireless sensor networks," *Computer Communications*, Vol. 30, No. 13, pp. 2560–2576, 2007.

H. Zhang and H. Shen, "Energy-efficient beaconless geographic routing in wireless sensor networks," *IEEE Transactions on Parallel and Distributed Systems*, Vol. 21, No. 6, pp. 881–896, 2010.

W. Zhang and G. Cao, "DCTC: dynamic convoy tree-based collaboration for target tracking in sensor networks," *IEEE Transactions on Wireless Communications*, Vol. 3, No. 5, pp. 1689–1701, 2004.

H. Zhang and J. Hou, "Maintaining sensing coverage and connectivity in large sensor networks," *Urbana*, Vol. 1, pp. 89–124, 2003.

H. Zhang and J. Hou, "On deriving the upper bound of α-lifetime for large sensor networks," in *Proceedings of the 5th ACM International Symposium on Mobile ad hoc Networking and Computing (MobiHoc '04)*, New York, NY, USA, 2004, pp. 121–132.

Y.P. Zhang, G.X. Zheng and J.H. Sheng, "Excitation of UHF radio waves in tunnels," *Microwave and Optical Technology Letters*, Vol. 22, No. 6, pp. 408–410, 1999.

F. Zhao and L. Guibas, *Wireless Sensor Networks: An Information Processing Approach*, San Francisco, CA: Morgan Kaufmann, 2004.

F. Zhao, J. Shin and J. Reich, "Information-driven dynamic sensor collaboration for tracking applications," *IEEE Signal Processing Magazine*, Vol. **19**, No. 2, pp. 61–72, March 2002.

F. Zhao and L.J. Guibas, *Wireless Sensor Networks: An Information Processing Approach*, Morgan Kaufmann, ISBN: 9781558609143, 2004.

J. Zheng and A. Jamalipour, *Wireless Sensor Networks: A Networking Perspective*, John Wiley & Sons, 2009.

J. Zheng and A. Jamalipour, *Wireless Sensor Networks – A Networking Perceptive*, Hoboken, NJ: Wiley, 2009.

J. Zheng and A. Jamalipour (Eds.), *Wireless Sensor Networks: A Networking Perspective*, Wiley, 2009.

R. Zheng, J. Hou and L. Sha, "Asynchronous wakeup for ad hoc networks," in *Proceedings of the 2003 ACM International Symposium on Mobile Ad Hoc Networking and Computing*, pp. 35–45, 2003.

T. Zheng, S. Radhakrishnan and V. Sarangan, "PMAC: an adaptive energy-efficient MAC protocol for wireless sensor networks," in *Proceedings of the 2005 IEEE Parallel and Distributed Processing Symposium*, April 2005.

J. Zheng and A. Jamalipour, *Wireless Sensor Networks: A Networking Perspective*, Wiley, 2009.

R. Zheng, J. Hou and L. Sha, "Asynchronous wakeup for ad hoc networks," in *Proceedings of the 4th ACM International Symposium on Mobile Ad Hoc Network Computers*, pp. 35–45, 2003.

L. Zhou and Z.J. Haas, "Securing ad hoc networks," *IEEE Network Magazine*, Vol. **13**, No. 6, pp. 24–30, Nov./Dec. 1999.

Y. Zhou, Y. Fang and Y. Zhang, "Security wireless sensor networks: a survey," *IEEE Communication Surveys*, Vol. **10**, No. 3, 3rd Quarter 2008.

S. Zhu, S. Setia and S. Jajodia, "Leap: efficient security mechanisms for large-scale distributed sensor networks," in *Proceedings of the 10th ACM Conference on Computer and Communication Security*, pp. 62–72, 2003.

Subject index

Entries for tables and figures are noted in bold typeface.

3-dimensional underwater localization, 333–334, **334**

A Line in the Sand application, 32
ABEE. *See* access-based energy efficient cluster topology algorithm
access control list (ACL), 63–64
access-based energy efficient cluster topology algorithm (ABEE), 184–185
active query forwarding in sensor networks protocol (ACQUIRE), 83
active sensor nodes, 20, 35
adaptive beacon placement localization algorithm, **140**, 141–143
adaptive local update-based routing protocol (ALURP), **269**, **270**, 269–271
adaptive periodic threshold-sensitive energy-efficient sensor network protocol (APTEEN), 89
Advanced Encryption Standard (AES), 10
aggregation points, 2, 80
aggregation routing algorithm (ARA), 277, **277**
agriculture
    WSN applications in, 44
    WUGSN, 349
air traffic control application, 2
algorithms. *See also* scheme
    ABEE, 184–185
    ARA, 277, **277**
    CODA, 161–162
    for performance modeling, 211
    localization, 140–158
    MCFA, 83
    Minimax, 252–253
    MIPS, 273–275, **274**
    topology, 173–190
    VOR, **251**, **252**, 251–252
Aloha MAC protocol, 338
analog to digital converter (A/D), 7, 19
anchor-free localization algorithm (AFL), 152–153
angle of arrival localization method (AOA), 139–140, **140**
ant colony optimization (ACO), 118–119
ant-based routing with congestion control (ARCC), 118–119
ant-based service aware routing (ASAR), 305
APIT algorithm (localization), 144–146, **145**, **146**, **147**

application functions (WSN)
    contour line detection, 36–37
    edge detection, 35–36
    target detection and tracking, 30–35
application-specific integrated circuits (ASICs), 19
application types
    agriculture, 44
    definition, 1
    health care, 38–39
    intelligent home, 39
    manufacturing, 39
    military, 44–45
    multimedia, 282–284
    security, 39
    underwater, 39–44, **42**, **43**, 322–323
applications
    layer in WSN OS architecture, 23
    memory is dependent on, 19
    PSFQ in, 125–126
    STCP in, **129**, 129–131
    WMSN, 310–313
    WSN, 2–3
    WUGSN, 349–351
APTEEN. *See* adaptive periodic threshold-sensitive energy-efficient sensor network protocol
ARA. *See* aggregation routing algorithm
architecture. *See also* design
    LEACH routing, 87
    PMAC, 66–67, **67**
    topology, 173
    UWSN, **42**–44, **44**, 331–332, **331**, **333**
    WMSN, 286–290
    WSN, **106**, **201**
    WUGSN, 354–357, **355**
asymmetric and reliable transport protocol (ART), 127–128
authentication service
    security requirement, 229, 235–236
    SPIN, 237
automatic repeat
    WMSN, 302
    ZigBee network, 9

405

Printed in the United States
by Baker & Taylor Publisher Services